INFLUENCE OF PSYCHOLOGICAL FACTORS ON PRODUCT DEVELOPMENT

Influence of Psychological Factors on Product Development

Lessons from Aerospace and other Industries

by

Eginaldo Shizuo Kamata

Springer-Science+Business Media, B.V.

Library of Congress Cataloging-in-Publication Data

ISBN 978-90-481-6096-9 ISBN 978-0-306-48079-9 (eBook)
DOI 10.1007/978-0-306-48079-9

Printed on acid-free paper

This book is an independent work by the author. It is not a result of his contractual activities with Contraves Space AG. In addition, no implicit nor explicit indication is made that the contents of this book are supported or endorsed by the Contraves Space AG management or its board of directors. Hence, it does not imply that the concepts and approaches described herein are (or were) used in programs in which this author is (or was) involved.

Originally published by Kluwer Academic Publishers in 2002
Softcover reprint of the hardcover 1st edition 2002

We shall not cease from exploration
And the end of all our exploring
Will be to arrive where we started
And know the place for the first time.
Through the unknown, remembered gate
When the last of earth left to discover
Is that which was the beginning;
At the source of the longest river
The voice of the hidden waterfall
And the children in the apple-tree
Not known, because not looked for
But heard, half-heard, in the stillness
Between two waves of the sea.
Quick now, here, now, always
A condition of complete simplicity
(Costing not less than everything)
And all shall be well and
All manner of thing shall be well
When the tongues of flame are in-folded
Into the crowned knot of fire
And the fire and the rose are one.

T. S. Eliot, *Little Gidding*, 1944

To my Origins

Contents

Acknowledgments

This book is the continuation of independent research conducted in pursuit of a PhD thesis at the Swiss Federal Institute of Technology (ETH) in Zürich, Switzerland. I therefore take great pleasure in thanking my thesis examiner Prof. Dr. Eberhard Ulich, then at the Institute for Work Psychology, who fostered my research with many fruitful discussions as well as with his continuous and expert support. Co-examiners Prof. Dr. Alessandro Birolini, then at the Swiss Reliability Laboratory at the ETH Zürich; Prof. Dr. Hans Dieter Seghezzi, then at the Institute of Technology Management in the Hochschule St. Gallen; and Prof. Dr. Paul Frauenfelder, then at the Institute of Management and Business Systems Engineering (BWI) in Zürich, also supported my thesis research in an (actually not) surprising convergence of fields. Many thanks for your comments and our discussions of reliability, quality management and technology management, respectively. I am deeply indebted to you for all your critical comments and suggestions that improved and strengthened this work.

Mr. Martin Landolt of the Dean's Office of the ETH Zürich is gratefully acknowledged for providing financial support, and Ms. Elisabeth Schniderlin helped unflaggingly with numerous solutions to general issues associated with my stay in Switzerland.

Prof. Dr. Wolfgang G. Weber, Mr. Cuno Künzler, both of the Institute for Work Psychology, and Prof. Dr. Rolf Kiechl of the University of Zürich contributed—especially in the first years of the research—with insight, productive discussions, and information. Mr. Markus Schmidt, Dr. Christina Hirsch, Mr. Toni Wäfler, and Mr. Jörg Bürgi, then at the Institute for Work Psychology, were always available for discussions and also contributed valuable information.

Dr. Olaf Böhme of IDEE-Suisse and Mr. Peter Schweizer of the CIM-Bildungszentrum Zentralschweiz CBZS generously allowed my participation in fora for discussions with practitioners from the Swiss technical management scene.

The ETH Library personnel, especially Mr. Robert Bühler, Dr. Karl Böhler, Mr. Hans Koch, Ms. Lone Imoberdorf, Ms. Christine Gross, Ms. Monica Bronner, Ms. Marita Barengo, Ms. Maria-Joerna Hodel, Ms. Maria Alice Schlatter, Ms. Elisabeth Lupart-Gasser, and Mr. Ivan Macukic, among many others, provided invaluable support not only by enabling access to difficult-to-find references, but also by their flexibility in providing a truly excellent library service. Your helpfulness reminds me of the words of Erasmus von Rotterdam, "My home is where my library is". You were not only customer-oriented; you made me feel at home.

With Mr. Marc Urben and Ms. Charlotte Fritz I had the pleasure of many interesting conversations about this fascinating field of psychology, with Ms. Talita Paula dos Santos Bonfils and Mr. Martin Landolt on juridical principles, and with Mr. Ivo Matthias on the problems of historical research.

Thanks to Dr. Ernst Berninger, then head of the Library of the Deutsches Museum in Munich, who contributed information on the history of technology, and to Prof. Dr. Yehuda Elkana of the Collegium Helveticum, who provided crucial insight into the philosophy of science.

My former supervisors Prof. Dr. Fernando Toshinori Sakane, Prof. Dr. Takashi Yoneyama of the Instituto Tecnológico de Aeronáutica—ITA, Mr. Eduardo Vieira Lima Magalhães Gondim, and Mr. Raul Wuo are also gratefully acknowledged for the many private conversations on research and development issues prior to the start of this research.

Many thanks as well as to Ms. Lilli-Marie Pavka, my tireless copyeditor, for her well-appreciated efforts.

And finally, to the first: Mr. Alfred Volkmer. This work is the result of a resonating, cyclical process that started concomitantly with the countless conversations on product implementation issues we held during the intervals between his lectures on digital circuit design back in 1981.

Foreword by Eberhard Ulich

The subject of Eginaldo Shizuo Kamata's book is the product development process, whose economic relevance on the international scale can no longer be overlooked, especially given its potential for providing a competitive edge. Development processes are to a large extent thought processes — some of them extremely demanding ones — the psychological aspects of which play an important role in this context. These aspects have generally been ignored far too long despite their often crucial importance for success.

With this work Dr. Kamata has succeeded in identifying such psychological parameters, in addition to providing a conceptional framework for their classification and showing their importance by means of case studies. The methodological approach is the result of an extensive investigative research, spanning the gathering of data from his own observations, interviews and questionnaires to the statistical meta analysis of the data, and ultimately to the search for, description and realization of actual case studies.

This last mentioned method is not less demanding than the other approaches listed, but it is less familiar in (natural) science and engineering disciplines. This is especially true when the analysis concerns problem cases and instances of failure, where it is still not common practice to publish them and to learn from them.

Also unusual are the scope and the discussion of historical development trends, which not only include Max Weber's theoretical analysis of bureaucracy but also Burn's and Stalker's distinction between mechanistic and organismic systems, as well as the socio-technical system concept derived from Kurt Lewin's work, and the psychological orientation of

Morton Deutsch in connection with the cooperation ansatz. Eginaldo Shizuo Kamata presents a breadth of knowledge that it is only rarely found in the literature on product development.

What sets this book apart is not only its particular methodological approach but also, and most prominently so, its wealth of facets and knowledge, which taken into account systematically are a substantial contribution towards improving the quality and speed of product development. Thus it is hoped that this book will not only receive acclaim in engineering as well as in work- and social-sciences communities but that it will be studied attentively and integrated into actual product development work.

Zurich, April 2002 Eberhard Ulich

Translated from German by Charlotte Bolliger

Summary

This book focuses on the interrelationship of social, technical, and organizational aspects of and related to the product development process. It originated from activities in practice in industry and research laboratories.

In order to ensure relative autonomy from the short-term economic interests of a given industrial branch or specific company, the research for this work was first conducted in pursuit of a PhD thesis intended to provide practice-oriented results. With the ansatz practice – theory – practice, a generalizable approach was achieved. It then evolved to cover additional issues brought forth by recent cases in the aerospace industry, among others.

A combination of scientific methodologies is used to focus on the psychological factors that influence the quality of technical product development processes. The basic framework is provided by the *grounded theory*—a qualitative approach, in which data was not only collected by this author but is extended to cases from the history of technology described in the published literature. The inclusion of historical cases is possible, thanks to the availability of sufficiently detailed descriptions for examination in terms of the grounded theory principles. They appear in the Appendix. Cases cited were verified by comparing them with contrasting viewpoints from various sources. I was able to examine and complement some cases by contacting persons associated with the programs analyzed. Quantitative studies are also integrated in this research to verify the elements resulting from the grounded theory integration. They also complement the framework by contributing elements that would be difficult to arrive at based solely on qualitative methods. Work psychology provides the main structure in which this research is evaluated, but elements from cognitive and social psychology are also very important for the integration.

A historical context is presented to provide an overview of the approaches that are and have been followed in technology management, together with a description of the evolution of the social science concepts relevant to product development issues. Moreover, a brief description of the pertinent scientific concepts is indispensable, owing to the direct application of the corresponding scientific methods in the field of engineering.

Several approaches to managing the quality of development, their potentials and limitations are integrated in the framework. Issues related to the individual, collective and leadership levels are examined. The *interaction theory* of leadership is applied by generalizing it to product development activities, particularly to the role of quality leadership. A discussion of the role of quality professionals is presented and a proposal of its application to existing quality approaches is made.

Seven research propositions formed the foundation of this work. This resulted in a broad investigation aimed at the possible factors influencing cooperation and participation among the diverse individuals and disciplines involved in product development. They are stated *a priori* for the research effort, but are also listed *a posteriori* for the reader in the chapter on General Conclusions. Interested readers, particularly from academia, might wish to study the propositions beforehand, in which case they are urged to become acquainted with the meta-scientific questioning of the research method in the Appendix, which provides the necessary scientific basis of this work.

The result is an interdisciplinary, problem-oriented framework for analyzing conditions under which technical product development could be improved. It also provides a reference for job design. Although focused on the aerospace industry, it remains generalizable for other engineering branches, especially for electronics and mechanical industries.

Chapter 1

QUALITY OF PRODUCT DEVELOPMENT

1.1 Introduction

It is commonly accepted that the product definition established in the course of a development phase influences the work and lives of many other people inside and outside of the enterprise. From marketing to product deployment, suppliers, customers and society are all strongly affected by the activities that take place during the product development phase. Schmelzer (1990, 39) estimates that 70% of the production costs have been determined by the time the product conception phase has been completed. Fleischer & Liker (1992, 254) cite estimates that "as much as 85% of total product costs are 'locked in' by the time the early concept is completed."

The costs associated with the development phase are one matter—whether the final product will be feasible is quite another. The way from a functioning prototype to a deployed functioning product is frequently very long. The often-heard success criterion "My design works!" does not do justice to the complexity of the issues involved. The "functioning" product might, for example, be difficult to manufacture, thus incurring high production costs; it might be impractical for the end user to operate, or it might function only under certain conditions or require a prohibitively expensive infrastructure ("Yes, it works, *but*...").

Often, significant resources are spent to rectify such issues *a posteriori*, with the typical consequence that considerable additional time is needed until product maturity is achieved. In many cases, it was found that after a functioning unit had successfully been produced, the development process had to be repeated at least partially in order to ensure the necessary product optimization.[1]

Generally speaking, it is difficult to define "quality of product development". Of the forty German companies in the electronics, aerospace and automobile branches surveyed by Specht & Schmelzer (1991), quality managers were asked to define the term "Quality of the Development". Only 26 of them proposed a definition. The results are summarized in *Table 1*.

Table 1. Definition of quality of the development as perceived by quality managers.

DEFINITION	WAS INDICATED BY
Fulfillment of customer requirements or related specification	31%
Observance of project-following measures that would ensure the achievement of these requirements	23%
Adherence to development standards and guidelines as well to project plans (general definition)	NA

The achievement of these goals is not as simple as it might appear. Indeed, it seems that the intricacies of the process cannot be assessed, thus limiting attempts to influence product quality from *outside* of the development group. This is not an entirely bad thing, because if those acting directly on the product development are conscious of the quality of their work, they already establish *internal* control loops to achieve this quality. How well they cover *all* relevant aspects related to the "world" *outside* of the development department demonstrates the need for a systematic approach. Here is where the quality department steps in. Hence, many quality assurance measures attempt to establish a link to this internal control loop or to establish intermediate *external* control loops. Such attempts usually end with very limited success, however, and many are questionable or ineffective overall.

It is known that one of the reasons problems are developed *into* products is that not all pertinent information is available to the persons working on the development process: facts they do not or could not know, facts that—although present—escape their *perception*. Another source of errors is a distorted perception of the available information. Owing to a lack of knowledge about what happens with the product in the field, or about how

---

1 The literature on Design for Manufacture (DFM) and Design for Testing (DFT) presents numerous cases of this kind. It is not the objective of this work to cover this aspect extensively. We will therefore examine selected cases of interest for discussion purposes.

the product is processed during manufacture, or even about how other developers are solving related problems, developers make certain assumptions, some of which may turn out to be wrong.

If comprehensive information were available *a priori*, it would be possible for developers to work independently, even isolated, as their internal quality awareness would drive them to optimize product characteristics. Unfortunately for the developers, this is not the case, and they and the development organization have to endeavor to obtain all the necessary information. Merely ensuring the generous flow of information is not sufficient either, because the fact that crucial information is missing does not necessarily mean that there is a scarcity of information available. Indeed, in order to cope with the normal flood of information or details, it is necessary to determine the relevance of incoming information. Those responsible for implementing a product have to decide what constitutes an important product characteristic and what could be omitted. If not done appropriately, this step can render the entire process too slow to develop (at least) one product for launching on the market in a reasonable time, or the result might be a product with quality deficiencies to be tolerated by the customer. Consequently, not every piece of information can be taken into consideration, so developers have to make certain assumptions and compromises. Coping with prioritization issues requires knowledge of the relative importance of a given piece of information to the persons involved. Whether developers perceive the problem as relevant or irrelevant obviously determines the solution proposed.

Box 1. Examples of different perceptions of the importance of product particularities

Take for example the case of the renowned M.I.T. Charles Draper Laboratory, which developed a space sextant for the Apollo spacecraft. The equipment worked well but could not be operated by an astronaut wearing a fully inflated space suit (Gray, 1992, 177).[2] In hindsight, it seemed obvious that an astronaut would have to operate the equipment wearing space gloves, but, for some unknown reason, the developers assumed that astronauts would not be wearing them inside the spacecraft. According to Gray, the development had to be partially redone. Whether this pertinent information was present in the specification is unknown, but the prototype did not accommodate this requirement.

Another case occurred in connection with the repairs to be done on the failed Hubble Space Telescope. Although the telescope was initially conceived to be (partially) repaired in space, the defective high-resolution spectrograph was not originally supposed to be replaceable. During the

[2] Excerpts from: ANGLE OF ATTACK by Mike Gray. © 1992 by Mike Gray. Used by permission of W. W. Norton & Company, Inc.

discussions on how to reestablish this feature, an engineering manager proposed (apparently in all seriousness) that one merely had to loosen seventy-two bolts (several of them epoxied), remove the equipment bulkhead, and jump-start a dead power supply (Chaisson, 1994, 343)—in an uncontained environment with zero gravity, performed by an astronaut wearing space gear, including gloves.[3]

Yet another example illustrates the tacit assumption that other people work in the same way as oneself (case #1).[4] In the early 1990s, a technician working on the CAD of a piece of aeronautical equipment included a feature in the product that he believed would facilitate production. He provided the assembly with a pin that served as a guide for the manual assembly operation. Manufacturing engineering personnel, however, were puzzled by the purpose of the pin, and considered it superfluous. Moreover, owing to production processes that were beyond the influence of the designers, the pin had to be specially fabricated, thus incurring additional costs. It was ultimately omitted from the product definition. The interesting aspect of this case was not only the erroneous perception of a potential problem and the subsequent initiative of the designer to propose a solution, but that it became an object of ridicule, including by some of the leading managers. The persons involved in the design were of course unhappy about this treatment because—after all—they had merely been trying to anticipate the difficulties potentially encountered by production personnel. Encouraging and channeling employee initiative of this is a nontrivial task that can depend on many issues examined in this book. The managers in this case obviously did not handle the situation in an optimal manner and risked discouraging the designers. After a mistake of this kind has been made, as we shall see, it takes several process cycles to rekindle this kind of employee initiative and reestablish it as part of a methodical solution to development problems. These process cycles will be described below in the section on achieving cooperation.

In the past, it was assumed that one way to solve such problems was to repeat at least portions of the development process. Hence one had for example an engineering prototype, a manufacturing prototype, a pre-series production, etc. This repetition was actually part of an iterative process that converged toward product maturity. First the development department worked out the product definition, then it was handed over to production. In

---

3 In all fairness, it should be mentioned that a brainstorming situation precludes any suggestion being rejected. Any conceivable possibility, no matter how "stupid" it might seem—perhaps even to the engineer who proposed it—should not be assumed invalid from the outset ("Why not?").

4 For details of this sequential case numbering, please refer to section on Notation, page 30.

this phase, any problems that arose could be addressed and a non-negligible amount of redesign was necessary. After that, production commenced and the development department was no longer involved, at least in theory. In fortunate cases, inherent product flaws and weaknesses could be discovered and eliminated or, in cases where elimination was not possible, at least compensated for. In less fortunate cases, a product that was not conceived for the conditions under which the customer needed it, or one that did not lend itself to efficient production or testing, or one that had inherent non-production-related development flaws was launched on the market, sometimes with catastrophic consequences for the user as well as the retailer. As product complexity has increased and development times shortened by market forces, the successive iterations necessary to bring a product to maturity can no longer be performed in a casual manner. If an iterative process of this kind does not produce a defect-free product, users are exposed to these flaws, including the attendant risks and consequences.

Quality and reliability organizations worldwide have made many attempts to overcome these issues from inside a company or economic sector, not only in the aerospace industry. At the same time, customers' perception of product quality has become more acute, and their acceptance of defects has diminished, factors that have clearly influenced the evolution of quality (Seghezzi, 1992b, factor #2).

The perception of the need to improve the quality of a product development process is not at all new. The following list shows a few of the approaches taken:

- network planning techniques (such as PERT/CPM),
- quality systems (such as ISO 9000 or its predecessors),
- centralization (e.g. for enhanced efficiency or economies of scale),
- decentralization (e.g. for agility and enhanced job satisfaction),
- improved communication,
- automation / simulation / visualization (such as CAD/CAE),
- simultaneous engineering,
- design checklists,
- freedom, distance, a "stress-free" environment,
- rapid prototyping,
- 3-D visualization rooms.

All these approaches have certain advantages because they all emerged in response to a given (e.g. economic) situation. Unfortunately, little is known about the context in which these approaches were applied, although much has been published about their content. This missing contextual information is actually the essence of what makes a method or technique work. As a consequence, many a "success story" has too often led to an attempt to reproduce the results without first examining whether the application makes sense in the new context. Clearly, there are many context-dependent factors that influence product development. Not surprisingly, the inherent flaws and weaknesses of these techniques remain unknown. The result is that in some situations, the disadvantages of their application outweigh their advantages, causing those "promising" results to remain elusive or even leading to a complete project failure. The most fortunate cases may well achieve partial success and hence appear to justify the investment made, thus allowing their creators to save face before their superiors or maybe even save their careers.

We can therefore pose the question as follows: Why is it so difficult to manage quality in product development? The problems already emerge at the specification phase. Imagine the following (true) story:

Box 2. The boy, the wine label, and the managing engineer

After many months of negotiations, the contract for the development of new airborne equipment was signed. All those involved went to celebrate with the partner company because, after all, this product could mean the future of both companies, and it was no easy task to harmonize the numerous product specifications, the critical engineering challenges necessary to achieve them, the responsibilities and work share, technology transfer, training, legal and financial aspects, and many other things... Now, imagine the festive meal in a small village restaurant known only by insiders and, of course, the fine wine that the very special occasion demanded. One of the managing engineers wanted to keep the label of this particular bottle for his collection, of which his colleagues had no previous knowledge, and so he asked a young boy who was helping in the restaurant: "Could you please remove this label for me?" Promptly responding to this request, the boy returned after a while with a thoroughly cleaned bottle, the cleanest bottle the engineer ever saw. "Where is the label?" the engineer asked. Surprised by the question, the helpful boy answered: "I rubbed it off, sir!" One of the man's colleagues, a systems engineer, then observed: "Well, you *do* have to be precise about exactly what you specify." So although the specification was 100% fulfilled, it did not produce the desired results...

This anecdote illustrates that an unambiguous specification could have avoided customer dissatisfaction. The particularly ironic aspect of this story is the fact that the customer himself had defined the specification. However, although it is true that a good, clearly stated specification contributes to product success, it should not be forgotten that a "perfect" specification is not always achievable, or would require too long an analysis or description to be feasible. In addition, one can seldom establish a comprehensive specification at the outset and expect it never to change. Some engineers believe that once established, a specification can proceed directly to the product implementation phase. They apparently feel that the subsequent steps are simply a rational, technical activity: the kind of thing that engineers are trained for. But this is not necessarily true, especially when complexity is added to accommodate the dynamic content of usage requirements, which may well change due to any number of technical reasons, mission revisions or usage environment changes.

The need for a detailed specification is confirmed, for example, by the data of McGrath, Anthony & Shapiro (1992, 98). They found in a survey[5] of product development that 42% of the total work performed is repeated because of an upstream change such as late customer input, erroneous specifications, or something that had been overlooked. Care has to be taken in interpreting this number. To avoid the prospect of changes being introduced in a late stage of development, which often causes the developers to lose focus or lengthens the completion time, some managers try to minimize this *number* by freezing such factors as specification parameters or even by closing down communication channels in an effort to avoid changes at all costs, regardless of whether they are necessary or not.

The importance of optimizing the development process in terms of achieving an equilibrium of the parameters quality—deliverability—costs or quality—time—costs, is increasing also in sectors other than the aerospace industry. The basic idea is that there is mutual tension among the three constraints within which product optimization occurs. This joint optimization is not achieved by specifications alone. Rather than overspecifying a product, one would be well advised to improve personnel interaction in order to determine a product's weaknesses before it is too late, or in order to be agile enough to respond to necessary changes, even last-minute ones. This optimizes the tension triangle instead of acting on the number. Ignoring the nature of task repetition may hinder the process that actually leads to product optimization. Taken to the extreme, this might yield the specified product, but the product success on the market may quite probably be hampered, if not jeopardized.

We have now strayed from the case of the wine label, but the connection will become clear below.

The following statement by Mason (1986, 32) concerning the development of the European Tornado Multi-Role Combat aircraft confirms the observation made above: "There is a tenet, subscribed to by many, which holds that if the duty and performance of an aircraft are defined by specification, only one ideal aircraft can result. Such a design is, of course,

5 No precise details about this survey could be found. The authors mention only that they "surveyed hundreds of people involved in product development at many companies spanning a broad range of industry" during the six years preceding their publication (op. cit., p. 97).

unattainable owing to cost and on-going research, which in the former instance conspires to limit the ultimate efficiency and in the latter introduces timescale definition. Therefore 'state of the art' and timescale parameters are vital criteria in the design". He goes on to say that "it must be emphasized that in the normal course of informal intercourse between members of design and research staffs of several European (and American) companies and government agencies the subject of the 'state of the art' had been under constant discussion, with general reference to the results of theoretical research, while the presentation of papers (commonly referred to as 'transactions') to the professional bodies ensured constant dissemination of ideas between engineers and design staffs." This situation still exists today.[6]

This optimization process is highly dependent on the quality of the interaction among the people working to fulfill the specification. In addition, the implementation of a product for a dynamically evolving market or for changing use situations is described by many practitioners as "aiming at a moving target" and should not be regarded as a blind execution of an *a priori* stated specification. Swiss entrepreneur Branco Weiss stated in 1990 that there is a tendency to place excessive emphasis on establishing the specification[i] instead of on the implementation of the product. In this context it should be kept in mind that it is better to leave room to evaluate, establish and modify a substantial part of a specification during the evolution of the activity of development.

In addition, unexpected problems as well as the interaction of individual parts might require a reassessment of the detailed specification. Such a case calls for flexibility. Clearly, the specifications of product characteristics should be set *as broadly as possible*. The key issues that contribute to accomplishing a customer's mission specification or fulfilling market needs are specifications that are *as detailed as necessary*, or alternatively, specifications that differentiate between *static content* and *dynamic content*. The establishment of product specification will not be covered in detail in this work.

Let us now turn to the implementation. Some people mistakenly argue that if the product to be developed is adequately specified, the only task left is to implement it, which is *merely* a rational, technical activity of solving technical problems, and which proceeds in a straightforward manner to product delivery.[7] But this is not necessarily true. Although product

---

6 The Tornado Conceptual Definition took place in 1968.

7 During the course of the research underlying this work, a technically-oriented engineer stated to me in an informal conversation that he did not see what the problem was that was being addressed by this research. In engineering, he said, problems are there to be solved and we (the engineers) solve them! (German original text: *"Die Probleme sind zum lösen da, und wir lösen sie!"*). This engineer had complained only moments before about how

development is essentially an activity of technical nature, the many non-technical factors that affect its performance should no longer be neglected.

The study by the American Association of Engineering Societies to determine the time utilization of engineers disclosed that the majority of their time goes toward documentation, planning, and communication activities (Morton 1987, 19). The following quip by Wernher von Braun supports these findings: "We can lick gravity, but the paperwork is overwhelming". Moreover, the above-mentioned dynamic content of specifications and the tension triangle enhance the need to consider such non-technical aspects of development as group interaction.

According to a report by A. D. Little, the R&D management of the western world concentrates on the technical competence of the individual for decisions regarding recruitment and professional advancement, and places less importance on one's personality and character. The complementarity of talents and personalities necessary for the optimal execution of tasks has not been adequately appreciated (Saad, Bohlin & van Oene, 1992, 218). Kolleck (1995, 47) found that the proportion of quality problems attributable to technical and organizational causes amounts to only 20 – 30%, whereas the remaining 70 – 80% are caused by human factors. Schmelzer (1990, 48) in turn attributed more than 50% of product development problems to behavior issues and not to technical factors. Neither of them provide further details concerning the industrial sectors from which the data was derived, nor about the population surveyed, but it is safe to assume that they culled their findings from their vast experience in and profound familiarity with this field.[8] In any case, their figures are indicative of the existence and perception of the importance of the problem.

Box 3. Effect of the separation of activities between disciplines or departments (case #2)

Another problem lies in the separation of activities or disciplines. Not seldom, variations of the statement, "I develop products—it does not concern me how to produce them" could be heard worldwide. One of these variations was particularly bewildering: When asked whether there were any built-in self-test capabilities of the product his group was developing, an (indeed very brilliant) development engineer declared he gave it no thought

difficult it was to work with his colleagues and that he ultimately had to do everything himself. So technical issues are apparently not the only difficulty encountered in the course of solving a technical problem after all...

8 At the time of their publication, Mr. Schmelzer was with Siemens AG in Munich. Mr. Kolleck has many years of experience in process technology in the chemical industry, especially regarding quality in plant construction (German: *Anlagebau*). He has also served as auditor of various companies in different industries and was active in the German Society for Quality (DGQ, *Deutsche Gesellschaft für Qualität*). Their present activities are not known to this author.

because that was the task of the quality department in charge of testability issues, as if the quality personnel would come at the end of the development and add a magic box that would render the design self-testable. His director subsequently impressed upon him that if he did not develop it into the product, thus *building* it *in*, who would? (early 1990s).

The arbitrary boundaries drawn around disciplines and areas of competence play a major role in this phenomenon. As a good development engineer, the man in the case above knew how to test *one* unit in the development lab with staff who are quite familiar with every module to a degree of detail that nobody else in the world could match. But could the people in the manufacturing hall test the item to the extent required to ensure the proper functioning of *all* units? Can a pilot sitting under the canopy of a waiting aircraft under the summer sun test it to decide whether the equipment is GO/NO GO? Can the maintenance people readily determine which unit to replace when impatient passengers are waiting inside the aircraft for take-off? Can mission control know whether the box is still working in low earth orbit?

Diverging perspectives and tendencies among professionals performing different activities or working in different disciplines can also hinder cooperation across departments for various purposes. Ziemke & Spann (1991, 48) state, for example, that a marketing department may value unique product design features, whereas the accounting office may value control. Wiebecke, Tschirky & Ulich (1987, cited in Ulich 2001, 507)[9] investigated cultural differences between R&D and marketing professionals and found the differences presented in Box 4.

Box 4. Cultural differences between R&D and marketing professionals according to Wiebecke, Tschirky & Ulich (1987)[10]

Relationship to the environment

"R&D and marketing have different views about the relationship of the whole organization to the environment. R&D considers the technological and scientific relationship to the environment as crucial: The scientific and technical quality of its products justifies the existence of the whole firm, and the provision of technically useful products to the environment as the fundamental task of the organization ... Marketing however regards the

[9] Excerpts from: Ulich, E. (2001). Arbeitspsychologie (5. Aufl.), Zürich: vdf; Stuttgart: Schäffer Poeschel. Reproduced here by kind permission of vdf Hochschulverlag AG an der ETH Zürich and the author.

[10] From: Wiebecke, G.; Tschirky, H. & Ulich, E. (1987). Cultural Differences at the R&D-Marketing Interface: Explaining Interdivisional Communication Barriers, Proceedings of the 1987 IEEE-Conference on Management and Technology, Atlanta, pp. 94-102. © 1987 IEEE. Reproduced here by kind permission of the IEEE.

firm's role in the economical environment as most important: By supplying products that suit the market demand the financial means are obtained, and the organization survives through its commercial activities ... "

Time scales

"Marketing has a shorter time perspective than R&D, is today-oriented and focusing on the rapidly changing markets. Cooperating with R&D, marketing people are impatient and do not appreciate the long development time. Instead they change their opinion and lose enthusiasm for projects during development. On the other hand, R&D has a long term perspective and has to do long-term anticipation into the future. Interacting with marketing, scientists are perceived as lacking the sense of urgency and not caring about delays. They give the impression of working at a leisurely pace at projects they always want to continue, believing that everything will one day work out."

Nature of reality

"R&D and marketing use different concepts of truth to determine the reality as a basis for decisions. On one hand, R&D uses experiments to determine the physical reality, resulting in the empirical proving or falsification of assumptions. The problems can be clearly defined, and to confirm the results extensive data collection is necessary. Marketing on the other hand is concerned with determining social reality. It deals with societal differences and changes, and has to cope with inseparable factors which can't be easily measured. In marketing's social reality, the value of information relates to the circumstances, and truth is determined by consensus."

Another diverging perspective is that of production engineers, who value the ease of production and inherent low variability of the design under production. Other designers may value an elegant solution, even if it poses certain difficulties to the production. Project controllers may attach utmost importance to the budget and schedule, and so on. All these differences not only hamper communication, they also render the focusing process difficult, especially if one person or group tends to impose his or their view on the others. It is a true challenge to integrate all these diverging perspectives into one picture.

In a study performed by Arthur D. Little (ADL) , Saad, Bohlin & van Oene (1992) state that, among other things,

- there are differences between the culture and education of researchers and those of business people,
- there is an absence of a common vocabulary,
- managers with technical training have a poor grasp of business management,

– managers with a nontechnical education do not appreciate the functioning of R&D and the meandering course of science and technology.

Some managers and R&D practitioners describe it as a modern Tower of Babel. Similar problems plague interdisciplinary research, as described by Perez (1995, 5): "Those who deal with 'interdisciplinarity' are—in all likelihood—occasionally tormented by the question of whether they have opened a Pandora's box".

Quality of development also runs in two distinct directions: On the one hand it comprises overall product characteristics reflected in the general product concept, the physical/chemical effect explored, the appropriateness of the algorithms used, the system architecture, general handling/operation characteristics, and overall costs, among other things. On the other hand it includes such details as the idiosyncrasies of a given part, the use of an electronic component or material, the particularities of the physical/chemical processes chosen, the implementation of the algorithms, a particular use situation, local costs, and much more. Therefore one can have, for example, a very good fly-by-wire architecture that prolongs aircraft life, and at the same time a computer that refuses to apply the brakes because it still detects lift on the wings, causing the runway to be overrun (Airbus A320-211 Serial No. 105 in Warsaw, 1993). One can have very nice winter handling characteristics of an aircraft but a latch-opening button that is not designed to be operated with winter gloves, causing a mechanic's ungloved finger to stick to the surface at extremely low temperatures, conceivably resulting in a considerable departure delay before the mechanic can be freed without injury (one of the cases influencing the design of the Boeing 757/767).

Alternatively, one can include design features that, from the outset of product development, could solvc a myriad of minor issues encountered when working with the product. In a low-cost program, one can specify cheaper materials that result in more scrap or one can chose a more expensive material that is easier to work with, thus yielding less scrap and hence reducing overall costs (V-22 Osprey cost reduction for mass production). One can have a satellite attitude control that includes redundancy, which turns out to be very useful in case of incorrect orbit injection by depleting almost all fuel for orbit correction but still having the means to stabilize and operate the satellite (Artemis). Or the system designer can consider using aerobraking in case of propulsion failure (Mars Express). Alternatively, one can eliminate entry, descent and landing telemetry data of a (low-cost) planetary probe in a program that was also meant as a learning experience for future missions (Mars Polar Lander). MPL used a novel pulse mode control of the descent engines.[11]

[11] Even if it did work, one cannot be sure it would work again.

Thus let us assume that a development organization is capable of achieving a functioning product, such as an aircraft that flies (even with suboptimal handling qualities), a car that runs (even if it overturns occasionally), a satellite that orbits the earth (even if on the wrong path), scientific space instruments that function well on earth (but not in the space environment, or cannot withstand launch), a planetary probe that is successfully deployed (but has a limited capacity to communicate its data with humankind), then the problem of development quality turns into: How good are its flight handling characteristics? How well does it handle cargo? How easily can you detect failures or defects and restore correct functioning? How fast can routine maintenance operations be performed? How well does it handle difficult traffic situations? How well is the suspension trimmed? How well can you perform daily tasks with it? Hence, the following two contrasting interpretations of the nature of details become valid: an English proverb claims that "The divine is in the details", whereas a German proverb holds that "The devil hides in the details". Although a thorough analysis of the differences between these two proverbs is beyond the scope of this book, let us assume that both are correct, depending on one's perspective—the choice is left to the reader.

Concerning autonomy, practitioners do not agree on what constitutes "best practice" to achieve development quality. Should a development group be granted extensive freedom to create or should they be kept on tight reins? Should a centralized or a decentralized structure be used? Should one emphasize control or initiative and stimulus? Should specific, detailed tasks with a full description of the intermediate steps be given to development personnel or would it be better to state a broad goal and a general direction to follow and let them find their own way? Some managers feel that one should not leave "undisciplined engineers" to do as they wish, whereas others claim that one has to leave people alone and let them do their work because, as experts, they know better than any manager what it is to be done. There are successful managers who argue that they have to put pressure on development staff to get them to work; at the same time unsuccessful cases abound of the negative effects of such pressure on cognitive abilities. Take for example the case of a development executive at Honda, who said: "It is like putting the team members on the second floor, removing the ladder, and telling them to jump or else. I believe creativity is born by pushing people against the wall and pressuring them almost to the extreme" (Takeuchi & Nonaka, 1986, 139). In contrast, Noyce (1987, 23) of the equally successful company Intel has declared "We can't stand over people with a whip and tell them, 'Create, or else.' It just won't work. The process of innovation depends on our ability to give our innovative people breathing room and leave them alone so that their ideas can eventually bear fruit." Blindly following either of these two *apparently* contradictory statements without evaluating their detailed characteristics would be misleading and perhaps even dangerous, as we shall see below. There have been companies that tried to cultivate their R&D by creating a "*lab-in-the-woods*" removed from daily stress. They eventually realized that they had to discontinue this somewhat utopian-sounding activity because employees were completely isolated from what was going on in the rest of the enterprise and, even worse, were not achieving the desired results. On the other hand, successful cases have been reported where isolation from the usual practices, culture and pressures was the main factor of a development's success.

Other companies worry that their R&D is not innovative enough, their managers claim that personnel is already overloaded, that they could not squeeze more out of their personnel. At the same time such companies often refuse to recognize that in reality their personnel complains about having insufficient time to work on new ideas.

Some companies believe that computer technology is the recipe for quality of development; others proclaim that human resources are the key to success. Every approach has its advantages as well as its disadvantages, its potentials, and its limitations.

Motivation is also a key factor for improving development quality. Juran (1992, 426-427) listed the following actions as being necessary for quality planning:

- bring planning for quality into the strategic business plan,
- train the planners in quality-related methodology,
- provide participation to all who are impacted,
- provide a structured approach to quality planning.

Moreover, he mentions that, during the 1980s when many companies tried to carry out this "list of actions [they] encountered various obstacles, some of which could be avoided through motivation". These obstacles are listed in *Table 2*.

Companies reported such issues as (i) task-related conflicts, (ii) interpersonal disharmony and rivalry, which caused team efforts to deteriorate, which in turn impeded decision-making, and (iii) unproductivity and/or a waste of personal energies. Some managers actually tried to intensify conflicts by reasoning that this would "set incredible creative energies free". Conversations with some of these managers revealed that they apparently considered it a good idea to "rock the boat" with no regard for the type of personalities involved, nor for the timing of a given conflict, nor—much more gravely—with any thought to a conflict-solving strategy.

Table 2. Some obstacles to quality planning according to Juran.

	DESCRIPTION
Unawareness	People are not aware that they are creating quality problems.
Competition in priorities	People are unaware of the need to give top priority to quality because other goals actually have higher priority, and hence get in the way.
Sub-optimization	Achievement of quality locally gets in the way of overall quality.
"Here comes another one"	Prior unsuccessful drives [can] result in a cynical reaction to any new initiative, including an effort to give top priority to quality.
Managerial myths	Such as "Quality will get top priority if the upper managers so decree".

1.1.1 Quality, soft factors and the aerospace industry

1.1.1.1 Paperwork reduction

The once-innovative aerospace sector that set world standards for high quality and reliability is changing.

One of the changes requires that the need for paperwork be reevaluated. At a workshop on ISO 9000 in the early 1990s, a member of a certification institution described to an audience of non-aerospace professionals an audit of a company that kept records for all the screws they used in aircraft altimeters. Their customer required such documentation. The auditors could not say exactly why, but the customer requirement was there, and it had to be fulfilled. This apparently justified the existence of the extensive records. Indeed, meeting this requirement was included in the price tag (or costs, if the reader prefers).

A joke that circulated during the Apollo moon program stated that when the paperwork reached the height (or weight) of the stack, the rocket would be ready to launch. Another laconic musing held that archeologists in future centuries who excavate sites of the aerospace and defense industry will conclude that a paper factory existed, and that the factory was so successful in producing paper that the people built aircraft just to transport it. Other variations were that it was so successful, they could even build rockets in their spare time, or that they made their own anti-aircraft cannons to protect it, or that they built their own armored convoy trucks to transport it, etc. Jokes aside, considerable paperwork resulted from such necessities as traceability requirements[12] and it was a means for coping with quality-related

[12] For example, one must know the exact configuration of an item that failed (for space: in that fireball), or to repeat a success that was unretrievable (for space: then in orbit), or to

issues. However, in an attempt to reduce the problems associated with mounting quantities of paper, the tendency has been observed that paperwork is readily eliminated with the simplistic argument that "paper does not fly". But when the hardware in fact does not fly either or demonstrates anomalous behavior, then the same persons who labeled paperwork a "waste" are often quite ironically the first to search for the non-flying documentation and cling to it as to a lifesaver. The solution to the problem posed by paper (or whatever support media the information is stored on) requires that the reasons for producing such documents be reexamined, i.e. whether the *activity* behind it is in fact essential, the paper being merely a result. In other words, one has to cure the disease, not treat the symptom. A complementary observation made by McCurdy (2001, 94) confirms the tendency to emphasize paperwork after failure in the cases of the Clark and Mars Climate Orbiter satellites, where "outside investigators typically recommend tighter oversight or better paperwork trails."

As trivial as it might sound, this is not obvious. In a recent low-cost space program, a project manager was budgeting quality activities with respect to materials and mechanical parts as comprising nothing more than what he called the elaboration of the declared materials and mechanical parts *list.*[13] What he explicitly meant was the trivial task of typing the list into the computer, printing the document, and then formally releasing it. He considered only the output of the WBE, in total disregard of the activities implied by the items on the list. These *activities* lead to the assurance that the materials are appropriate for the given space application, not the list itself, of course. So it becomes clear that omitting the paper (which does not fly) does not omit the activities it represents (leading to a good flying object). Those tasks would have to be performed elsewhere in the process, or distributed among various functions. Attempting to solve the paperwork problem in such a simplistic manner only creates a time bomb—unaddressed issues have the potential to significantly reduce confidence in the final product. Companies should refrain from "throwing the baby out with the bath" in an effort to reduce the costs of such mundane activities as paperwork.

1.1.1.2 Myth of the cool, rational engineer

Other changes require critical scrutiny of any technical activity that generates the desired products. Due in part to a bias introduced by basic

---

investigate how a failure could affect subsequent launches or missions, or to command recovery sequences from earth, or to guarantee that the part used was a part with sufficient quality for the intended purpose.

13 For non-space readers, the document "Declared Materials and Parts List" is a form of reporting to the prime contractor and the customer the results of the activities of materials and parts compliance to the space environment.

educational systems, there is the myth that the activity of product development is solely a technical venture, determined only by calculations and rational argumentation.

The awareness of product development as an activity strongly influenced by social processes, not only by the "hard" sciences, is growing in all sectors of the technical community as stated by Weyrich[14] (1998, 73): "In order to fulfill the demands for interdisciplinarity and innovation placed on future Research and Development (R&D) personnel, non-technical subjects as well as teamwork skills and risk affinity should be included in basic professional training. The personality profile of engineers and scientists should be cultivated not only by means of early transfer of responsibility but by exposing them to challenges."[ii]

Engineering students are already feeling the influence of such so-called soft factors. The topic was discussed intensively during a workshop on "Success factors in cooperative ventures" during the International Juniors Forum at the 42nd Salon International de l'Aéronautique et de l'Espace, in Le Bourget, Paris, 19 June 1997 conducted by the then Daimler-Benz Aerospace AG. There was a very noticeable tendency among the German-speaking students to acknowledge that soft factors—not technical competence—are the major determinant to success.

This trend has a longer history, as described in the early 1990s in the study conducted by Arthur D. Little (ADL) (Saad, Roussel & Tiby, 1991, especially pages 161-168): The third-generation management of R&D defines itself as an activity to integrate R&D personnel into the corporate culture by allowing them, on their own initiative, to develop their own identification with the company goals or, better yet, by actively taking part in the evolution of these goals.

Breaking the isolation of R&D within a company is not easy to accomplish, for it has cultural roots in such things as different perspectives, established ways of thinking and performing, diverging views of other functions and even starkly differing technical jargons. In the past, this isolation was taken for granted. Saad, Roussel & Tiby, (1991) say that the wise management of R&D personnel requires that they not be characterized by their specific technical-scientific knowledge but primarily by their individual personality and motivational profiles. The removal of this isolation barrier has not been easy. As recently as 1991, recruitment was still conducted by discipline. A semiofficial hierarchy of technical functions prevailed, as did a stereotypical image of R&D personnel.

---

[14] Mr. Weyrich is with the Central Division for Technology at Siemens AG, and is a member of the board.

Since the publication of the ADL study, much progress has been made, but there is still a long way to go before far-reaching changes will have been achieved in the workforce.

The aerospace industry, like every other technical branch in the western world, must step up to these challenges.

1.1.1.3 Aerospace in socio-technical crisis?

The study on *People Management in Aerospace* conducted by the Society of British Aerospace Companies (SBAC) in 1997 (Rivers & Thompson, 1998) surveyed more than 350 establishments. As shown by the selected results in *Table 3*, a strong awareness of the importance of human nature in solving technical problems was found.

Table 3. SBAC study on people management in aerospace (selected results).

HUMAN RESOURCE PRIORITIES	PERCENTAGE OF ESTABLISHMENTS DECLARING PRIORITY AS BEING 'VITAL' OR 'VERY IMPORTANT'
Raising employee skills	>80%
Encouraging innovation	>80%
Increasing employee flexibility	>80%
Improving understanding of business	>60%
Empowering workforce	~60%

The study was conducted to disclose the companies' attitudes toward personnel strategy, not what they actually do in practice (op. cit., p.5). Nevertheless it also assessed the adoption of innovative working practices for more than two-thirds of non-management employees, see *Table 4*. The authors of the survey regard the penetration of the analyzed innovative practices as "relatively high" but, considering the potential effects of the practices mentioned, the numbers suggest a very broad possibility for improvement.

Table 4. Innovative working practices for more than 2/3 of non-management employees in the SBAC study (selected results).

WORKING PRACTICE	PERCENTAGE OF ESTABLISHMENTS
Responsibility for own work quality	53%
Formal teamworking	25%
Self-managed work teams	~15%
Problem solving groups	~12%
Job rotation	~5%

Another finding of the study is the slow adoption of certain innovative working practices, such as: "two-thirds of establishments have no form of team or group-based rewards, despite that nearly three-quarters have formal work team arrangements" (op. cit., p.6).

The study was not limited to product development. Recall that product development requires involvement of other areas, and as such, it would also benefit from the company-wide successful implementation of better personnel strategies.

Another interesting finding is that "nearly four in ten establishments do not think that they have sufficient managerial talent for their future business needs" (op. cit., p.12) and three key management development challenges were identified as "vital" or "very important".

Table 5. Key management development challenges identified by the SBAC study (selected results).

MANAGEMENT CHALLENGE	PERCENTAGE OF ESTABLISHMENTS
Improving communication skills	76%
Developing leadership skills	75%
Developing change management skills	60%

Also surveyed in this study was the perception of skill shortage as being "almost exclusively in the engineering area", especially CNC machinists, software engineers, skilled technicians and design engineers (op. cit., p. 8).

During the same time period, but on the other side of the world, a series of articles published in *Aviation Week and Space Technology* pointed out problems in the American aerospace industry associated with "people issues" (Scott, 1999d). In particular, they cited the focus on management incentives (Scott, 1999b) and called attention to the obsolescence of the hire-and-fire personnel policy (Scott, 1999a). The articles were published in the June 21 issue. Readers' reactions were immediate, prompting explicit mention in the Editorial of the July 5 issue. Evidently, several high executives denied the "crisis". This was noted in turn in the Nov. 15 editorial, which also mentioned one CEO who "did acknowledge that the stories were on target" but asked AW&ST "to stop printing them". The articles were based on a five-year multidisciplinary study of management practices and the environment faced by aerospace companies (Scott, 1999d, 65).

At the same time, an independent review team focusing on Lockheed Martin and the American launch failures of the preceding two years revealed that (i) too much emphasis was being placed on costs at the expense of mission success, (ii) there were weaknesses with regard to accountability of individual managers and engineers, (iii) there was a need for better management of subcontractors and suppliers, as well as (iv) a need for workforce quality enhancement (Scott, 1999c, published Sept 13). Furthermore, the problem of loss of expertise was addressed by Scott (2000,

March 13) in connection with the "crisis" discussed above, as well as by Velocci (2000a, April 24), who confirms the prevalence of such "denial" behavior. In a speech at a U.S. Space Foundation National Space Symposium, J. Albaugh (2000), president of Boeing Space and Communications Group, stated: "If aerospace companies are to compete with the dot.coms for people, capital and investment, we can't do using the old business paradigm...I'm particularly concerned about our ability to attract and retain the best people". Velocci (2000b), also in connection with the loss of "intellectual capital", quotes M. Clark's—vice-president of human resources of Rockwell Collins—explanation of improving management training: "We want to minimize the chance that good engineers will leave because they feel they may have been treated unfairly". Dornheim (2000, Sep. 25) states that "recruiting and retaining engineering talent, particularly in the software field, remains a concern in many, but not all, areas of the space industry."

This mosaic of articles concerning the situation in the American space industry may raise questions whether it is perhaps caused by factors that are not typical in other countries (e.g. the common hire-and-fire policy), or whether it is a result of globally adopted organizational recipes. But regardless of the competition for technical talent posed by the so-called "dot.coms", and regardless of whether a "crisis" can justifiably be denied, action is clearly required in the mid and long-term—not a "quick fix". A panel of aerospace CEOs[15] unanimously agreed that "to attract and retain skilled employees, the aerospace/defense industry must fix basic problems that kill motivation and taint perceptions" (Scott, 2001, 33).

These factors mean a potentially high strategic advantage for companies that do not merely want to mitigate the difficulties of recruiting. Recruiting bottlenecks may prevail periodically, but alleviating them alone would not avoid the reoccurrence of the problem. What is needed is an investment in the intrinsic and extrinsic characteristics of the jobs offered.

Returning to the SBAC study, a second survey performed in November of 1999 revealed similar findings that there were "not enough suitable skilled people" and a "lack of practical skills" compared to the situation surveyed in the 1997 study. Moreover, the confidence in the current skill level to meet future business needs had fallen considerably: from 64% ("very confident" or "quite confident") to 36% (Thompson, 2000, 16). It seems that companies in the U.K. aerospace industry are becoming increasingly aware of the changes necessary to accomplish these goals and that merely raising

---

[15] Presumably at the 17th National Space Symposium, 19-12 April 2001, Colorado Springs. The source is not completely clear from the article by Mr. Scott.

technical competence will not be sufficient to cope with the challenges faced.

As the study states: "Failure to create workplaces that can retain these valued employees could be very costly in the long run." (op. cit., p. 17). The second survey included 90 establishments that had also been respondents in the 1997 study. In this group a statistically significant increase was found in the use of the human relations practices mentioned above. The study authors maintain that "high performance work systems bring with them the need for greater collaborative working in teams, more information sharing and greater adaptability to deal with change and uncertainty in production [and therefore would expect] more effective firms to shift the balance of their training strategies from job-based to behavioural-based training" (op. cit., p. 19).

These non-technical factors should be considered when designing jobs and the organizations that encompass them, at least for the sake of resource efficiency. Although it is to some extent understandable that "aerospace in crisis" is sometimes regarded a taboo topic, the phenomena it describes are not new in technical companies throughout the western world. It is not wise to deny the existence of the problem, especially when it could still be manageable, including by the workforce itself.

The SBAC study reveals primarily the perception of the problem and the trends within British industry to cope with it. It was not conceived to assess how well companies are coping with the perceived problem. It is assumed that neither the activities proposed by the SBAC nor the recovery from the "crisis" will happen overnight or even within a couple of years—a quick perusal of the tables published in the SBAC study confirms this statement. It requires the mid- and long-term consideration of numerous factors in order to cure many of the "illnesses" or, if the reader prefers, alleviate many "defects" currently inherent particularly in technically oriented companies.

1.1.2 Objective of this book

The findings of the SBAC study, the perception of "aerospace in crisis", and the results of research supporting this work (initially released in 1998) converge in the present book. It is intended as a discussion of selected concepts and will attempt to provide some answers to meet the challenge of (re)considering the significant, non-technical aspects of technical activities in and around product development, including aerospace products.

1.2 Information for the reader

1.2.1 Learning from failures

Companies and laboratories throughout the world have been trying for many decades to improve the quality of product development processes. Especially in the second half of the previous century, many techniques and methods were developed in an attempt to ensure the success of activities that are directly and indirectly related to the definition of product characteristics and features.

Most methods described in the product development management showcase success stories and/or highlight the positive experiences of the authors as practitioners or researchers in the field. This of course seems only natural, as one understandably wants to emulate the winners. However, the repeatability of success is not straightforward. Nevertheless, methods such as project scheduling, resource planning, quality systems, design reviews, and information technologies, particularly in the context of simultaneous engineering or computer-supported cooperative work, have been introduced to varying degrees of success. Although these methods may certainly provide solutions to *some* of the problems encountered in product development, practitioners today face many issues not dealt with in textbooks, yet they often attempt to apply textbook "recipes" and have a greatly exaggerated faith in the methods established in the conventional literature.

Another problem associated with the difficulty of reproducing a success story is that it is not always sufficient to jump prematurely to the final "solution" without going through the inevitable learning process associated with the introduction of any new method. Paradoxically, the ultimate success can be a result of transient effects that occurred in the course of the learning process, and not necessarily a result of the method itself. Unfortunately, however, there is little opportunity to learn from the errors and constraints that led to a failure because such cases are typically not widely published or analyzed.

Hence many unsuccessful cases are swept under the rug, so to speak, instead of being analyzed for the useful lessons they might contain. This tendency is by no means exclusive to the industrial context—it is quite generally the case that success stories are celebrated for reasons of public image, stock values, customer opinion, positive effects on one's career, to name but a few. This phenomenon can also be observed in academic publications, where "successful" research results are more apt to be published than those that run amiss of what was intended or desired. But the fact is that much can be learned from well-conducted "failed" studies. In this respect, product development is no different: it is instructive to study the factors of or conditions in which a given technique did not produce the intended outcome. Furthermore, it is valuable in itself first to expose the *problems* faced in the implementation of real cases in order to be able to develop better instruments to help the *persons* working in product development to achieve better results.

This approach implies a critical appraisal of whether a case is a "success" or a "failure" in dealing with the problems encountered by practitioners. The cases reported here are primarily of the latter kind.

This might give rise to the false impression that these techniques are being attacked or being declared to be without merit. *This is not the case.* All these techniques have positive features—this has been established in the published literature and need not be reiterated here. By leaving the description of their potential to the published literature and focusing in this book on their limitations (due to restrictions of space and scope), it will be shown that it is not necessarily the techniques themselves but a misguided *implementation* that can render a technique useless, particularly if its *implementors* continue to ignore the problems that the technique cannot solve, or may even create as a byproduct.

All the examples presented here are true cases and were assessed either by means of interviews, personal experience, observations, or based on documentation.[16] Care has been taken to omit personal and company names as well as technical details that might be considered confidential information. Exceptions to this are well-known, publicized cases. As McCarthy (1995) stated in his book on software development, however, many of the examples he depicted evidently bore a sufficiently strong similarity to entirely other cases that individuals unrelated to the case of interest thought they were the persons being described. I hope this does not happen in response to the cases and situations described herein.

1.2.2 From the past to the future

The American philosopher George Santayana stated that those who do not study the past are condemned to repeat it. The value of studying past cases, particularly in the history of technology, is to learn from the mistakes made, to recognize error patterns, and to avoid the myths that surround both successful and unsuccessful R&D programs. To consider the past in the

---

16 In the published cases, care was taken to select primarily *real* cases. Cases indicated by the original authors as fictitious, or those suspected of being so, were included only for illustrative purposes. This is indicated explicitly.

context of the evolution of social science theories and science theories is to provide a historical perspective for approaches to product development. It also provides concrete, documented cases including their technical, social and political for elaboration or confirmation of theoretical efforts to explain, model or orient the product development process.

Cases depicted in this work were chosen to represent typical occurrences of the "phenomena" described. Past cases are also useful to illustrate how long these issues have been grappled with, as indeed this shows that the source of these issues lies in the nature of human beings.

Therefore, although computers are no longer developed today as they were at the time of the ENIAC, it is not surprising that many similarities in the development process still exist. Let us examine the case described by A. P. Speiser, upon returning to Switzerland from the USA in 1950. He reported having been surprised by being given the task of installing and operating the Zuse Z4 computer, which called for electromechanical relays. He had originally expected to work on the original assignment of designing and building a computer with the then modern technology of electron (vacuum) tubes. As he later commented,[iii] however, this detour turned out to be a valuable experience because it led to considerable improvements of the later design of the originally proposed electronic computer[17] (Knauer, 1990, 39').

Or let us consider the problems associated with designing cockpits for modern aircraft (and piloted spacecraft). In many cases, instances of pilot error appear to have been induced by the design of the cockpit. Although it would help considerably if engineers were themselves pilots in order to truly appreciate the user's situation, it is hardly realistic to expect design engineers to have the comprehensive skills of an air transport pilot. Therefore a different approach is necessary to ensure that the user experience is properly accommodated by the design. Although much progress has been made in coping with such issues, it is still a common quip that an engineer no longer has to be a pilot to design an aircraft system, but the pilot now has to be a design engineer to fly it. Ironically—though fortunately, only occasionally—this is sometimes true.

But let us put this issue aside for the moment and return to the design process similarities between older and more recent cases: we shall see that the same kinds of design process errors or tendencies are repeated over and over again. Clearly, therefore, there is a need to enhance our awareness of problems by analyzing cases from the past. A good example of this is the "historical" case of the Z4 computer mentioned above. With such an

[17] The ERMETH was a success also because the two-year delay brought better component technology with considerably higher reliability (Speiser 1981, 152). The pioneering machine is displayed at the Technorama Museum in Winterthur, Switzerland.

approach, not only the discussion is presented of the topic itself, but the history of the problem is couched in an engineering context that is still valid today.

The reoccurrence of "old" problems is not peculiar to the type or the current state of technology and suggests that there are elements missing in a modern engineering education that should be covered in basic technical training. Failure to consider certain issues in future products may result in the embarrassing repetition of past errors. We will return to the case of the Z4 in our discussion of a recent space project failure.

But not all is amiss. Several creators of product development methods used today have long emphasized the importance of bringing the user's experience into the designer's environment, and others have subsequently rediscovered this necessity. Unfortunately, some practitioners still ignore its importance, or neglect it under pressure (e.g. time or cost constraints). Other methods evaluated in this book suffer from the same problem.

Similarly, social science studies conducted as long ago as the 1950s are considered here, as well as concepts originated in the 1920s, not because of their "classic" character, but to illustrate how long psychological factors affecting product development have been the subject of intense investigation.

Applying lessons from the past requires an understanding of the historical context in which the situation occurred. Having said that, the purpose of citing cases from the Apollo moon program is *not* to propose a return to the context in which it took place: e.g. the frequent use of a cost-plus-fixed-fee type of contract, the technocratic orientation prevalent in the 1960s, the status of space flight as a thing of national prestige, the tensions of the Cold War, the *Sputnik* "surprise", among others. Instead, we shall examine true cases of applying the methodical concepts that were used in the Apollo era. Some of these concepts are not only still applicable today, they make a meaningful contribution to the implementation of successful products in the present context of cost-time-performance optimization and international cooperation, to name but two common situations in today's business world. Some of the techniques used back then are actually being "reinvented" today, with the obvious result that much time and resources are being unnecessarily spent relearning how to use them. The key to learning from others' experience lies not in merely repeating what has already been done, but in understanding the reasons for success or failure and transferring this insight to the pertinent context. One does not have to go far into the past to confirm this. Take for example the notion in the western world in the 1980s that Japan's economic success could be reproduced merely by copying such Japanese methods as extreme factory automation, without a more profound analysis of whether they are applicable in the western context. This produced some very embarrassing results.[18]

Moreover, there are myths to be deconstructed in technology as well as in science (let alone in management). Exploring the past requires finding ways out of what is termed here "historical advertising": the phenomenon of some trendy notion having been "sold" to posterity.

Some readers may challenge the validity of analyzing cases from the 1960s or even earlier, or object to the citation of studies conducted more than ten years ago. However, it is the author's belief that one should not dismiss valid sources solely on the basis of their publication year, but to use them as a frame of reference. The chosen studies provide a sound insight into the issues described. This insight will be applied in turn to challenges currently faced in the field of product development.

1.2.3 Introductory note for engineering professionals concerning data from social studies

Although mathematics is an attempt to model reality, its use in the social sciences is slightly different—not in terms of theory, but in the criteria used. A social science system is much more complex than an engineering system: its intricate nonlinearities and variables are very often coupled. In addition, when dealing with statistics, a social scientist copes with a considerable level of uncertainty, whereas an engineer is frequently interested in eliminating the lower-level uncertainty. This is one of the reasons why an engineer working on a noise cancellation algorithm for a communications channel might be dissatisfied with a correlation of 0.95, whereas a social scientist might well be very pleased by a correlation of 0.37. Furthermore, in the social sciences, a researcher does not have the engineer's option of reducing the complexity of the issues examined. Even if this might seem possible, other factors might mask the results. A reader from the field of engineering might prefer to think of the noise level as being much, much higher than in engineering cases. As this book is not intended to deal with the differences between the sciences, this will not be treated in more detail.

[18] Details will be trated in the CIM-Ruins case described on page 345.

1.2.4 Nomenclature

The following points should be clarified in advance to avoid misinterpretations due to the "common" meaning of certain terms used.

1.2.4.1 Cooperation and participation

There are various levels of cooperation and participation, and sometimes the concepts overlap.

Hornby et al. (1974) define cooperation as "working or acting together for a common purpose" and participation as "the act of having a share, taking part in".

For the purpose of the present work, it is useful to think of these two terms in logical-complementary terms as in, what *non*-cooperation means, and what *non*-participation in the product development process means.

A *non-cooperating* person or group would then *not* be working or acting together for a common purpose. This does not mean they are bad performers. Even though they may perform their own activities well, it does not follow directly that they might truly be contributing to the overall success of the project. Such behavior might seem illogical, but a person (or even a group) might simply be working on his/her/their own task and as such, although perhaps present during the activities, not truly act together with others. They might be following only their goals or have another, conflicting agenda. An extreme example would be the case of a person intentionally withholding pertinent information in order to prove at the end that he or she had known better all along.

Although maybe cooperating, a person *non-participating* in a process would *not* have a share or take part in the development process if he or she were not contributing actively to the definition of solutions or the search for alternatives to the issue at hand, even though he or she might well be actively providing information about the issue itself. At this point, it must be emphasized that determining *what the problem is* constitutes a very important part of the solution. The aspect to be underscored here is not whether it is positive or negative to be non-participating but to focus on what standpoint the person or group assumes or can assume and its effects on the quality of the final product.

Turning away from this definition-by-negation of cooperation and participation, we can see that one of the main differences between the two is that by cooperating, one allows others to do things, supplies them with helpful information, or one covers complementary aspects of the product. Cooperation can occur even if the activity is performed independently. By participating, one takes part in the process of making active contributions. For instance, one might contribute interconnected, complementary skills or complementary activities to achieve a common goal, even if this goal commonality is only partial. The point is that the responsibilities of defining product characteristics and the decisions affecting the actions to be taken are shared.

The differentiation among forms of cooperation proposed by Hacker (1978, cited in Ulich, 2001, 72-73) illustrates this aspect further:

1. *Work in spatial grouping*: several persons working in the same room, but the tasks to be fulfilled are not dependent on each other (i.e. work without cooperation).
2. *Cooperation in successive grouping*: every member working on an object provides continuation to a preceding activity in a different, individually conceived task.
3. *Work in integrated grouping*: work of several employees on the same object at the same time.

Many real cases of alleged participation or collaboration concerning product development also illustrate the need to define these concepts more clearly.

Here is an example of the difference between participation and cooperation from the concept of the "customer's share"[iv] established by Seghezzi: if a patient goes to a dentist but refuses to open his or her mouth or to stay seated, the patient cannot expect to receive quality service. The customer has to *cooperate* with, albeit not *participate* in the process.

For the purposes of this book, however, such behavior is not necessarily sufficient to qualify as "cooperative". The information exchange between customer and service provider may, however, be on a level that may be considered cooperative in the sense developed herein.

In the description of the activity *Preparation of a Quality and Reliability Assurance Program*, Birolini (1994, 319) also differentiates between three kinds of task interaction among the functions marketing, development, production, as well as quality and reliability assurance. They are (i) primary responsibility, (ii) *active* cooperation, and (iii) information [emphasis mine].

To reinforce the difference between participation and cooperation, the description of participative leadership by Ulich, Baitsch & Alioth (1987, 15-16) is adopted here: "In contrast to the imperative leadership, the supervisor sees the process of setting goals as an effort to be jointly achieved. He would then involve his personnel in the planning and would give them a more or less large degree-of-freedom for initiatives (in designing the job). Appropriately, he would refrain from continuous and detailed control."

1.2.4.2 Control and verification

The term *control* can be defined as (Hornby et al. 1974):

- the power or authority to direct, order, restrain,

- the means of regulating, restraining, keeping in order,
- to check,

whereas the expression *to control* can carry the meaning of

- having power over,
- verifying.

Completely different results are to be expected if the word control is meant as a *verification* or in the sense of a person or group *taking control* or exercising *power* over another person or group. These different meanings are also familiar in other languages and, depending on company or group culture, the word control is used in very different senses.

Examining the uses of the expression "quality control", for instance, we find measuring mechanisms, trend plotting, histograms, standardized levels of quality to be met, established standards. Semantically speaking, these are methods to verify, check; regulate, keep in order. In practice, however, the expression conventionally means some person or group that directs, orders, restrains, or has power over the work of others who have done a given job, i.e. the notion of evaluating someone else's work. Consequently, the associated techniques assume a character of directing, ordering, restraining, having power over others.

Solving quality problems, however, involves much more than controlling (in the narrow sense). This is illustrated by the evolution of the discipline's name throughout its existence.[19]

The different flavors of these words are significant for this work. For reasons that will become clear below, the following terminology will be adopted, unless an expression appears in quotation marks:

- *Control in a broad sense*: general term for all concepts of control.
- *Regulate*: the means of regulating, restraining, keeping in order, constraining the variables to a desired value in keeping with control systems theory.
- *Check or verification*: an assessment of a situation as a first step toward taking appropriate action.
- *Control in a narrow sense*: the activity of comparing an object with a known "good" reference in order to evaluate the correctness of the work performed.

---

[19] One example illustrating the distorting effect of these meanings is described by Lerner (1988, 29) in that the quality method called "control chart" was translated in German as "Kontrollkarte". The English term was meant as an instrument to regulate the quality level of a given item at desirable levels, whereas the German term emphasized the activity of checking instead of actively regulating the levels. Owing to this misunderstanding, the German standard DIN 55 350 (quality terms) suggests changing to the more exact term "Qualitätsregelkarte." The English translation for that would be something like Quality Regulation Chart.

1.2.5 Notation

One of the limitations of citing numerical results obtained by objectively measured studies lies in the fact that the mere presentation of numbers can easily lead to a misinterpretation of the data. Work satisfaction, for example, can be distinguished in several different forms, leading to very different conclusions (cf. Ulich, 2001, 261-262). To the extent possible, a brief description is provided of the studies cited but, due to space limitations, emphasis was placed on describing the scopes of the methods used in order to provide a sufficient context for the findings discussed. In many cases, additional numerical data was available but was not cited in this work because the necessary explanation would have burdened the work with excessive explanations. In these cases, it was preferred instead to limit the discussion to the conclusions drawn in the studies. This is an implicit limitation of this work to an integration of results.

Boxes were used to highlight empirical observations in order to more easily identify the cases that support the theoretical part. The texts inside and outside the boxes were composed such that they constitute a cohesive text. A suggestion to the reader is simply to ignore the text boxes when reading the text or, alternatively, to use it as a reference when browsing the text or taking a closer look at the specific cases described or the concepts presented in the boxes.

As the theoretical and empirical parts do not appear as separate chapters, the cases and situations that constitute the empirical part of this present work are explicitly marked as follows: *"case # "* followed by a number.

Care has been taken to avoid the annoyance of excessive quotation marks. Nevertheless, the correct citation of the sources required many quotes throughout the text. Translations, whenever intended as citations, were included in quotes to better indicate their origin. Summaries of findings make spare use of quotes because of the shifted responsibility inherent to summarizing, but the source is clearly indicated nonetheless. In some cases, boxes were used to set off the conclusions or to depict structured descriptions of the original author(s). Sometimes, they are used to highlight an essential study. For clarity, quotes were not used in these cases because the original sources are always cited.

Secondary literature is listed only as a reference to primary literature if it was a part of the concept illustrated by the main author. The following cases are exceptions: (i) when the cited literature is a part of the development of this research and its citation by the new author confirmed that it is a convergence of individual studies, and (ii) when the secondary literature was further processed to arrive at a more complete or detailed description not present in the primary work. Whenever text is set in a box assigned to a single cited work, it is implicit that further references are secondary literature and therefore were not explicitly marked as such.

Owing to the unavailability of the original English text of certain translations, or of authorized translations into English of non-English texts, texts were translated by this author into English or back into English as appropriate. In some cases, the source appears in the end notes for enhanced clarity.

Unless a precise use of quotations is necessary, the convention for reprinted passages is that the original documents are cited in the list of references.

Within quotations, text inserted to explain or add emphasis appears in brackets. In order to add emphasis or highlight an already emphasized text, *italics* formatting was applied, together with an appropriate comment.

In order to provide readers with more than just the author's view of the issues associated with product development, particular importance was attached to the "scientific validity" of the development you are about to read. The interested reader will find in the appendix a clarification of the scientific approaches upon which this book is based.

Chapter 2

HISTORICAL CONTEXT

2.1 Introduction

This section presents the historical context relevant to the quality of product development processes.

2.2 Social sciences

As described by Ulich, in the first phase of the development of work psychology, the generally accepted notion of a "man", or *image of man*,[v] assumed the average person to be an *economic man* who "avoids responsibility, is oriented toward maximizing his profit, and is primarily motivated by monetary stimuli. The consequences for the organizational structure include: extensive division of work and authority, and a small span of control and individual incentive systems, where at the same time, agreements about limitation of performance should be made difficult. The company is understood as a technical system to which persons had to adapt" (Ulich, 2001, 7-8).

The approach of F. W. Taylor (1911) to use scientific concepts of objectivity to optimize working methods in production activities followed three main principles as pointed out by Hill, Fehlbaum & Ulrich (1992, 410):[20]

- Detachment of the work process and the skills[vi] of the workers (objectification[vii] through depersonalization and radical simplification of tasks).
- Full separation of intellectual and manual work (objectification through thorough planning and structuring of work by management).
- Complete objective control (objectification through scientifically justified standardization of the work object, the workplace, and the work time).

The consequences of *Scientific Management* were analyzed by Mankin (1978, cited by Ulich, 2001, 10) as being the paradigm that characterizes the three first decades of work psychology and having the following elements:

- Focus on the individual worker as the object of analysis.
- Assumption that the "average" worker is almost exclusively motivated by monetary rewards.
- Tendency of rarely dealing with the worker as a person but rather as a machine whose inefficiencies can be compensated by the appropriate conception of tools and procedures. This is also known as the *mechanistic view* of workers.
- Consideration of the worker as a factor to be manipulated to increase productivity.

Max Weber's idealized *Bureaucracy* is an approach that embodies the maximally efficient organization as the most rational formal form of exerting control and power over[viii] the organization. Hill, Fehlbaum & Ulrich (1992, 415) describe the effect of Weber's approach on organizations as having the following characteristics:

- Division of labor based on functional specialization.
- Exactly fixed hierarchy of authority.
- System of rules in place prescribing the rights and obligations of the persons occupying each position.
- System of procedures to cope with work situations (standard procedures).
- Impersonal nature of relations between people.
- Promotion and development based on technical[ix] competence. Career opportunities with bottom-up progression of salary.
- Principle of work process is "according to the book".[x]

[20] Excerpts from: Hill, W.; Fehlbaum, R. & Ulrich, P. (1992). Organisationslehre 2 - Theoretische Ansätze und praktische Methoden der Organisation sozialer Systeme, 4. Aufl., Bern: Verlag Paul Haupt. Used here by kind permission of Verlag Paul Haupt.

The same authors take issue with Weber's approach as follows (op. cit., pp. 416-417):

- "It neglects the conditions in the organizations. Weber uses the old, non-situational end–means approach. His alleged instrumental rationality may be valid for a static environment, for mere routine tasks, and for a constellation comprising one authoritarian and one less-motivated person. In a dynamic environment of problem-solving tasks and socially motivated people, this situation is inverted and corresponds to the popular connotation of the word bureaucracy: the excessive overhead of a clumsy, inefficient organization."
- "It neglects social-emotional rationality. The aspects of motivation and work satisfaction are not taken into consideration. This is not dependent only on the *image of man* predominant at that time but also on a one-sided emphasis on the aspect of control/power. The latter resulted in the notion of hierarchical subordination. In addition to ethical issues concerning the authoritarian principle of control/power, every bureaucratic structure has the tendency to create a sensation of helplessness towards the all-powerful 'system'."
- "There is a danger of exaggeration. Bureaucratic techniques change very easily from being an instrument to an end in themselves. Precision turns into pedantry, stability into stiffness, written documentation into red tape, etc. The same elements that should render a bureaucratic organization efficient introduce instead the principles of inefficiency."

March & Simon (1958)[21] provide a viewpoint on theories of bureaucracy after Weber. They describe the proposition by Merton (1940) that can be summarized as follows: "Merton's system of proposition begins with a demand for control made on the organization by the top hierarchy. This demand takes the form of an increased emphasis on the reliability of behavior within the organization" (accountability and predictability). "Standard operating procedures are instituted, and control consists largely in checking to ensure that these procedures are, in fact followed." Among the consequences of this model, the authors point out that:

- "There is a reduction in the amount of personalized relationships. The bureaucracy is a set of relationships between offices, or roles."
- "Internalization of the rules of the organization by the participants is increased. ...Rules originally devised to achieve organizational goals assume a positive value that is independent of the organizational goals."
- "There is increased use of categorization as a decision-making technique" associated with a tendency to "restrict the categories to a

[21] Excerpts from: March, J. G. & Simon, H. A. (1958). Organizations, New York: Wiley, 9th printing 1967. Reproduced here by kind permission of Blackwell Publishing.

relatively small number and to enforce the first formally applicable category rather than search for the possible categories that might be applied and choose among them. An increase in the use of categorization for decision making decreases the amount of search for alternatives."

The combination of these three consequences leads to a rigidity of behavior.

The depersonalization of relationships enhances the sense of commonness of purpose, interests and character. This in turn increases the propensity of organization members to defend each other against outside pressures, which again solidifies the rigidity of behavior. "The rigidity of behavior increases the amount of difficulty with clients of the organization ...and complicates the achievement of client satisfaction. ...In addition, client dissatisfaction may in itself reinforce rigidity." The Merton model is depicted in *Figure 1*.

The same authors discuss another bureaucratic model proposed by Selznick (1949): that the need for control leads to an increased delegation of authority. This delegation increases the amount of training in specialized competences, which improves the ability of the worker to deal with problems and tends to decrease the difference between the organizational goals and achievement, thus stimulating more delegation. However, at the same time, "this delegation leads to departmentalization and an increase in the bifurcation of interests among the subunits in the organization. The maintenance needs of the subunit goals dictate a commitment to the subunit goals over and above their contribution to the total organizational program." The bifurcation leads further to "increased conflict among organizational subunits". As a consequence, "the content of decisions made within the organization depends increasingly upon considerations of internal strategy, particularly if there is little internalization of organizational goals by participants." *Figure 2* presents the simplified Selznick model.

In contrast to the above, a contemporary of Taylor and Weber took a completely different approach to organizations. The political scientist Mary Parker Follett successfully applied her concepts in the field of social work by pioneering the organization and management of vocational guidance centers in America in the early 1900s. She presented her concept of "power" to a Bureau of Personnel Administration conference group in January of 1925, but it was apparently not published until 16 years later by Metcalf & Urwick (Eds.) (1941). Follett basically makes the distinction between *power-over* and *power-with*. *Power-over* means the conventional implementation of power exerted by some person or group over some other persons or group. *Power-with* is a jointly developed power, a co-active, not coercive power. This view of power led directly to Follett's concept of executive leadership in which the leader combines the "different wills" in an organization to create "a driving force", a "group power" that reconciles disparate, often conflicting parts into a "team."

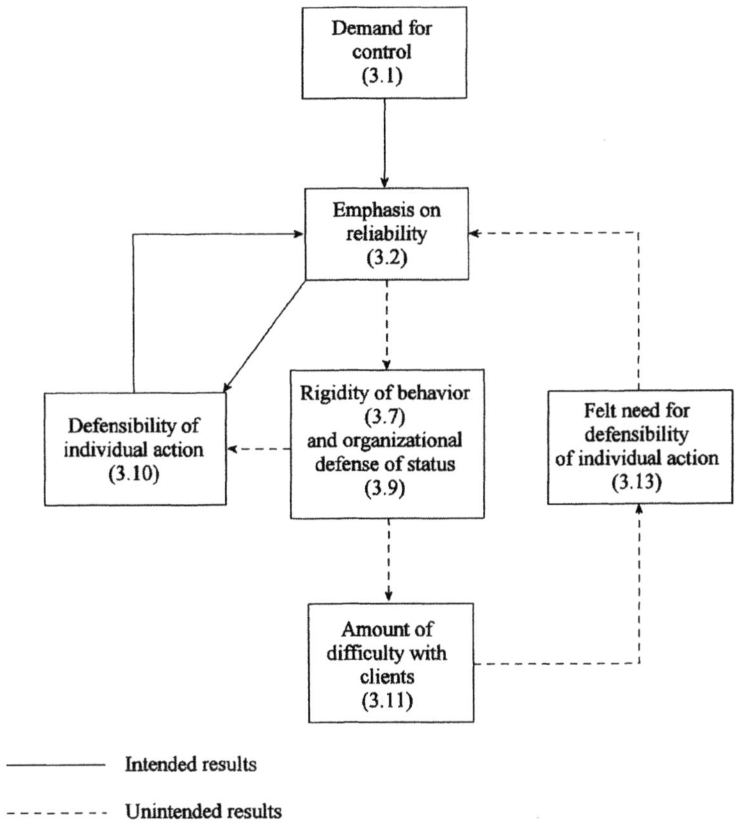

Figure 1. Simplified Merton Model[22]

22 From: March, J. G. & Simon, H. A. (1958). Organizations, New York: Wiley, 9th printing 1967. Reproduced here by kind permission of Blackwell Publishing.
Numbers in parentheses are references to variables in the original text. Although unnecessary for this work, they were kept for copyright reasons.

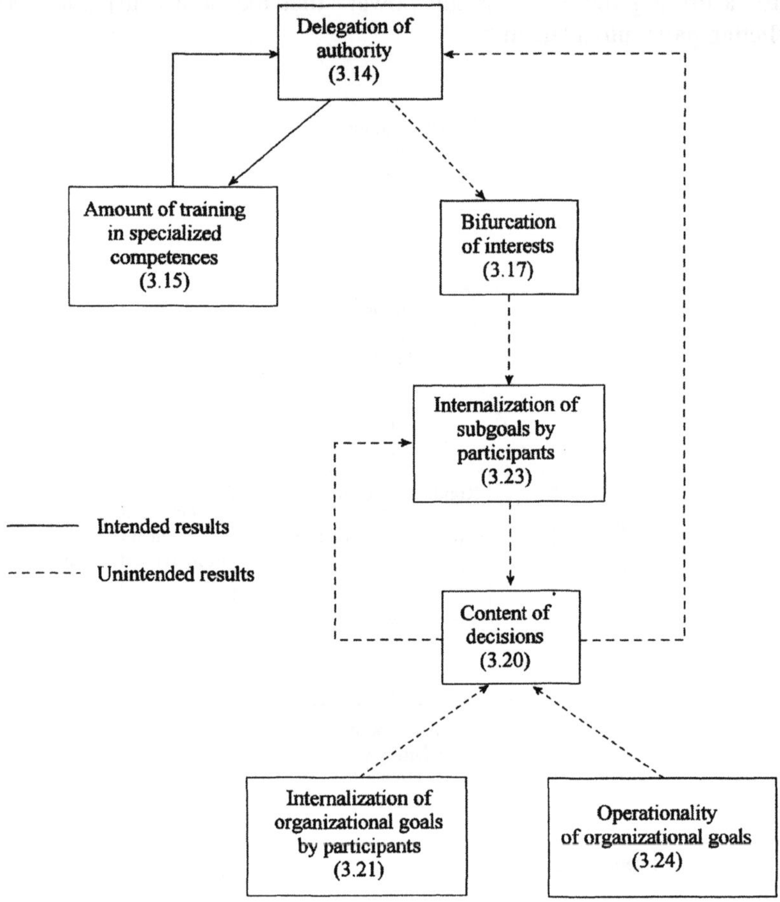

Figure 2. Simplified Selznick Model[23]

Follett also proposed concepts of "control", which we shall examine later in this book, as well as the involvement of workers in the process of coordination. Her view of using conflicts constructively is based on the notion of conflicts as a means to *understanding*. This is done not by

[23] From: March, J. G. & Simon, H. A. (1958). Organizations, New York: Wiley, 9th printing 1967. Reproduced here by kind permission of Blackwell Publishing.
Numbers in parentheses are references to variables in the original text. Although unnecessary for this work, they were kept for copyright reasons.

determining who or what is "right" in a conflict situation. Instead, it is assumed that both parties are likely to give correct answers to different questions. As such, each participant is led to ask herself/himself what would be *the question* behind the other's answer. After becoming acquainted with the other person's question, the point is to find a new answer that satisfies both questions.[24] The result of this approach is neither "victory" for one side (e.g. the strongest) nor "compromise", but an integration of interests.

Follett "emphasized human beings and their interactions as the bedrock of all organization. Rather than Taylorism's 'one best way' to manage, this flexible iconoclast wrote of the 'law of the situation' " (Lambright 1995, 24). This *law of the situation* means determining what is to be done in a *given situation* or to achieve *the task* at hand, instead of being defined by the will of the superior (or by a procedure defined by management). Hence this constitutes a better approach than merely giving orders and expecting them to be obeyed without question (Follett 1926), which creates a system of obedience and control in that the order-givers must ensure compliance with the orders given. This control-oriented approach: (Follett, 1949a)[25]

- loses "what we might learn from the man actually on the job" (op. cit., p. 19),
- "ignores one of the fundamental facts of human nature, namely the wish to govern one's own life" (op. cit., p. 20),
- neglects that the "worker is often...equally interested...in [doing] the best of which he is capable" (op. cit., p. 21, brackets maintained), and
- "decreases the worker's sense of responsibility" (op. cit., p. 21).

She proposed an organizational form that, according to Nohria (1995, 157-158), "stands in sharp contrast to the Weberian bureaucracy" as she "believed that in any situation, reasonable people could come to an agreement on the demands of the situation. Orders that emanate from the law of the situation, Follett argues, are less likely to be resented and more likely to be followed. Furthermore, since the orders are based on the logic of the task and incorporate the input of all those whose expertise is relevant to performing the task, the task can inevitably be performed more effectively".

Follett's ideas, which she introduced in the 1920s, were not heeded even throughout the 1930s and 1940s, not only because of the mechanistic view of an enterprise accepted at that time, but because of the interpretation of

---

24 Follett uses a trivial case to illustrate this. Two individuals want to orient a truck docking bay their own way. Instead of quarreling over whose is the "right" solution, the two experts can examine the reasons underlying their approach (i.e. their question), including their context. As a result, a third solution arises that satisfies both parties.

25 Cited in Nohria (1995, 158).

power. Peter Drucker (1979)[26] describes an event in the late 1940s when General Motors attempted to introduce what was later called *quality circles* as a partnership between management and workers based on workers' responsibility for their own tasks and work. A practice—Drucker emphasizes—"we now credit the Japanese with". General Motors surveyed workers to determine which issues workers considered important. The response was overwhelming, but "the United Auto Workers (UAW) protested to the National Labor Relations Board (NLRB) that even asking workers about their jobs was an "unfair labor practice". The survey implied the cooperation between company and workers, that there were areas in which both had the same interests—in other words, that there could be harmony. According to the UAW, this violated the axiom of conflict on which both the country's labor law and the union contract were based" (Drucker, 1995, 5).

But let us return to work psychology. Ulich has stated that "among the central results arising from the work of Mayo (1930, 1933) and Roethlisberger & Dickson (1939) is a new *image of man* known as the *social man*. This image places social motivation in the foreground and assumes that the workers' behavior is to a great extent determined by the social norms of his work group. The company is essentially considered to be a social system where information and communication processes deserve special consideration. The perception that group membership as well as the type of group relationship can more strongly influence job performance than the financial incentive system, working hours or pause timing served as starting point for the *human relations movement*. Its goals were primarily to improve the interpersonal relations inside of working groups as well as between superiors and subordinates" (Ulich, 2001, 43-44).

Only after the human relations school of management thought was established did Mary Parker Follett's ideas regain influence on the management of complex organizations. James Webb, then personnel director at Sperry, experienced the focus shift from Taylorism to human relations. He had previously trained to become an educator and particularly liked to take talented people, train them and place them where they would be needed most. He acted as a teacher in a fast-growing organization. In this process, he became "particularly influenced by [Follett's] writings, ... especially on how groups could be made to operate at an optimum level. Many committees at Sperry, he had noted, wound up compromising at the lowest common denominator. Follett argued that such groups could instead yield decisions based on their highest common denominator" (Lambright 1995, 24). Webb

[26] Cited in Drucker (1995).

was to become the NASA administrator during the successful Apollo Moon Program of the 1960s.

At the same time, the work of Maslow (1954), Herzberg, Mausner & Snyderman (1959) and Argyris (1964) "placed the needs of self-actualization and psychological growth in the first line of interest. The *image of man* seeking self-actualization and autonomy (*self-actualizing man*) constitutes the third paradigm of the work psychology" (Ulich, 2001, 44-45). The work of Herzberg et al. led to the *two-factor theory,* where work satisfaction and dissatisfaction are not understood as the extremes of *one* measurement dimension but are separately influenced by different aspects. As such, work dissatisfaction is not the result of the absence of *satisfiers* (or the factors that cause work satisfaction) but are caused by *dissatisfiers*, and the converse. Generally, satisfiers are related to the activity itself and its possibilities of growth—the job content—and are consequently called *content factors* or *motivators*, whereas dissatisfiers are primarily related to the work environment and are also called *context factors* or *hygiene factors.* The term "hygiene" refers to the preventive character of actions in response to these factors (to correspond to workers' needs), thus avoiding unpleasant situations (op. cit., condensed from pp. 47).

Another development of special interest is known as the *socio-technical systems* developed in the Tavistock Institute of Human Relations in England and represented by the work of Emery & Trist (1960), among others. As Ulich describes, "the understanding of the company as a socio-technical system makes it clear that the relationship between production system and organizational design, between organizational structure and organizational processes are extraordinarily connected and that (i) the use of technology, (ii) the work organization and (iii) the development of human resources can only be jointly optimized" (Ulich, 2001, 51). From the assumption—characteristic of this period—that workers strive for self-actualization, autonomy and self-control, persons in the enterprise become active. They become the subject of their own actions, the results of which they can control themselves. In terms of the self-image of superiors, this means they should not regard their tasks as 'instruct-and-control' but as 'stimulate, support, and promote' " (op. cit., p. 53).

In 1961, following a study of the British electronics industry, Burns & Stalker launched the contrasting forms of mechanistic and organic systems of management. The first was intended for stable conditions and the second for changing conditions, which constantly give rise to fresh problems and unforeseen requirements for action (Burns & Stalker, 1994).[27] *Table 6* summarizes the predominance of the organizational style in various forms.

Table 6. Differentiation between mechanistic and organic systems of management. (condensed from Burns & Stalker, 1994, 119-121).

	MECHANISTIC	ORGANIC
Global task	Broken down in specialized functional differentiation	Contributive nature of special knowledge and experience to the common task
Individual task	Abstract nature, pursued with techniques and purposes distinct from those of the whole. Agents pursue improvement of means, rather than accomplishing the ends	"Realistic" nature, regarded as being determined by the overall situation
Individual task integration	By reconciliation of distinct performances by the immediate superiors	Adjustment and continual redefinition of individual tasks through interaction with others
Responsibility	Precise definition of rights and obligations and technical methods attached to each functional role	Shedding of "responsibility" as a limited field of rights, obligations and methods. Problems may not be posted upwards, downwards or sideways as being someone else's responsibility
Centralization – decentralization	Translation of rights and obligations to functional positions	Spread of commitment beyond any technical definition
Control, authority, communication	Hierarchical structure	Network structure
Sanction of individual	Derived more from a contractual relationship between himself and a non-personal corporation, represented by his immediate superior	Derived more from presumed community of interest with rest of the working organization in the survival and growth of the firm
Locus of knowledge	Reinforcing the hierarchical structure by location of knowledge of actualities exclusively at the top	Omniscience no longer imputed to the head of the concern, may be located anywhere in the network
Direction of interactions	Tendency to be vertical in the organization (superior - subordinate)	Tendency to be more lateral than vertical in the organization with communication between people of different rank; also resembles consultation rather than command

[27] © Tom Burns and G. M. Stalker 1961. Reprinted from The Management of Innovation by Tom Burns and G. M. Stalker (revised edition, 1994) by kind permission of Oxford University Press.

	MECHANISTIC	ORGANIC
Character of exchange	Tendency of superiors to issue instructions and decisions to govern the operations and working behavior	Content of communication which consists of information and advice rather than instructions and decisions
Membership	Insistence on loyalty to the concern and obedience to superiors as a condition	Commitment to concern's tasks and to the "technological ethics" of material progress and expansion
Importance & prestige	More attached to internal (local) than to general (cosmopolitan) knowledge, experience, and skill	Attached to affiliation and expertise valid in the industrial and technical and commercial milieus external to the firm

Burns (1971, 49, cited in Mark & Levine, 1984, 177) describes the difference in the span of responsibility in the two systems as follows: "Mechanistic systems tell (the individual) what he has to attend to, and how, and also tell him what he does *not* have to bother with, what is not his affair, what is *not* expected of him, what he can post elsewhere as the responsibility of others. In organic systems, such boundaries disappear. The individual is expected to regard himself as fully implicated in the discharge of any task appearing over his horizon. He has not merely to exercise a special competence, but to commit himself to the success of the concern's undertaking as a whole." (italics maintained).

Following the historical development of *image of man*, interpersonal differences are considered in the concept of *complex man*. In this approach, the following points should be considered:

- changes in human needs in the course of a person's life,
- individual differences concerning the level of importance attached to these needs,
- interaction between motives and needs,
- combination of values and goals, and
- evolution of motives throughout one's working life.

In this case, no general solution is possible. What is necessary are flexibility and the opportunity to tailor the work design to the individual.

In agreement with the work of Ulich, *Table 7* is an attempt to present an overview of the development of the design of organizations. It is of course a simplification because the reciprocal relationships between the elements cannot be shown. Nevertheless, the effects of the initial presupposition to derive an organizational form are clear.

Ulich observes that the image of the *economic man* still cannot be considered overcome[xi] as it continues today to determine the business culture in diverse organizations with work-sharing structures (op. cit., pp. 62). It must be emphasized that the achievement of productivity gains made possible by modern concepts may be hindered by the use of such anachronistic approaches.

Table 8 summarizes current schools of thought in work psychology.

Table 7. Connections among various aspects of the development of work-psychological concepts according to Ulich (2001, 61).

IMAGE OF MAN	ECONOMIC MAN	SOCIAL MAN	SELF-ACTUALIZING MAN	COMPLEX MAN
Reflected in the				
UNDERSTANDING OF ORGANIZATIONS	Technical system	Social system	Socio-technical system	
Affecting				
THE SHAPING OF WORK	Tayloristic rationalization	Human Relations	Job enrichment	Individualization
Changing				
ORGANIZATIONAL STRUCTURES	Central/ bureaucratic, on individual basis	Central/ bureaucratic, on group basis	Decentralized/ flat, on individual or group basis	
and				
EVALUATING CRITERIA	Profitability, absence of damage	Satisfaction, psychosocial well-being	Personality promotion	

Table 8. Current schools of thought in work psychology (condensed from Ulich, 2001, 62).

CURRENT SCHOOLS OF THOUGHT	MAINLY REPRESENTED BY	CHARACTERISTICS
SOCIO-TECHNICAL SYSTEM DESIGN	Trist & Bamforth (1951) Rice (1958) Emery (1959) Emery & Trist (1960) Lewin, Kurt	Joint optimization of the prevailing social and technical system (organization and technology)
JOB DESIGN	Hackman & Oldham (1976)	Indications of motivation-promoting design of work activities
PERSONALITY PROMOTING JOB DESIGN	Hacker (1978) Ulich (1978) Volpert (1979)	Work activities are among the relevant life activities that play a significant role in the development of an adult's personality.

2.3 Scientific methodology in engineering

The following section is intended to clarify the often misunderstood and hence not fully appreciated engineering sciences as well as the consequent application of scientific practices to product development. Once again, the main goal is not to describe the science itself but to provide a basis for the description of product development practices.

The (over)utilization of Cartesian methods in product development and decision-taking as well as the claim of many engineers to be (only) Cartesian-thinking confirms the need to clarify certain concepts of science from the start. In addition, emphasis placed on the pure rational aspect of such reasoning also needs to be clarified.

It is commonly accepted among scientists or people using scientific methods that the basis for their reasoning is based on Renè Descartes' work *Discours de la Méthode* originally published in Leyden in 1637.[28] The four Cartesian principles are stated as follows (Descartes, translation by Heffernan, 1994, 35):[29]

1. Never "accept anything as true that [one does] not evidently know to be such; that is to say, carefully...avoid precipitation and prejudice; and ... include in [one's] judgments nothing more than that which would present itself to [one's] mind so clearly and so distinctly that [one] were to have no occasion to put it in doubt."[xii]
2. "Divide each of the difficulties that [one] would examine into as many parts as would be possible and as would be required in order better to resolve them."[xiii]
3. "Conduct [one's] thoughts in an orderly manner, by beginning with those objects the most simple and the most easy to know, in order to ascend little by little, as by degrees, to the knowledge of the most composite ones; and by supposing an order even among those which do not naturally precede one another."[xiv]
4. Perform everywhere "enumerations so complete and reviews so general that [one would be] assured of omitting nothing."[xv]

Descartes himself could not resolve one issue with his own method: that one would need to prove the absolute truth of the points described in step 1 above. The result for engineering sciences is that one has to (1) divide the problem, (2) analyze the elements, (3) perform the integration of results, and finally (4) review, in this order. At the end it is then "proved", "demonstrated", or "verified" that the device works, always. Unfortunately (for engineers), this "always" is not always true.

In his work "The Logic of Scientific Discovery" initially published in 1934, Karl Popper (1959, 32) describes four lines along which a theory can be tested deductively:

---

[28] Original french text in the endnotes from Descartes (1969, 30-32) correspond to page 20 in the original work.

[29] Text in brackets denote the conversion from the first person discourse by Descartes to an impersonal form, more adequate to this work. Additionally, ellipses indicate suppressions of the infinitive form with "to".

1. Logical comparison of the conclusions among themselves (internal consistency check).
2. Investigation of the logical form of the theory (character of empirical, scientific theory or logical deduction).
3. Comparison with other theories (determination of potential scientific advance).
4. Testing of the theory (empirical application of its conclusions).

The test procedure turns out to be deductive as well. Hence we take a decision by comparing the derived statements with the results of practical applications and experiments. If the decision is positive, the theory is *verified* and if the decision is negative, the theory has been *falsified* by these results (op. cit., p. 33). Popper proposes *falsifiability*, not *verifiability*, as the criterion of demarcation (i.e. the means to distinguish between empirical sciences on the one hand and mathematics and logic on the other). He emphasizes that it is not necessary that a "scientific system ... be capable of being singled out, once and for all in a positive sense" but that its logical form "be such that it can be singled out, by means of empirical tests, in a negative sense: *it must be possible for an empirical scientific system to be refuted by experience*" (op. cit., pp. 40-41, italics maintained).

Popper thus founded the notion of Critical Rationalism from which Albert (1971) derives the idea of *Critical Testing* as a "methodological principle that assumes that our thinking and actions are inherently fallible. As a result, anyone with a sincere interest in the truth will have an interest in acknowledging the weaknesses and difficulties of his thoughts and solutions, will be willing to entertain counterarguments, and will welcome alternative arguments in order that he be given the opportunity to compare, modify and revise his position. Only those concepts will prevail that have been subjected to critical debate. The advantages and disadvantages of a given concept can only be evaluated against a backdrop of alternative opinions."[xvi]

Another frequent misinterpretation of science present in engineering activities was approached by Kuhn (1962). He pointed out that science suffers from the "myth of agreement" in that textbooks perpetuate the notion that science is pursued in a step-by-step, impersonal manner, an uninterrupted, patient accumulation of impartial facts. Thus textbooks distort the way science is perceived by neglecting the processes inherent in the sometimes bitter controversies surrounding important science achievement. Similarly, maybe consequently, engineering management textbooks often distort the way product development is accomplished, especially when innovation is part of the goals.

One important psychological aspect of science touched upon by Maslow (1966)[30] was his idea of self-actualization (cf. page 41) in terms of the psychology of scientific activity. He differentiated between cognitive activities instigated by anxiety and those performed in the absence of fear or having overcome fear, a state he calls *courage*. Maslow found that "cognitive impulses seem to function under conditions *either* of fear or of courage, but they will have different characteristics under these two different conditions" (op. cit., p. 20, italics maintained). His focus on problem-centering versus method-centering, his investigation on the anxiety-allaying mechanisms and the coping-related mechanisms in cognition led him to differentiate between the defense-motivated and the growth-motivated scientist, between "safety science" and "growth science". In an earlier work, he asserts that "sometimes safety needs can almost entirely bend the cognitive needs to their own anxiety-allaying purposes. The anxiety-free person can be bolder and more courageous and can explore and theorize for the sake of knowledge itself. It is certainly reasonable to assume that the latter is more likely to approach the truth, the real nature of things. A safety philosophy or religion or science is more apt to be blind than a growth philosophy, religion or science" (Maslow, 1962, 62). He also identified the need for a certain "love" of the object of study, in contrast to an impersonal relation to the subjects. He went even further regarding a point only touched upon by Fromm (1956) concerning a personal bond with the object created in any human activity.

Personal relationship with the object was also studied by Mitroff (1974). In his investigation of the subjective side of science, he analyzed interactions among scientists conducting research on the material samples collected during lunar expeditions. Among his findings is the conclusion that "the greater the scientist, the more likely he is to belie the myth of the 'disinterested, uncommitted scientist'." (Mitroff & Mason, 1974, 1509). Mitroff was influenced by the work of Churchman (1971) on inquiry systems. In Churchman's work, the conceptualization or modeling of a problem basically amounts to conducting an inquiry into its nature. It follows that what one knows about a given issue is dependent on the "style" with which one obtained information about that issue and its nature, i.e. on the *inquiry system*, with the characteristics described in *Table 9*. In other words, *the style affects the results* with the contrasting differences listed in *Table 10*. The work of Churchman is based on philosophical archetypes and

[30] Excerpts from: Maslow, A. H. (1966). The Psychology of Science - A Reconnaissance, New York, etc.: Harper & Row. © 1966 by Abraham H. Maslow. Copyright renewed 1994 by Bertha G. Maslow. Reproduced here by kind permission of HarperCollins Publishers, Inc.

his classification is not intended to be complete and exhaustive. Both tables are derived from the work of Mitroff.

Among Mitroff & Mason's astonishing findings about the community of scientists investigating the lunar samples is that, instead of relying on a Leibnizian inquiry system, they formed a *Lockean* inquiry community, thus documenting the "tremendous role that 'irrational' factors play in the attainment of scientific objectivity" (op. cit., p. 1501).

Table 9. Main characteristics of inquiry systems (IS).

INQUIRY SYSTEM	ARCHETYPE OF	EMPHASIS ON
Leibnizian	Formal-deductive systems	The purely formal, the mathematical, the logical, and the rational aspects of human thought
Lockean	Experiential, inductive, consensual systems	The purely sensory, the empirical aspects of human knowledge
Kantian	Synthetic multimodel systems	Both the formal and the experiential—the integrative—aspects of human thought. Synthetic because it attempts to reconcile (synthesize) the demands of philosophic rationalism (Leibnizian-IS) with those of philosophic empiricism (Lockean-IS); multimodel because it requires inquirers to produce at least two models
Hegelian or dialectic	Conflictual, synthetic systems	The antagonistic and the antithetical, the conflictual aspects of human thought
Singerian – Churchmanian	Synthetic, interdisciplinary holistic systems	The richly diverse modes of human thought: the scientific, the ethical, and the esthetic

Another aspect of scientific work that is either neglected or misunderstood is the value of serendipity. Hornby et al. (1974, 777) describe it as "(a talent for) making fortunate and unexpected discoveries by chance".[31] The myth that in science nothing happens by chance, that everything is discovered only by intentional hypothesizing, then by the planning and performance of experiments to test the hypotheses, resulting in the development of theories is misleading. Many consider the element of luck to be anathema to good scientific as well as engineering practice. One engineering manager in a space science project declared that he did not like the idea that things happen by chance: it gave the impression that things are not carefully planned and executed, that one is not responsible (case #3).

The reason behind this reaction is mainly because people cannot frame the role of chance in a methodical system, nor in a management system.

[31] Parentheses as in the original.

Even though discoveries *are* made by means of meticulously organized and rationally conducted experiments and the linear evolution of knowledge, this clearly is not the only way scientific progress is made, for which there is no scarcity of examples. Alexander Fleming's chance discovery of penicillin is one such example that springs to mind, as does the (failed) experiment of Hans Christian Oersted to demonstrate electromagnetism, the discovery of the photoelectric effect by Heinrich Hertz, the X-ray by Wilhem Conrad Roentgen, the radioactivity of uranium by Henri Becquerel, pulsars by Jocelyn Bell Burnell, and the Cosmic Microwave Background[32] by Arno Penzias & Robert Wilson, to cite only a few famous instances.

Table 10. Summary of strengths and weaknesses of inquiry systems that the Singerian/Churchmanian IS tries to correct.

INQUIRY SYSTEM	MAIN STRENGTHS	MAIN WEAKNESSES
Leibnizian	consistency rigor logical coherence precision little or no ambiguity in the use of terms conditions of proof	Extremely difficult to defend why they chose to solve a particular problem and why they chose to represent it in the manner used; has been accused of being empty; rich in analytic or formal content but extremely poor in experiential or empirical content.
Lockean	Ability to sweep rich sources of experiential data	Experience is undoubtedly a rich source of knowledge, but it can be extremely fallible and misleading.
Kantian	Best suited for handling issues of moderate ill-structure where the alternative views of problems are absolutely essential for problem definition	There is no guarantee that the multiple models will include the correct model; danger that the decision maker will be more overwhelmed than aided by the multitude of models; more costly to operate.
Hegelian or Dialectical	The decision maker is actively involved in the information creation process. Unlike most scientific information systems that speak in the dry language of facts and symbols, Hegelian IS speaks in the more active language of drama (using conflict as intellectual and dramatic device to induce decision maker to take an active interest in the system	May create conflict where there is none or inappropriate; not all decision makers can tolerate conflict or learn from it optimally; costly and time-consuming because it involves the creation and use of at least two equally credible theories or experts.

[32] CMB is a piece of evidence supporting the Big Bang theory.

The term serendipity comes from the ancient Persian tale entitled "The Three Princes of Serendib"[33] about fortuitous discoveries. Serendipity also happens in product development, as illustrated by the cases of the invention of cotton harvesting, rubber vulcanization, stainless steel, and microwave ovens. Another example is the (failed) development that led to the sticky glue used on Post-it™ (3M) notes—theory books indicated that a glue with such properties is not possible, or Teflon™ (DuPont). Serendipity also happens when problem areas lead to quality improvements in product design. Among the elements that make this "chance", "luck", "random", or even "chaotical" factor work are (i) the alert mind, (ii) the ability to think laterally, (iii) the ability to mentally connect observations, (iv) the opportunity to discuss with peers, (v) insight, and (vi) the initiative to look closer. All of these elements can be blocked or even completely disabled by an inappropriate organizational context.

2.4 Quality and reliability systems

In spite of its critical consequences for the working world, Taylor's undisputed contribution was the introduction of concepts taken from the exact sciences to establish an "objective" measurement of individual performance. Taylor also introduced the notion of minimally training workers to perform their (minimal) activities as well as the idea of controlling production processes based on predefined standards. This mechanistic view of work, however, was contradicted by unavoidable fluctuations of an individual's performance, so statistical techniques were introduced. This occurred in 1924 with W. A. Shewart's quality "Control Chart" (cf. footnote 19, page 23) and its implementation at Western Electric Co. As a result, test personnel was reduced from 5000 to 2000 persons (Lerner, 1988, 29).

In the 1930s, the concept of quality of manufacture was introduced and became the predominant approach to ensuring the quality of technical systems. Tests were usually conducted in a department in the production area. During the 1940s, it was recognized that quality is not introduced in the product by means of control and therefore must be built into an item. The concept of "quality of design" was beginning to gain importance (Birolini, 1994, 16).

Although quality systems achieved popularity and wide geographical coverage through the ISO-9000 standards, their actual origins go further back. Organizations such as the American Department of Defense (DoD),

---

[33] Alternatively, *Serendip*: an ancient name for Sri Lanka, also *Sarandib* or *Taprobana.*

the National Aeronautics and Space Administration (NASA), the North Atlantic Treaty Organization (NATO), and other institutions connected with the aerospace and defense industries have applied institutionalized quality systems to their acquisitions. Compliance is required of their suppliers and contractors as a precondition of being allowed in the procurement process. Such policies were originally intended primarily as a means to establish a common base amongst the participating industries in order to deal with frequent instances of poor equipment quality.

To illustrate the quality issues encountered in the 1940s, let us examine a character created by the British author Roald Dahl, a former RAF fighter pilot who served as an air attaché at the British Embassy in the United States. Dahl wrote stories for children about the "infuriating little creatures called 'gremlins' that during wartime created the problems causing equipment to malfunction and planes to crash". They were cited again in 1981 by John Yardley, NASA's associate administrator for space transportation systems, to explain the rationale for the extensive tests and flight trials of the space shuttle to "uncover unanticipated problems that might be lurking in the shuttle's unique flight profile" (Bilstein, 1994, 317-318). In its investigation of the Challenger explosion, the Rogers Commission[34] found that the solid rocket booster O-rings were one of the 748 "gremlins" that were "individually capable of causing the loss of the shuttle" (Larson & LaFasto, 1989, 16). For those who regarded the space shuttle as a perfect machine and the result of a perfect system, the fate of the Challenger was a shock that forced them to rethink this notion.

Up to the founding of NASA at the end of the 1950s, there were two basic approaches taken by government defense procurement. The first is termed the "arsenal concept" , derived from the centuries-old approach by militaries to research, develop, and build their own warfare tools. Arsenals possessed nearly comprehensive facilities to equip itself. The second is called the "contractor approach" because it relies on existing industries to perform R&D and manufacture activities. The pure application of one of the concepts did not occur frequently in the previous century: mixed approaches were more typical.

The predominance of one approach is the actual point of interest in this discussion. One of the basic differences between these two approaches in terms of quality is that, similar to Medieval guilds, the arsenal concept encouraged internal quality standards to be developed and followed. Advocates of this principle argue that conventional industrial quality standards are unsatisfactory or that the available hardware is not adequate to the specific purpose. The main arguments in favor of this principle, however,

---

[34] Named after Senator William P. Rogers, who presided the commission.

especially in the mixed forms referred to above, are that in-house development brings the necessary technical know-how to bear on the real implementation problems encountered, and that an essential "measure of reference" is established as a result of the experience gained. The implication of this point will be treated in Chapter 4. Until the founding of NASA in the late 1950s, rocket development by the Army Ballistic Missile Agency (ABMA) under von Braun followed the arsenal concept. This "centralized" approach had the advantages of being cost-efficient, flexible and fast, but its effectiveness apparently also depended on the staff at hand (cf. Medaris & Gordon, 1961). The strategy was nevertheless changed to the contractor approach following the Space Act, which incorporated the ABMA into NASA, and created the Marshall Space Flight Center.

The following case reported by Launius (1994, 74-75)[35] illustrates the reasoning behind the arsenal approach.

Box 5. Intensive knowledge of the object under the arsenal approach.

The Saturn second stage was built by North American Aviation at its plant at Seal Beach, California, shipped to NASA's Marshall Space Flight Center, Huntsville, Alabama, and tested there to ensure that it met contract specifications. Problems developed on this piece of the Saturn effort and Wernher von Braun began intensive investigations. Essentially his engineers completely disassembled and examined every part of every stage delivered by North American to ensure no defects. This was an enormously expensive and time-consuming process, grinding the stage's production schedule almost to a standstill and jeopardizing the presidential timetable.[36]

When this happened, Webb[37] urged von Braun to desist, adding that 'We've got to trust American industry.' The issue came to a showdown at a meeting where the Marshall rocket team was asked to explain its extreme measures. While doing so, one of the engineers produced a rag and told administrators that 'this is what we find in this stuff.' The contractors, the Marshall engineers believed, required extensive oversight to ensure they produced the highest quality work.

Engineers in the same program sometimes actually ended up knowing more about the problems in the implementations than the people who had built them.[38]

---

[35] Excerpts from: Launius, R. D. (1994). NASA: A History of the U.S. Civil Space Program, Malabar, FL: Krieger. Reproduced here by kind permission of Krieger Publishing Company.

[36] This refers to the commitment of President J. F. Kennedy to put a man on the moon and bring him safely back to earth by the end of the 1960s.

[37] J. Webb, NASA Administrator.

[38] Cf. page 269.

In contrast to the arsenal approach, the contractor approach relies on suppliers having the necessary know-how. The customer is merely the user. This is not, as it might appear, necessarily a result of confidence in the capabilities of industry, but is associated with the in-house lack of expertise and facilities to achieve the desired implementation or an attempt to reduce costs by avoiding the effort to develop one's own expertise and facilities. The U.S. Air Force, for example, chose to apply this principle as an alternative to building its own intensive knowledge infrastructure, as the arsenal approach would have required. This by no means indicates incompetence of any kind. The Air Force used its expertise to evaluate the products outsourced to contractors, and it was very rigorous in its evaluation. If necessary, it would even suggest modifications or improvements, which implies a certain technical competence in judging the end product. Nevertheless, even though the contractor approach might solve the problem by "acquiring knowledge" of the processes necessary to develop a product (e.g. by hiring experienced developers), it is difficult to keep this knowledge current because the former "experts" become too far removed from the practice. Adding to this situation the rapid evolution of technical areas and support tools, the capabilities of discussing implementation details with the designers is limited at best. Even if people from the development process are brought in to improve a product by assuming the role of the customer, the pure contractor approach contains no implicit and sustainable way to have or maintain profound expertise to evaluate technical solutions on an equal level with developers. Even though the customer's knowledge as a *user* can potentially contribute essential information to the development process, one of the reported problems lies in *discussing* implementation problems in detail. Another problem is discussing a product's feasibility when the customer assumes the role of the *developer*. The contracting agencies reportedly had to rely on the contractors' estimates because they lacked the experience to evaluate critically and discuss such issues.

With both approaches, development testing exposes the difference between a good and a bad product. What is known about the true characteristics of a product that may function well "on paper" or in the simulation is dependent on how extensively it has been tested. Testing is what turns a failure into a success, because it affords developers the opportunity to trace the causes of malfunction and correct them, even in times of increased computer simulations. Testing also contributes strongly to the proficiency of a development group. Asked years later whether the 120 scientists and engineers brought to the United States to advance the development of rocket technology were technically better than their American counterparts, Ernst Stuhlinger, one of von Braun's closest co-workers, declared that they were not better. He said, "the Americans were on their training, their commitment, their interest as good as we were but we have had in Germany twelve more years to make errors and learn from them"[xvii] (Piechowiak, 1989, 6'30").

Muirhead & Simon (1999, 140-142) pointed out the similar importance of intensive testing to "[reduce] the risk level of Pathfinder". Mars Pathfinder is the main advertiser for NASA faster – better – cheaper space programs.

From the late 1950s on, many attempts were made to improve the reliability of military electronic equipment as reflected by the report of the Advisory Group of Electronic Equipment (AGREE, 1958). AGREE's technically-oriented approach concentrated on development testing, although it also served to influence environmental stress screening in the production phase as a means of filtering out products with so-called "childhood diseases" introduced by the manufacturing process. It introduced the concept of *burn-in* or *environmental stress screening*. Organizations were also urged to approach the problem of quality systematically by the military standard MIL-Q-9858 "Quality Program Requirements" (for contractors) first issued in 1959 (cf. DoD, 1981a).

In contractor-based product development, testing and the reporting of test results were also in the hands of the contractors. This appears to have worked well until the scandal reported by Vandivier (1972)[39] involving the development of improved brakes for the USAF A-7D—the "D" version of the Corsair II fighter aircraft originally conceived for and already in the service of the U.S. Navy—caused the DoD to make "sweeping changes in its inspection, testing, and reporting procedures." Falsified test data of a brake conception that exhibited catastrophic failure even during bench testing allowed its integration in the aircraft. This almost ended in disaster when, during ground test runs of the aircraft, the brakes disintegrated and the wheels blocked, rendering the aircraft uncontrollable. These findings raised serious doubts about the reliability of the self-reporting approach taken hitherto.

Consequently, control mechanisms had to be introduced, albeit not overtly. A person from the contracting agency was assigned to the premises of the main contractors and subcontractors. Although the presence of a resident government representative was basically intended to promote the government–industry interface, "discomfort" was reportedly caused by the

[39] Excerpts from Vandivier, K. (1972). Why Should My Conscience Bother Me? The Aircraft Brake Scandal, in: R. L. Heilbroner et. al. (Ed.), In the Name of Profit, pp. 3-31, Garden City NY: Doubleday & Co. Reproduced here by kind permission of Random House, Inc., Doubleday.

suspicious nature of the relationship.[40] Amongst subcontractors in the aerospace industry, one effect still noted decades later was the obligation, sometimes stipulated by contractual clauses, to advise the customer when certain tests were to be performed and to invite the customer in advance to witness the tests. Indeed *witness* is the very word still used in such situations.

Standards for ensuring the quality of new developments and their associated practices evolved, and many variations thereof emerged. NATO, for instance, established its "Allied Quality Assurance Publication" (AQAP) series in 1968. In NATO member countries, national variations and industry-specific standards followed.[41] One of the characteristics of these standards is the formalism in the structure to "systematically" ensure quality.

As a result of this formalism in dealing with quality issues, companies had to document standard operating procedures in order to demonstrate compliance and thus achieve the status of a qualified supplier. However, many complaints concerning the content and effectiveness of such an approach were heard throughout the organizations involved. Formalized procedures were frequently not resource-efficient, some were error-prone and in many cases not updated appropriately to reflect real practices. Very often, the structure called for by these procedures closely resembled the internal bureaucratic organization. Management and especially the supplier evaluators from the customers expected these procedures to solve quality issues independently of the persons applying them. In fact, instead of preparing a framework in which the work of a good professional can make a difference, such systems actually impose this independence from a specific worker. Quality problems inherent in such an approach were reportedly solved most efficiently when repetitive activities such as production steps were performed. The poorest efficiency was reported for non-repetitive, intellectual[42] activities such as in the product development and evaluation processes.

Formal implementations of bureaucratic systems are scorned worldwide, and sometimes precondemned without true justification. The idea of developing a set of matured procedures to cope with a problem or with a kind of problems is not entirely bad, at least in theory (cf. page 34ff.). It

40 In some cases, it was merely a communication interface and in other more elaborate cases, the representative also had some technical competence and was authorized to make certain decisions. It must be said that there were cases in which this interface functioned very well without the suspicion character, whereas in others the representative acted as a veritable spy. Sometimes, the "spies" overreacted to rumors causing energy-consuming trouble.

41 Repercussions were felt even by non-NATO members. For instance Brazil introduced the "Requisitos Brasileiros para a Indústria Aeroespacial" (RBIA) series, which had a one-to-one correspondence to the NATO requirements.

42 Intellectual, i.e. relating to the intellect.

could liberate the persons involved in an activity to concentrate on more important things other than devising solutions for problems that have already been solved. Thus, if a solution has already been provided, a person trying to cope with the given problem would be well served. The idealized bureaucracy is a measure of the organization of a system, thus leading to efficiency in coping with problems. In practice, however, the bureaucratic system that should help workers cope with the issues they encounter often hinders them. For example, bureaucratic systems frequently impose additional obstacles or insist on a predefined "one and only way" to solve a problem, even when workers are sure this is not the case.

The formalism of the Quality System Procedures is in theory an interesting idea—as Max Weber's bureaucracy (cf. page 34) still might be. It also carries the potential of solving at least part of a given quality problem in technical ventures. Nevertheless, complaints about the effectiveness of formalized procedural systems like those required by Quality Systems were heard in many countries. Among these complaints were: (i) excessive bureaucracy, (ii) the inefficiencies inherent in the formalized procedures, (iii) the nonapplicability of compulsory measures, (iv) the existence of the system that exists only "on paper", (v) the inability of the system to really help the person confronted by the problem, (vi) the inadequacy of handling exceptional conditions, and, especially in the case of product developers, (vii) that the system limits their freedom to create. In addition, due to the demarcation of accountability, (viii) there is a tendency in bureaucratic systems for people to do only what is defined in the procedure AND not to do what is not defined in it, even when they know they could or should. Finally, the approach (ix) eliminates the option of considering the very important tacit knowledge[xviii] of experienced workers because, no matter how valuable, tacit knowledge cannot be assessed in procedural terms (cf. Ulich, 2001, 302-311).

More or less parallel to the aerospace industry, other industries also found that "expanding" the quality department from one that controls the activities of the production department to one that has responsibilities across the entire company was hindered for several reasons. One of them is the tendency observed to extrapolate control techniques (in the narrow sense) to encompass other activities, including those of product development.

Until the mid-1980s, there were two basic approaches: quality control and quality assurance; they were generally regarded as completely different schools of thought. The term quality control (QC) emerged from statistical quality control techniques developed during the 1940s. In essence, QC catches problems after they have occurred or after predefined levels of tolerance have been exceeded. One unexpected consequence of this approach is that the objective was no longer to produce error-free products but to "pass the control" (Seghezzi, 1996, 5). This shift in objectives resulted in cases where manufacturers reported situations in which a rejected lot was resubmitted without any further correction and was subsequently accepted by the same quality-control entities, sometimes even by the same personnel. In one case, production staff made a veritable sport out of deliberately introducing bogus production items to see whether the items would be detected by control personnel. Bogus items that escaped notice were subsequently removed, but of course provided a rich source of "entertainment" (case #4).

As quality efforts began to concentrate on preventing errors in the first place and subsequently implementing quality improvements, the term quality assurance (QA) gained significance. As a result, numerous techniques were developed to incorporate this task into manufacturing processes.

In the 1980s, it was obvious to which school of thought a person in the field of quality subscribed. This became especially apparent at workshops, where issues associated with improving quality were discussed in detail. Those accustomed to catching the errors of others tended to favor imposing a mechanism of control. Those whose focus was to improve quality per se acting on the sources of problems. When the differences between these two approaches were the greatest, strong resistance was the result. Very often, it boiled down to the simple dilemma that "QAs" and "QCs" did not speak the same language, so to speak. From the practitioner's point of view, it took a while before this conflict was resolved, at least partially. Even as total quality management (TQM) or company-wide quality control (CWCQ) was attracting much attention in the second half of the 1980s—at a time when responsibility for product quality was diffused within the enterprise instead of concentrated in the quality department—those serving in quality functions still suffered from a significant negative influence of their professional background on their performing styles. Having those who execute a given activity assume responsibility for the quality of its outcome was a notion that encountered formidable obstacles, and not only from within the quality department. Some people still subscribe to the dogmatic belief that persons who perform a given activity should not be allowed to control the quality of its results, irrespective of the applicability of an independent control or self-control.

The International Standardization Organization proposed the first total quality standard in 1987. Called the ISO-9000 series, it was proposed as an integral approach to quality issues. The standard applicable for companies developing products was the ISO-9001 (ISO, 1987a). Although the applicability and the level of detail observed by NATO and ISO differed in terms of their implementations, both entities adhere by and large to the requirements established in the standards. MIL-Q-9858-compliant companies did not comply as closely, but the requirements placed on them were essentially equivalent.

Wolf (1992) describes two contrasting cases of ISO 9001-certified companies in Germany. The first company did not achieve the intended quality improvements within one year of certification. Although an evaluation revealed that the procedures and guidelines were known and had been implemented in the day-to-day activities, employee motivation was the greatest detriment to quality improvement. Employees felt overwhelmed by the abrupt introduction of the quality system. Particularly counterproductive was the perception that management had tried to force employees to accept the new quality assurance measures.

Unlike the first company, the second one enlisted the participation of employees from all sectors and hierarchy levels in setting high-level goals (i.e. goals of overriding importance). An essential supportive measure was the company's policies regarding all professional and employee-oriented information. In addition, the second company implemented a comprehensive training concept at all hierarchy levels. The latter company achieved a high level of competence in quality assurance issues within a short time. The two contrasting cases illustrate that quality is dependent on the people using the system.

It should be noted at this point that the factor of employee motivation was taken into consideration in the ISO 9004 guidelines (ISO, 1987b, 15-16) from the outset, but this standard is still not obligatory for certification.

Similar problems occurred with the quality assurance of software development. Starting with self-criticism within their own field and the problems associated with designer-dependent quality, software practitioners faced pressure to evolve their craft from an "art" to a more disciplined approach. The term software engineering appeared. Now the main problem with software quality turns out to be the human inability to conceive perfect complex logical structures. It was common wisdom among software engineers in the early 1980s was that "software is never debugged". Sometimes correcting a bug introduced a series of others, and it was hypothesized that, after debugging, the quantity of bugs in a complex system remained constant, independent of the effort to correct them. Quality and reliability organizations, on the other hand, reacted in substantially different manners. Some regarded the problem as being development-oriented in the sense of determining suitable techniques to support the engineering aspect of software development. One example of this line of thought was followed by the German working group of the Telecommunication Society of the Association of German Electrotechnical Professionals—NTG[xix] and the German Society for Quality—DGQ.[xx] The initial meetings took place in 1982 and 1983. The resulting publication entitled "Software Quality

Assurance" (DGQ/NTG, 1986) clearly set the focus on development, with its comprehensive listing and description of the available techniques to develop software and prevent errors—not on finding and correcting errors. In the description of the principles underlying software quality assurance, these entities proposed the following maximal constructive quality assurance principle: "errors that cannot occur do not need to be corrected"[xxi] (op. cit., p. 23). This terminology emphasizes that work's character of preventing errors instead of detecting and correcting them. With this principle, the need for software testing is not eliminated because the probability of error cannot be completely zero. But at least the focus is on contributing constructively to improving quality. The authors of that publication did not require their guidelines to be followed, but focused instead on gaining the information that each of the techniques represents as well as their implications for error prevention.

On the other side of the world, with preliminary meetings occurring in 1979 and discussions being held from 1983 through 1985, the American Department of Defense and three major industry associations established the military standard DOD-STD-2167 (DoD, 1985). Although the considerations of software engineering are also clearly present in that document, the focus was placed on the documents to be generated during the software development process. These documents had a formal character, especially from the perspective of the procuring agencies. The prevailing sentiment was that software development was "the work we have to show the customer". But the DoD went too far in this endeavor, and the standard had to be revised. The resulting version 2167A was released in 1987.

Many problems were experienced in using this approach: in one situation (case #5), the standard was primarily intended for mission-critical software. But in this case there was the tendency on the part of the contracting agency to apply it to *every* software item associated with product development. The developers were worried about the amount of presentation-quality paperwork they would have to generate to fulfill the requirements. The designers and quality personnel as well as company technical directors were aware of the importance of the techniques, but could not argue with the imposition of the standard in spite of their concern that it would block the development process. Customer representatives countered with a simple "you must comply with it."

Other industries experienced the same problems. Coad (1988, 48) describes why the standard had to be revised. Among other reasons, the intention to provide "visibility and control" for the contracting agency regarding the development process went too far and "too much time and money [had to] be spent on the contractually required documentation and the multitude of minute details called for by the 2167".

In addition, it predetermined what was considered good practice for software development in 1985, and although the standard left room for companies to propose alternate approaches and methods, it provided default documents that implicitly obliged use. Even the term "development phases" unintentionally prescribed the use of the phase model of software development. Coad also reminds us that "no process or method is appropriate for all development projects in all organizations. These defaults have inhibited the introduction of the latest (and possibly proprietary) software methods, which are different than the defaults. To be safe, the defaults are taken as law. Producing every document according to the 'letter of the law' has significant economic and engineering effects ... For larger projects, the burden of documents is so high that the entire engineering organization becomes document-driven. More effort is expended to produce the documents than to engineer the system under consideration. Furthermore, when customer reviews focus on format compliance, rather than on engineering content, the dilemma is only amplified. The data requirements and default tables sometimes call for descriptive attributes sooner than necessary in the development cycle. At times, to be compliant, attributes are added with little meaning or engineering value, simply to fulfill a data requirement." (op. cit., p. 48).

The 2167 standard was an attempt to establish a framework for developing software with consistent process and product standards, but it placed too strong an emphasis on defining requirements, preliminary design, and detailed design. This "consistent approach is incompatible with alternative development processes, notably object-oriented development, evolutionary development, rapid prototyping, and multiple build (design a little/ code a little/ test a little) approaches" (op. cit., p. 48).

The revised 2167A had to reconsider all these aspects. It subsequently required simpler, shorter, fewer documents and allowed for more flexibility. The revision laid the groundwork for "tailoring" by having companies propose for each program a list of pertinent documents appropriate to the situation. Defaults were minimized and development phases were called development "activities" in order to avoid bias toward a certain process. Contractors were encouraged to "readily define and utilize" alternative development processes (op. cit., p. 48).

Nevertheless, this did not solve the problems completely. One contracting agency (case #6) complained that the new, essentially slimmer version did not provide the visibility that the earlier version had, owing to the documents it called for. This contractor tried to retain the first version as a contractual obligation, even when the applicability of the standard had been corrected by its originators. The negotiations over the application of that standard were very difficult to conduct in this case. The contracting agency declared its intention to retain control (in the narrow sense) of the process.

After examining all these hurdles of assessing and improving software development quality, it is important to mention the approach proposed by Humphrey (1995) in which, according to Demarco (1996, 116), the author "proposes that we end two decades of attempting to make measurement and estimation of software effort an institutional responsibility, and transfer those tasks back to the individuals who build the software".

Similar issues had been faced before by reliability organizations in the aerospace industry. The basic standard that governs reliability issues is MIL-STD-785, originally issued in 1965. In revision B of this standard (DoD 1980), explicit mention was made that "increased emphasis has been placed on reliability engineering tasks and tests. The thrust is toward prevention, detection, and correction of design deficiencies, weak parts, and workmanship defects" (op. cit., p. iii).

One major difference in reliability activities compared with development activities is that, on the one hand, one has an engineer trying to make a product that will *always* work, and the reasoning behind the activities were directed toward this objective. On the other hand, one has a reliability engineer trying to determine the conditions under which a device would *not* work and estimating *when* it would not work (cf. the *falsifiability* concept of Popper, page 46). The resulting analyses are complementary by nature, but contrast sharply. They require different types of analysis that are not easy to understand from both perspectives at once, yet they require an intensive exchange of information that is common to both. They must influence each other in order to solve reliability problems effectively. These differences are also present in quality assurance activities in general, but are much more accentuated in reliability activities. To deal with these issues, some organizations have established a division of responsibilities by having the development people on the one side trying to demonstrate how something works, and the quality/reliability people on the other side, trying to demonstrate the contrary. This dichotomy evidently still exists today.

The characteristics of product maintainability are an issue in product development quality that also deserves consideration in this section. The problem is of importance whenever a high value or a not easily disposable item has to resume its function promptly after a failure. The American Department of Defense established standards for guiding the development of maintainable products: the MIL-STD-470 "Maintainability Programs for Systems and Equipment" initially published in 1966 (cf. Birolini 1994, 506). However, problems were reported in achieving this goal and the *standard* was changed to a *handbook* to mitigate the compliance obligation and give it more of a guiding or orientational character. In 1996, a draft of the handbook was released that had been combined with another handbook (which in turn had also originally been a standard). This produced "a handbook that is as useful to industry as it is to the Department of Defense" (DoD, 1996).

2.5 Project management

The origins of project management go back to the 1950s, when the U.S. Navy Special Projects Office (SPO) developed the Polaris ballistic missile. According to Sapolsky, because high-level officials were accountable for the failures of their subordinates, the former use a number of bureaucratic controls to limit the initiative of the latter. Such controls include special audits, detailed reports, and staff assistants. In extremely complex and risky programs such as the Polaris, these officials have the tendency to limit even more drastically the independence of their subordinates by intervening directly in management duties. Thus SPO sought to avoid being required to justify each decision in detail by building official and public confidence in its general managerial abilities. This introduced the concept of project management and related tools such as the project evaluation and review technique (PERT), PERT/COST, Program Management Center, Reliability Management Index (RMI), weekly program review meetings, and the invention or refinement of managerial graphics (Sapolsky 1972, 58-59).[43]

In order to deploy a force of submarines in the shortest possible time, the achievement of organizational autonomy was the basic objective of the Polaris proponents. From its extensive experience, the SPO was "aware of the many ways in which projects fail because of their dependence on cooperation among government agencies, each having independent bases of power and conflicting goals and interests." As the uncertainty associated with the development of the Fleet Ballistic Missile (FBM) program was considerable, many of the entities and persons allocated to the Polaris program would "have been tempted to monitor and direct its progress." This was further reinforced by the emergence of the Cold War. To achieve the necessary autonomy, two subobjectives had to be reached: (i) there had to be a broad base of support both inside and outside the Navy and (ii) the rest of the Navy and the rest of government had to be prevented from interfering in the project management tasks. To accomplish these subobjectives, four bureaucratic strategies were used, namely: *differentiation* and *co-optation* for

---

[43] Reprinted by permission of the publisher from THE POLARIS SYSTEM DEVELOPMENT by H. M. Sapolsky, Cambridge, Mass: Harvard University Press, © 1972 by the President and Fellows of Harvard College.

the first subobjective and *moderation* and *managerial innovation* for the second (op. cit., pp. 41-43).

The *differentiation* strategy is defined as the "attempt of organizations to establish claims on valued resources by distinguishing their own products or programs from those of their competitors" (op. cit., p. 43).

The *co-optation* strategy is described by Selznick (1952, 135-137) as the attempts of an organization to assimilate new elements in its structure as a means of "averting threats to its stability or existence. ... Depending on their political strength, potentially disruptive elements may be given control over decisions vital to their interests in order to gain their support in other matters, or they may be given symbolic participation in decision making in order to add legitimacy to established policies".

Moderation is the attempt to build long-term support of programs by sacrificing short-term gains. By following this strategy, SPO refrained from establishing a base with the specialized groups of rocketry in Huntsville, AL, or in California. Instead, it stayed at its Washington location, which, although an aging facility, allowed it to guarantee the survival of the program. Although having the ability to leverage its new priority status, it chose to stay geographically close to the Navy (Sapolsky, 1972, 54-56).

Finally, the strategy of *managerial innovation* can be defined as "the attempt of an organization to achieve autonomy in the direction of a complex and risky program through the introduction of managerial techniques that appear to indicate unique managerial competence" (op. cit., p. 58). The context in which it appeared is illustrated by the management system described by Downs (1967) as the "means by which the highest-level officials, presumably the most responsible officials, in an organization monitor and control the behavior of their subordinates". Sapolsky goes even further by stating that "even in programs that do not involve the technological uncertainties of the Polaris, there is a concern on the part of the organizational leadership that the lower-level participants in the organization are not dedicated to their work or honest in their reporting". He also includes this interest as one of the possible motivations for the managerial innovation in the Polaris program (op. cit., p. 95). In addition, "the traditional means of controlling costs, schedules, and quality of technological projects—experience, bargaining skills, and double-entry ledgers—were being depicted in numerous congressional and executive reports as being patently inadequate". He goes on to say that "the failure at that time of many large-scale technological projects to meet their cost and schedule projections was being attributed to deficiencies in management methods" (op. cit., p. 96). Maybe this contextual situation explains why "defense officials have tended to emphasize the efficacy of the managerial techniques and depreciate the importance of the organizational autonomy of the agency that produced both techniques and the Polaris" (op. cit., p. 59).

Following the program's success, many governmental and non-governmental institutions became interested especially in PERT and associated techniques such as the Critical Path Method (CPM). Within ten months of publicizing PERT in mid-1958, SPO received some one hundred requests for further information and detailed briefings. A Naval public relations officer declared that by the time of the first submarine launch of Polaris in 1960, the press coverage of PERT was almost as great as that devoted to the Polaris itself (op. cit., p. 111). In 1962, the U.S. government alone had issued 139 documents and reports on it (Poletti, 1964, 44), and by mid-1963, the bibliography compiled by the U.S. Air Force (1963) cited 702 books and articles published in English on the topic. In 1964, the list was still growing.

Several myths surrounded the SPO management innovation strategy. Based on interviews, Sapolsky concludes: "An alchemous combination of whirling computers, brightly colored charts, and fast-talking public relations officers gave the Special Projects Office a truly effective management system. It mattered not whether the parts of the system functioned or even existed. It mattered only that certain people for a certain period of time believed that they did. Thus, a former Secretary of the Navy was convinced that the Special Projects Office never missed a scheduled commitment (it did) and a former Chief of Naval Operations and a former Assistant Secretary of Defense were certain that PERT always caught impending errors (it did not). The Special Projects Office won the battles for funds and priority. Its program was protected from the bureaucratic interference of the comptrollers and the auditors. ... The very few that took a closer look discovered the fabrication, but also its value. The British Admiralty, for example, examined the operations of the Special Projects Office carefully in order to construct their own FBM force. Its representatives were initially surprised that the Special Projects Office's documented management philosophy did not match its management practice. Nevertheless, they too recommended the adoption of the entire management system as advertised since they recognized the advantages in terms of organizational independence and resource support that such a system could provide" (op. cit., pp. 129-130).

The apparent success of PERT helped it become a contract requirement of the American Department of Defense, the National Aeronautics and Space Administration and the Atomic Energy Commission. A special room called the "control room" was created to demonstrate management's supposed supervision of the program. *Figure 3* illustrates an artist's conception of the control room for the X-20 program[44] (Sadow, 1964).

Figure 3. Control room for the X-20 program at Boeing (early 1960s).[45]

A similar control room was implemented for the Apollo/Saturn moon landing development program. Bilstein (1980, 286) describes this room as being "rimmed with recessed, sliding status charts and double picture screens for comprehensive, up-to-the-minute briefing[s] on progress and problems in the far-flung program." In the back of the room there was a large PERT chart of the program as well as extensive communication facilities and presentation equipment (*Figure 4*). To illustrate the extent of the Moon program, for example, "the PERT network was broken down into 800 major entities, and summarized 90,000 key events taking place around the country" (op. cit., p. 286). It was estimated that the Apollo/Saturn "stack"[46] comprised some 6 million parts.[47] The Saturn V launcher alone,

---

[44] The X-20 was a U.S. Air Force space plane cancelled in 1963 because there were too many technological advances to be solved before such a program could be pursued. This was not possible until a decade later with the Space Shuttle.

[45] From: Sadow, R. M. (1964). How PERT Was Used in Managing the X-20 (Dyna-Soar) Program, IEEE Transactions on Engineering Management, Vol. EM-11, December 1964, pp. 138-154. © 1964 IEEE. Reproduced here by kind permission of the IEEE.

[46] As the composition was called.

[47] This estimate considers all bolts, nuts, washers, transistors and circuit boards. The Apollo Command and Service Modules had two million parts; the Lunar Excursion Module had one million parts (Bilstein, 1980, 288).

which was coordinated and partially developed by MSFC, had three million parts and involved some 400,000 people (including all contractors).

Figure 4. Apollo Program control center at NASA Marshall Space Flight Center in Huntsville. (Bilstein 1980) photo: NASA

Some people argue that PERT played a major role in the Apollo Lunar Landing Program. However, Tompkins (1977)[48] qualitatively investigated in 1967 the communication processes at the Marshall Center under von Braun during the development of the Saturn moon rocket and reported the so-called *formalism-impersonality syndrome:* "It was refreshing to hear from people who had made important contributions to the development of computers and management-technology a stern denouncement of the sterility, formalism and impersonality of many of the management tools they used. The specifics they condemned included computers, control centers, reviews and PERT; it was expressed as a grave concern that *things* rather than *people* were in control" (op. cit., p. 16, italics retained). He noted that the skepticism of the interviewed Marshall managers was caused by their interest in the quality of management and the details of the results, and summarized their complaints as follows: "Realizing that PERT was a 'GIGO' device ('garbage in, garbage out'), they often knew that information posted in the center was inaccurate; but their most acrimonious complaint was the intimidating effect on the center during briefings and meetings there. 'You can't argue or compromise with a control center or with PERT,' they said repeatedly to me. The result seemed to be a classic case of goal-displacement. Things that were devised to serve as management tools, as means, had become ends in themselves"

[48] Excerpts from: Tompkins, P. K. (1977). Management Qua Communication in Rocket Research and Development, Communication Monographs, Vol. 44, No. 1, pp. 1-26. Reproduced here by kind permission of the National Communication Association.

(op. cit., p. 17). Furthermore, the program director frequently discovered discrepancies between what he knew to be true and what was visible in the control center.[49] Tompkins concluded that the control center was "more for show than for utility" (op. cit., p. 18). Indeed, congressmen and VIPs were brought to the center to witness the agency's contributions to management communications. Tompkins also contrasts his observations with the statement by Drucker (1974, 595, cited in Tompkins 1977, 18) that "when NASA first started, the scientists who dominated it then believed that controls, and specially, of course, computer-based information, would run the system." He adds that, "It should be clear from these observations of the sterility of such beliefs. The Marshall Center, most important in the NASA system, worked well because the managers had a healthy skepticism for such controls. In the end, it was the people, not controls, who got the job done" (op. cit., p. 18).

Regarding the Navy SPO, Raborn[50] later observed: "We used a philosophy of utter communication...There was no such thing as hiding anything from anybody that had an interest in it. And there was nothing that got a person into trouble quicker than to delay reporting potential trouble. And, boy, if he waited until he had trouble, then he really had trouble." (Heppenheimer, 1997, p. 137).[51] The account continues: "His colleague Captain Levering Smith, a propulsion specialist, adds: 'Much of what was done was contrary to current government procedures and regulations. They were set aside.' This meant that Raborn could expect at all times to have the information he would need to make a decision. He then could do so, rather than order further studies." (op. cit., 1997, p. 137).

By 1972, PERT was no longer mandatory in DoD practice (Sapolsky, 1972, 125). Moreover, the limitations of managerial innovation techniques introduced by the Navy SPO were widely known and their influence was fading (op. cit., p. 246).

The defenders of the technique, however, state that the problem with project network techniques was the technology supporting it. Zerega (1976) provides historical data showing that the use of PERT in NASA's Goddard Space Flight Center[52] reached a low in 1971 and, after introduction of computer graphics network software in 1972, the number of projects utilizing PERT started to increase. His argument that the new program

---

49 For a complete account on the problems found, please refer to pages 17-18 of Tompkins' article.

50 Admiral William Raborn, head of the Navy SPO at the time of Polaris.

51 Excerpts from: Heppenheimer, T. A. (1997). Countdown - A History of Space Flight, New York: John Wiley & Sons. © 1997 by T. A. Heppenheimer. Reproduced here by kind permission of John Wiley and Sons, Inc.

52 (unmanned flights)

reduced the cost and time of implementation of the network is partially valid. Indeed, one of the reasons pointed out by Bilstein (1980, 287) for phasing out PERT during the Apollo Moon program is that it was "tremendously expensive". However, Zerega's data is inconclusive because he does not provide information about the total number of projects pursued at Goddard during the period in question. Nor does he provide information about the effectiveness of its use or details of the project complexities or interconnection of the activities. His summary indicates that the computer graphics program: (i) "*permits* increased use of PERT"; (ii) "*permits* more effective use of manpower" (he only cites the time-consuming drafting work eliminated) and (iii) "produces highly presentable, streamlined(!) networks" (parentheses and italics added). The latter point deserves closer inspection. A contemporaneous account by Weinberg (1971, 111) illustrates its "need":

Box 6. Case illustrating the "show" effect of project management network graphics.

"One of the most beautiful of such demonstrations (to casual visitors) consisted of displaying a PERT chart of some project function on the CRT.[53] When the manager pushed a button, the critical path in the chart was illuminated with greater brightness than the other paths. Then, by pointing the light pen[54] at one of the nodes, a deletion was made and the new chart was displayed. One more button push and the new critical path was illuminated. *This demonstration never failed to impress visitors with the greater control the manager had over his project.* Of course, it was a complete fake—all four patterns were prestored in the system, and they were the *only* four patterns available". (italics and parentheses added)

With the availability of the microcomputer, PERT and related network planning techniques began to return to the industrial scene in the early 1980s. Many advocates of these techniques argued again that the ability to perform calculations and display information was the limiting factor in using such techniques successfully before this period. Interviews with users of these "modern" programs reveal the curious fact that problems encountered in the 1960s were still being replicated from the mid-1980s through the mid-1990s.

53 CRT (Cathode Ray Tube) is the denomination for the computer terminal that at that time was starting to replace batch processing of the 1970s. Until then, the main form of data visualization was on paper and data input was mainly by means of cards. Weinberg lists it as one of the many status symbols of programming managers.

54 The light pen was a positioning device that the operator pointed at the desired position directly on the CRT screen.

Box 7. Case illustrating the repetition of the (in)visibility problem when using network techniques (case #7).

At the end of the 1980s, an electronics company had allocated an engineer to "manage" a program based on one of the microcomputer-based project management programs. His office was practically wallpapered with computer printouts of the PERT/CPM networks. The technical director once remarked that, in spite of all the work this engineer may have performed, he did not know more about the program than the director himself was able to achieve merely by interacting with the appropriate technical people during his "management by walk-around".

To reinforce this observation, take for example the statement by Ziemke & Spann (1991, 49) that "PERT charts are enormously time consuming. Managers often complained that they spend more time on PERT charts than on getting the job done".

Illusions concerning the use of PERT persist. The following arguments of a development manager at a convention on methods for product development in 1997 illustrate this situation. After mentioning engineers' fear of PERT as a monolithic system (supposedly caused by the implied limitation of their freedom to create), he attributed it to the observation that engineers are not accustomed to thinking in terms of schedules (sic). Thus he proposed a "drag-and-drop" tool for project scheduling, with which the engineer would be familiar. Whether this is a sensible way to cope with fear is left to the reader to decide[55].

Currently, some people argue that networking simplifies the gathering of data and as such, the use of networking techniques can be boosted again by technology. However, the many other essential issues associated with non-technical techniques of this kind remain untouched.

Another illusion stems from the fact that development is essentially an information flow process, not a physical flow as assumed in the PERT/CPM model (McGrath, Anthony & Shapiro, 1992, 113).

Project management is actually an application of systems engineering and is not limited to PERT diagrams. Yet many project managers build their activities around PERT instead of using it as a means. Or, in an attempt to

[55] In all fairness, it must be said that this was said in response to a publicly posed question and that he himself might have been driven by fear when giving this answer. Possible reasons for his fear are the realization that his software could not solve the problem, that he would have to admit publicly his lack of a better answer, or that he was not willing to attack an essential feature of his own product. On the other hand, perhaps he truly believed in it. None of the hypotheses can be eliminated in this case. The reader is advised to avoid an over-interpretion of his statement but to consider only his argument (overcoming the fear of engineers by providing a "drag-and-drop" software tool). This case serves *merely* as an illustration.

make a complex task manageable, some project managers adhere to the work breakdown structure to an exaggerated extent.

But this does not solve the problem of interaction described by Frosch (1969, 25), cited in Pill (1971, 24): "One of the key misassumptions in modern systems engineering and systems analysis is that the total problem can be, and frequently is, decomposed into subproblems; the subproblems can be solved more or less independently, and the total solution can be synthesized by combination of the subsolutions treating the interactions of the parts as 'interfaces.' The real world is, however, highly nonlinear, and unless real attention is paid to this fact, the linear decomposition treatment will fail catastrophically, because the interaction terms be as large as the subproblems and not reducible to simple interfaces. The result may well remain decomposed." Frosch was Assistant Secretary of the American Navy in a time when systems engineering was a euphorious field and, in his account, "the fundamental difficulty is that we have all become so entranced with the technique that we think entirely in terms of procedures, systems, milestone charts, PERT diagrams, reliability systems, configuration management, maintainability groups and the other minor paper tools of the 'systems engineer' and manager. We have forgotten that someone must be in control and must exercise his management, his knowledge, and his understanding to create a system. As a result, we have developments that follow all the rules, but merely fail." (op. cit., p.12-13).

Although system engineering has its merits and has certainly progressed since then, this assumption of the validity of the Cartesian approach (cf. page 45ff.) is still made in practice. The problem reported by Frosch is regaining importance today as product complexity and interconnection continue to grow, regardless of whether this occurs in conjunction with interdisciplinary product development.

Furthermore, as evaluated by Sapienza (1995, 142-143), project management, its structure, the multiple disciplines and functions it encompasses, and the resource constraints under which a project has to be accomplished "entails formal collaboration" (understood as: to work together).[56] She argues that "informal collaboration arises when people want to work together". In this scenario, one person determines that the input of another person is needed, and approaches that person. Sapienza goes on to say that "just putting people together in collaborative units like project teams does not mean they will collaborate. In fact, they generally will not." Informal collaboration is at least as important as formal collaboration to achieve development quality.

[56] The word "collaboration" is retained here because the author does not specify what form of "working together" is meant.

2.6 Skunk Works

A concept was originated at the end of World War II by Clarence "Kelly" Johnson at Lockheed as a way to arrive rapidly at high-performance, state-of-the-art aircraft. It was called "Skunk Works", and was originally composed of Lockheed's 23 best engineers and some 30 shop mechanics.[57]

The group was intentionally kept relatively small so that interaction would remain high (Kelly's rule #3: "The number of people having any connection with the project must be restricted in an almost vicious manner. Use a small number of good people", op. cit., p. 51).[58] As Ben Rich, Johnson's successor, explained: "inside the Skunk Works, we were a small, intensely cohesive group consisting of about fifty veteran engineers and designers and a hundred or so expert machinists and shop workers" (Rich & Janos, 1994, 7). Its small size was supported by Lockheed's size, which provided a pool of highly qualified personnel who could be "borrowed" as necessary (op. cit., pp. 318-319).

As Rich describes: (in 1976) "we had our own very unique method for building an airplane. Our organizational chart consisted of an engineering branch, a manufacturing branch, an inspection and quality assurance branch, and a flight testing branch. Engineering designed and developed the Have Blue aircraft[59] and turned it over to the shop to build. Our engineers were expected on the shop floor the moment their blueprints were approved. Designers lived with their designs through fabrication, assembly, and testing. *Engineers couldn't just throw their drawings at the shop people on a take-it-or-leave-it basis and walk away. Our senior shop people were tough, experienced SOBs and not shy about confronting a designer on a particular drawing and letting him know why it wouldn't work.* Our designers spent at least a third of their day right on the shop floor; at the same time, there were usually two or three shop workers up in the design room conferring on a particular problem. That was how we kept everybody involved and integrated on a project. My weights man talked to my structures man, and my structures man talked to my designer, and my designer conferred with

---

[57] Their greatest successes (among many "failures") up to now have been the first operational American jet fighter P-80, the U-2 and SR-71 spy planes, the F-117 Stealth aircraft, among others.

[58] Excerpts from: SKUNK WORKS by Ben Rich © 1994 by Ben R. Rich. Reproduced here by kind permission of Little, Brown and Company, Inc. SKUNK WORKS: A PERSONAL MEMOIR OF MY YEARS AT LOCKHEED © 1994 by Ben Rich and Leo Janos (Little, Brown, Boston). Reproduced here by kind permission of the authors.

[59] Internal development denomination of the aircraft that turned out to be the F-117 Stealth.

my flight test guy, and they all sat two feet apart, conferring and kibitzing[60] every step of the way. We trusted our people and gave them the kind of authority that was unique in aerospace manufacturing. Above all, I didn't second-guess them" (op. cit., pp. 46-47, italics added).

Rich continues: "The shop manufactured and assembled the airplane, and the inspection and quality assurance branch checked the product at *all stages* of development. That was also unique with us, I think. In most companies quality control reported to the head of the shop. At the Skunk Works quality control reported directly to me. They were a check and balance on the work of the shop. Our inspectors stayed right on the floor with the machinists and fabricators, and quality control inspections occurred almost daily, instead of once, at the end of a procedure. Constant inspection forced our workers to be super-critical of their work before passing it on. Self-checking was a Skunk Works concept now in wide use in Japanese industry and called by them Total Quality Management [sic] " (op. cit., pp. 47-48, brackets added).

They also intentionally limited the formality of the drawings release: "Very simple drawing and drawing release system with great flexibility for making changes must be provided in order to make schedule recovery in the face of failures" (Rule #4, op. cit., p.51).

With this flexible concept, they achieved a much faster learning curve than with conventional approaches within the industry sector, even though they built fewer airplanes per month (op. cit., p. 89).

Skunk Works also featured major differences in management control. One of the principles established by Johnson stated that: "The Skunk Works program manager must be delegated practically complete control of his program in all aspects. He should have the authority to make quick decisions regarding technical, financial, or operational matters" (Rule #1, op. cit., p. 51).

The application of this rule ensured the necessary circumstances for the team to perform, namely its flexibility and autonomy. This was not standard in the management principles of the time. In agreement with Johnson, Rich insisted that "the bottom line is that most managements don't trust the idea of an independent operation, where they hardly know what in hell is going on and are kept in the dark because of security. Don't kid yourself, a few among our own people resent the hell out of me and our independence. And even those in aerospace who respect our work know damned well that the fewer people working on a project, the less profit from big government contracts and cost overruns. And keeping things small cuts down on raises

---

[60] *Kibitzer* – an onlooker (at cards, etc.) who gives unwanted advice, an interferer (North American slang of Yiddish origin). *To kibitz* - to give unwanted advice, to meddle, comment out of turn (source: The Chambers Dictionary, Larousse plc).

and promotions. Hell, in the main plant they give raises on the basis of the more people being supervised; I give raises to the guy who supervises least. That means he's doing more and taking more responsibility. But most executives don't think like that at all. Northrop's[61] senior guys are no different from all of the rest in this business: they're all empire builders, because that's how they've been trained and conditioned. Those guys are all experts at covering their asses by taking votes on what to do next. They'll never sit still for a secret operation that cuts them out entirely. Control is the name of the game and if a Skunk Works really operates right, control is exactly what they won't get" (op. cit., pp. 287-288). Rich also points out that one reason for the limitation of widespread use of the concept "is that most other companies don't really understand the concept or its scope and limitations, while many others are loathe to grant the freedom and independence from management control that really are necessary ingredients for running a successful Skunk Works enterprise" (op. cit., p. 317).

The reward structure had to be modified. Kelly's Rule #14 stated that, "because only a few people will be used in engineering and most other areas, ways must be provided to reward good performance by pay not based on the number of personnel supervised" (op. cit., p. 53). This was not a conventional approach in the 1940s. Equally unusual at that time was their relationship with the customer: "there must be absolute trust between the military project organization and the contractor with very close cooperation and liaison on a day-to-day basis. This cuts down misunderstanding and correspondence to an absolute minimum." (Kelly's Rule #12, op. cit., p. 52).

Rich reported the later tendency of customers to increase control and bureaucracy, practices that go against the autonomy and agility achieved with the Skunk Works concept (cf. op. cit., pp. 51, 317).

With this approach of an agile functional organization, they established a parallel organization that was able to react much faster than the mother organizations they were part of.

Rich knew (1994) about the existence of about 54 similar groups (op. cit., p.318). He cites, for example, Ford's success in the beginning of the 1990s with a Skunk Works-like team that developed a new car in "25% less time and [with] 30% less money".[62] He also felt that "the French aerospace company Dassault-Breguet probably has the best operation in Europe" (op. cit., p. 319). Lazard (1987) from Dassault-Breguet cites how multi-functional teams consisting of design, quality and manufacturing personnel were able to optimize the Rafale fighter (after the flight of the Rafale demonstrator) such that, in one case, a different breakdown in the fuselage

---

61 Northrop apparently proposed that Rich create similar activities.

62 Concrete figures for this case: 1 billion US$ reduced to 700 million US$.

resulted in a cost reduction of 10% with the same geometry and reduced the assembly cycle from 41 to 28 days. In another case, the Rafale foreplane was analyzed for use of alternative technologies, resulting in an implementation cost of about 30%. It should be noted in connection with these examples that the learning effect gained from experience with the first demonstrator prototype is not compensated for. It is important to note that Dassault's creation, the Computer-Aided, Three-Dimensional Interactive Application (CATIA), was to become famous only after its use in the development of the Boeing 777 in 1995. CATIA had been mentioned 8 years before as an *element* of the development process, but the reason for Dassault's efficiency was the "harmonious integration of techniques and men", as remarked in an American RAND report issued 20 years before that (op. cit., p. 26F-4).

Quinn (1985, 79)[63] refers to Skunk Works as a style with which every highly innovative enterprise in his research sample "emulated small company practices with small teams of engineers, technicians, designers, and model makers with no intervening organizational or physical barriers to develop a new product from idea to commercial prototype stages" by eliminating bureaucracy, allowing fast, unfettered communication and rapid turnaround times for experiments, and instilling a high level of group identity and loyalty. He also describes it as a possibility for big companies to emulate "the highly interactive and motivating learning environment that characterizes successful small ventures".

2.7 Simultaneous engineering

2.7.1 Successes and failures

In a radical implementation of W. E. Deming's proposition of "breaking the barriers between staff areas," simultaneous engineering (SE) is increasingly appreciated as a solution to problems caused by anachronistic differences in organizational forms stemming from the advent of the industrial age. At the same time, it is considered a meaningful approach to the increased complexity and interconnection of technology areas. In order to anticipate and correct product problems introduced by product development in an effort to avoid wrong assumptions by "elite" development engineers, companies now involve various sectors concurrently in the

[63] Reprinted by permission of Harvard Business Review. Excerpt from: Quinn, J. B. (1985). Managing Innovation: Controlled Chaos, Harvard Business Review, May-June 1985, pp. 73-84.

product development process. Obviously, problems should ideally be detected and corrected before an idea is implemented.

SE enjoyed increasing popularity in the early 1990s. Successful cases with very impressive figures of what was achieved were widely publicized:

Box 8. Two cases of recent successes in developing aircraft.

With the revolutionary Boeing-Bell V-22 Tilt-rotor aircraft, years after the evaluation prototypes were flying, integrated product teams were in charge of product optimization for the pre-series. The prototypes were too heavy and too expensive, costing US$ 41.8 million each (Schwarz, 1997, 15). The 75 teams involved identified 2000 problems. Having eliminated the need of a mockup meant a savings of 150,000 man-hours and the use of more expensive materials associated with less work led to a savings of US$ 350,000 per aircraft (Schwarz, 1995).

With the F/A-18E/F Super Hornet, the then McDonnell Douglas Aircraft and Northrop/Grumman developed a scaled-up version compared to the earlier models of this strike fighter. Selected results: the first flight was more the one month ahead of schedule and within budget; the aircraft was 1,000 lb. under the weight target; 350 independent design teams working with distributed computer tools reduced defects by 50%; the cost of jigs was reduced by 25%; engineering change orders were reduced by 70%, representing a savings of about 300,000 manufacturing hours (Deitz, 1996). Although the basic aircraft layout is the same as that of the C/D version, there are so many significant differences than merely the bigger size that some say it can be considered a completely different aircraft.

Very impressive figures, without a doubt.

Table 11 summarizes the benefits of SE based on the literature cited by Lawson & Karandikar (1994).

Some say that SE is not essentially different from what every big company that started as a "garage enterprise" did. People were then not only close together but also had to have an integral view of the issues associated with the product, meaning that engineering personnel was aware of problems it was potentially causing in other sectors. This is considered one of the main reasons why many small startup enterprises are much faster and more efficient than larger industries. In comparison, the latter, with their departmentalization, specialization and mature infrastructure were *supposed* to be more efficient in all areas, including product development, but that was not necessarily the reality. The departmentalization common in big companies tends to enhance, and at the same time justify, the division into project phases.

A major departure from pursuing sequential project phasing was proposed by Takeuchi & Nonaka (1986, 140-141): a partial overlapping of development phases, which they called the *sashimi system.*[64] It is in contrast to the sequential phased program planning, where one can proceed to the next phase only after the requirements of the preceding phase have been met. The idea behind the sashimi system is that, by achieving the necessary integration of various viewpoints, the traditional phase method—though providing checkpoints between the phases—create a bottleneck that can slow down or even halt the entire development process. They compare this creative process of diverging points of view and convergence toward the implemented solution as a kind of "noise" generated throughout the product development process. When a bottleneck occurs, the level of noise increases. Overlapping phases allow the group to absorb the noise, but the process does not come to a halt. In the phased approach, "the crucial problems tend to occur at the points where one group passes the project to the next" (op. cit., p.141). The advantages and disadvantages are summarized in *Table 12*.

Table 11. Benefits of simultaneous engineering reported in the literature according to Lawson & Karandikar (1994, 4)[65]

GOAL	ACHIEVED RESULTS
Development time	30–50% less
Engineering changes	60–95% less
Scrap and rework	75% reduction
Defects	30–85% fewer
Time to market	20–90% less
Service life	100% increase
Overall quality	100–600% higher
White-collar productivity	20–110% higher
Return on assets	20–120% higher

Table 12. Summary of overlapping product development model.

ADVANTAGES	DISADVANTAGES
Greater speed	Problems managing an intensive process
Increased flexibility	Problems communicating with the entire team
Enhanced shared responsibility	Problems maintaining close contact with suppliers
Enhanced cooperation	Need for preparing several contingency plans
Stimulated involvement and commitment	Need for a strategy for handling surprises
Sharpened problem-solving focus	More tension within the group
Encourages initiative taking	More conflict within the group
Develops diversified skills	
Heightens sensitivity toward market conditions	

64 "Sashimi" is a Japanese dish of sliced raw fish arranged on a plate in an overlapping fashion.

65 The authors cite the sources as being: Business Week, IDA, ARPA.

Simultaneous engineering enhances the same aspects by almost piling the "slices," while increasing the advantages listed in the table above. However, its disadvantages are also amplified, thus creating a particular challenge to organizations that strive to explore its full potential.

Lawson & Karandikar (1994) attempted to confirm the success of SE as published in popular business journals (cf. *Table 11*). A total of 200 surveys were mailed to key individuals in product development organizations. The response rate was 35%, with multiple responses by certain organizations, reflecting the existence of multiple functional groups. The aerospace and defense industries were predominant in the survey population, but there was a broad representation in other industrial sectors. Although respondents, who were primarily from middle management (70%) and the engineering and manufacturing organizations (48%), could not provide hard data of the results, the qualitative data they provided "painted an extremely favorable picture", but the respondents nevertheless "did not report tremendous improvements". The best results are listed in *Table 13*.

Table 13. Effect of simultaneous engineering.

SE EFFECT	% OF RESPONDENTS
Greatly reduced engineering design changes	15%
Greatly improved customer relations	13%
Greatly increased collaboration between different stakeholders in the development	24%
Great increase in the quality of communication between and within product-development teams	20%

Compared with *Table 11*, which summarizes the alleged benefits of SE, *Table 13* lists the factors that possibly limit the successful application of SE. This is due to several misinterpretations, which will be discussed in the following sections.

2.7.2 Learning effect

One of the problems in using numerical data such as those in *Table 11* is that the learning effect gained from experience with previous versions of the same aircraft and that achieved by SE skews a simple comparison of the numbers. For example, take the following comparison: "The evidence suggests that the packages ('build-to' packages—complete packages of data and information for manufacturing teams) are doing their intended job, as defects per unit were reduced in the F/A-18E/F by almost 50%. Moreover, the forward and aft fuselages were assembled in just 22 hours, compared with several days on the last version of the aircraft" (op. cit., p. 72, parentheses added). Both The V-22 and the F/A-18 SE cases are evolutions from previous versions. The presence of success data from completely new designs such as the JSF actually contributes to the skewed comparison, but such programs also profit from the experience gained with many other aircraft, including the F-117A, the B-2, the AV-8B, to name just a few. It is not questioned here *whether* SE can accelerate the process, but *how much* it can accelerate it.

Skewed comparisons are a general problem tied to benchmarking: it is very difficult to filter contextual data from quantitative performance data. Saying, for example, that one can reduce development time depends on how it is measured: in terms of number of days-to-prototype or man-hours in development. Some managers tend to emphasize reduced man-hours, instead of reduced completion time, even if this might actually mean an increase in individual activities that pay out later in decreased costs due to optimal design.

Saying, for example, that 42% of development work is repeated due to upstream changes is misleading, especially if this repetition is considered "bad" or a "waste of time" to be cut to zero. Repetition could be the effect of an iterative process that is sometimes necessary to optimize product characteristics. Even if development processes are not iterative, changing circumstances (e.g. evolving customer requirements that require a better instrument resolution, slight corrections in mission profile causing the data transfer rate to be increased, changes on instrument case dimensions or weight, changed hole positions due to payload evolution, changes of production facilities or even changes due to the dynamic nature of the market) would make the repetition *necessary*. Consequently, it would be better if *more* repetition could be allowed in the product development, at least to some extent.

An even worse situation would ensue if one were to follow the rather tempting strategy that, if repetition were eliminated, productivity would increase by 72% (i.e. 58% → 100%) without adding any human resources. In fact, this is a very common reasoning among managers. But let us examine its flaws. First, it implies that repetition takes the same time as the first pass. In many development situations, repeating a step takes less time than the first execution because the step is better understood. For the sake of illustration, let us examine the case depicted on page 91, which took about two full weeks on the first pass. A subsequent cross-check to verify the generated data revealed that the round-up errors due to the double integration loop in the program (necessary because of the added convolution integral) caused an excessive loss of resolution, rendering the data invalid. The program had to be rewritten in another language, but the entire operation, including the cross-check, took half a day to be accomplished. This difference in duration is because a *learning process* took place during the first iteration: the problem was characterized, various experts were consulted, program implementation problems solved, etc. These were processes that did not have to be repeated. Moreover, the second implementation had a much more efficient structure than the first. If it had not been allowed to repeat the work, the resulting device in this case would have been useless. It would be more meaningful to calculate a repetition in terms of the number of activities repeated, which would produce completely different results.

Similarly, changing the position of a hole does not call for the entire development of the associated parts to be repeated. In other cases, repetition prompts changes in other parts in a chain reaction, which might sometimes be undesirable, but sometimes utterly necessary to achieve quality.

The same considerations are applicable to the number of modifications or engineering review changes. As SE assumes that activities should be started before all characteristics have been defined, e.g. by involving manufacturing or procurement personnel, it is potentially necessary to perform *more* changes than with the usual sequential approach. The supporting organization should thus allow faster changes and manage the risks of performing manufacturing, procurement, or training tasks that might turn out to be unnecessary or have to be redone if the design evolves differently than expected. As a consequence, the evaluation metrics should be coherent with these considerations.

2.7.3 Focus on technical aspects

Browsing the literature on SE frequently gives the impression that most of it is based on design for:

- manufacturing
- assembly
- test
- maintenance
- reliability
- cost or with
- CAD/CAE/CAM/CIM.

The emphasis of many of the books on SE is simplistically put on these technical aspects of product development, which can actually also be achieved with the conventional phase approach. Indeed, this has been pursued for a long time, not only in the aerospace industry, and partially with very good results.

In terms of these aspects, however, SE gains significance in the following contexts:
- interdisciplinarity
- multi-disciplinary
- elimination of interdepartmental barriers
- involvement of related non-development experts from the outset and keeping them in the daily routine
- solving communication problems (different orientations, viewpoints, language, styles, etc.)
- cross-fertilization among different disciplines
- overall product optimization, not departmental optimization
- doing as much as possible in parallel
- interconnecting activities from the beginning of product definition in order to achieve rapidly the product characteristics listed above.

The second list is not of a technical nature.

2.7.4 Role of information technologies

It is incredible how many persons in the aerospace as well as non-aerospace industry associate the success of the development of the Boeing 777 with the intensive use of computers for simulation and visualization, including the CATIA software.

In reality, one of the key success factors was the application of the design-build team (DBT) strategy, an approach that involved 250 small, multidisciplinary teams. This factor is very frequently overlooked or neglected.

The prevailing misconception results from the misleading emphasis of SE on information technologies. Consequently, it is frequently believed that implementing SE is merely a matter of buying technical installations such as computers and software.

It would be too naïve, however, to attribute the success of SE cases exclusively to such computer tools as CAD with 3D visualization capabilities, or to an engineering database, or videoconferencing, or computer simulations.

In reality, success could be achieved with non-computer-based methods, provided that sufficient attention is paid to the usage of these methods. In much of the popular literature citing these success cases, there is a lack of data to justify attributing the real reason(s) of success to information technologies.

The following early examples of SE and SE-like development of aircraft are called upon to illustrate this.

Box 9. Three cases illustrating the problem of design interaction and the need for changes with the simultaneous engineering approach.

In 1939, the Glenn Martin company had so much confidence in their design that in the prototype fabrication of their B-26 Marauder bomber, they invested in manufacturing tools such that the first prototype was actually the first production prototype. At that time, the interaction of product characteristics frequently called for considerable changes—including changes of the dimensions—after the prototype evaluation, thus incurring a loss of the work invested in production facilities. The resulting aircraft was in fact very successful, but, due to the high wing loading assumed in its design, its takeoff and landing speeds were too high, making the plane unforgiving of pilot errors during these phases. The Marauder was nicknamed "Widow Maker" and "Flying Prostitute" because of these unpleasant characteristics, which prevailed in all versions of the 5,157 Marauders built (Birdsall, 1970).

In Germany, Willy Messerschmitt heard about the B-26 and wrote in a report to Ernst Udet that "Glenn Martin smelt a commercial success, made the aircraft ready for series production during the design and built stages and at its own risk [and] embarked on large-scale production ... [In] about 16 months, not only could the first aircraft be test flown by the Air Force, but at the same time the production aircraft could be delivered. There was no actual prototype aircraft, no pre-production run, but the first aircraft was the first production aircraft. Modifications which were necessary were built in at that time. Non-essential modifications were simply not done. From the beginning of the design to the delivery of the first production aircraft took 16 months. That could be a record. Nevertheless, such achievements do not remain individual cases in the USA, but become the order of the day." (Vann, 1993, 167)[66]. Messerschmitt proposed doing the same with the twin-engined fighter-bomber Me-210, whose manufacturing facilities were also developed simultaneously with the first prototype. Orders of 1000 aircraft were placed before the first flight. But even during its first flight, "the aircraft displayed undesirable handling characteristics. The longitudinal stability was a cause for great concern and the tendency of the aircraft to go into a spin did not endear it to the pilots. It was reported that, after the first flight, the test pilot told Messerschmitt that the fuselage needed to be lengthened by at least one

[66] Excerpts taken from: Willy Messerschmitt by Frank Vann, published by Patrick Stephens Ltd, an imprint of Haynes Publishing, Sparkford, Yeovil, Somerset BA22 7JJ, UK. Tel + 44 (0)1963 442080. © by Mrs. F. Vann. Reproduced here by kind permission of J. H. Haynes & Co. Ltd.

metre. Messerschmitt is supposed to have replied that such a modification would mean the scrapping of millions of Reichmarks' worth of jigs. ... As a result no fundamental changes were made to the geometry of the aircraft" (op. cit., p. 78). After many crashes occurred, pilots refused to fly the aircraft (op. cit., p. 83), and production had to be stopped. The consequences of the Me-210 failure were disastrous for both the Luftwaffe and Messerschmitt. Vann states that Messerschmitt "had had failures before in his career, as every aircraft designer has, but none as humbling as this disaster" (op. cit., p. 86). After pursuing many possibilities to resolve the problems, the modifications called for a lengthened fuselage (initially refused by Messerschmitt) and stronger engines. This resulted in an effective aircraft renamed the Me-410, but it came too late.

Another instance of aircraft development SE during World War II occurred at General Motors in the case of the XP-75. General Motors had a good track record of manufacturing aircraft developed by other companies, and it had experience manufacturing parts for the aeronautical industry. It hired Don Berlin, the "thoroughly experienced aeronautical engineer" who designed the famous Curtiss P-40 to conduct its efforts to fulfill Air Force's requirement of developing a fighter in a "remarkably short time by using, as far as possible, structures, controls, and accessories already in full production for other aircraft in order to obviate the need for long delays in tooling up for production. The ingenious scheme called for the use of the wings of the P-40E, tail surfaces from the Douglas A-24, and so on. Only the main fuselage itself would have to be developed anew" (Holley, 1987, 587). A long-range escort fighter was needed, and if the Air Force had to wait until flight tests on the XP-75 were done, a delay of several months would have resulted. General Motors tried to cope with the task by increasing the number of prototypes to accelerate flight testing and issued a letter of intent for 2,500 units, conditioned to the fulfillment of the requirements by the prototypes such that GM could start placing orders and design jigs and fixtures. Like every new aircraft at that time, it was expected to undergo modifications to correct undesirable flying characteristics: "The test pilot reported that the plane lacked stability and displayed a distressing tendency to stall and spin when making tight turns, the maneuver most essential to fighter aircraft" (op. cit., p. 589). To correct for this, the tail surface had to be increased, thus precluding plans to use the off-the-shelf tail of the A-24. The heavier tail had to be compensated for with a longer nose, which involved redesigning a number of production tools. Then the engine did not perform as well as intended, thus disabling the speed needed to engage German fighters. Furthermore, to correct for the inherent instability, the ailerons were extended over the wingtips, thus leading to a modification of the existing wings. The longer ailerons called for hydraulic boosters to reduce the forces to be exerted by the pilot to a manageable level. After all modifications, the intended rate-of-climb of 5,600 feet per minute was reduced to 3,000 fpm. Bradley (1974), who flew the prototype, remarked that the plane that was supposed to protect bombers was not even able to protect itself. The project was canceled.

It must be said here that in the above-mentioned cases, the problem did not lie with the technical competence of the designers; indeed they are still recognized as very competent professionals. The many interactions and uncertainties in product development that could not be completely known in advance, even with today's simulation capabilities, pose a special challenge to the SE approach. Although the main points of interest in these historical examples are not of a technical nature, and although modern computer simulations could certainly help solve many of the technical issues faced during World War II,[67] it is not necessarily the case that new issues arising from advances in technology can be avoided.

The problem occurring with the "J" version of the Lockheed C-130 illustrates the point. Although the basic design of the Hercules was proved more than four decades ago, a new, undesirable stall characteristic was detected on the flying prototype. It was attributed to an interaction between the new 6-blade propellers and the wings. The manufacturer decided not search for the causes of the interaction but to deal with the effect by warning pilots of its occurrence (Warwick, 1997). This is not a bad practice in aeronautical design, yet it illustrates the difficulties of incorporating "mature" components into a new product concept, even if the new concept is an evolution of old designs. Other example is that of the F/A-18 E/F mentioned above. The prototypes exhibited a so-called wing drop problem, which was considered by R. Harney, the flight-test team leader, to be the "toughest problem to date, at the heart of the tactical manoeuvre envelope" (anon., 1997). Wing drop of up to 30° occurred at between 0.8 and 0.9 Mach with an angle of attack of 8 – 10°, making target tracking difficult, something that is "most essential to fighter aircraft". The cause was pinpointed to the airflow over the fairings covering the wing-fold hinges, causing an asymmetric stall. Three possible solutions were devised in an effort to avoid an expensive major wing redesign (Warwick 1998a). One solution was quickly disqualified (Warwick 1998b), but a "Blue-Ribbon Panel set up to review efforts to cure …the problem identified a 'national need' to understand the phenomenon. Concern [had] been raised by the failure of wind tunnel testing and computational fluid dynamics analysis to

[67] Adopting 6-axis flight simulators with wind tunnel data to gain insight into aircraft handling is not a new capability, and advances in computational fluid dynamics have also been achieved since then.

predict accurately how potential solutions [would] perform in flight." Interesting to note is not the technical problem itself, but the concern of not foreseeing the problem in spite of modern simulation technology.

The cases above emphasize the interaction of product characteristics, some of which cannot be known beforehand. A strategy involving parallel activities may not reveal such risks, which will require later changes of the design and may render useless some of the work done. Some work may then have to be repeated "downstream"[68] (e.g. such as expensive manufacturing jigs in production facilities), or product characteristics may be jeopardized. Note that in the cases of the C-130 J and the wing drop of the F/A-18 E/F, a solution was devised without incurring a major redesign, but this is not always possible.

It is not intended here to doubt the technical competence of the development personnel involved nor to doubt that the new version of the aircraft has superb qualities. Equally, it is not intended to deny the influence of computers on improving the development process. Computers are unquestionably a potentially very valuable tool. The point is that by executing the technical aspects of simultaneous engineering to an exaggerated extent and at the same time not actively confronting the processes underlying the non-technical aspects that support the interactions necessary for the successful application of SE, it would not be surprising for very modest results to be the outcome.

2.7.5 Influence of human issues

In addition to the interaction of technical parameters, the interaction of the *participants* is an issue that the utilization of modern tools alone, e.g. computer simulation, cannot necessarily solve or even help solve. A multi-disciplinary team obviously cannot be made to function merely by bringing people from different disciplines together and providing them with sophisticated tools. Nor can computer communication, integrated databases, and visualization rooms alone bring the advantages of the SE approach, where group interaction is a key success factor.

For the purpose of an analogy, let us examine an approach based on a proposal by T. J. Allen (1970). Implemented in the early 1990s by a quality manager of an aerospace company, the approach involved improving office communication by changing the office layout. The quality manager, after studying communication patterns among his personnel, introduced substantial layout changes. One engineer asked whether the manager really

[68] Downstream in the sequential approach.

thought people would communicate better as a result of the changes made. His answer was simple but effective: "*if* you guys want it, yes!" Indeed, in another department of the same company, two engineers assigned to different machinery had, although seated at adjacent desks, established two conflicting sets of requirements for electronics design. Unspoken, but as important as the message above was a series of measures undertaken at the individual and group levels to improve not only internal group communication, but also to improve individuals' action within the company. The layout change was merely the elimination of one of the many barriers, not the active driver of improved communication (case #8).

Similar to SE, it is not the availability of the tool that makes the difference. An action only becomes effective if its users are considered in its design and implementation, as well in the organizational design. The key to the focus of SE activities was also asked in the survey conducted by Lawson & Karandikar (1994) mentioned above. The results are summarized in *Table 14*.

Table 14. Key to simultaneous engineering (Lawson & Karandikar 1994)

	STRONGLY AGREE	AGREE	NEUTRAL	DISAGREE	STRONGLY DISAGREE	NO RESPONSE
Teamwork	**63 %**	30 %	1 %	3 %	1 %	2 %
Information Technology	9 %	40 %	31 %	12 %	6 %	2 %

The authors conclude that their "figures clearly indicate that the product-development community recognizes the need to improve their manual processes *first* before spending much time, effort, and money on technology solutions" (op. cit., p. 5) and also that "while top organizations see the potential advantages of information technology, it is not considered of primary importance in realizing SE."

The respondents to this survey clearly indicated that the most pressing need was to foster a teamwork environment. Information technology will certainly play an increasingly important role in creating a competitive advantage in product development processes, but the problem (and the greater leverage) lies in teamwork and process improvement. "*SE is first and foremost a human interaction challenge.*" (op. cit., pp. 5-6, italics added). It has repeatedly been recognized that simultaneous engineering is a people and communications issue, not primarily a question of computer technology.

2.7.6 One more aspect to consider

In 1957, the launch of the Sputnik satellite caused the alarming concern to Americans that the Soviets could (or would soon) have the capability of deploying nuclear weapons in ballistic missiles. The so-called "Sputnik shock" not only prompted many changes of strategies and tactics in military doctrine, it also had many consequences in the American technological, scientific and educational worlds. In the urgency to achieve operational capability, task phases that had formerly been conducted in a sequential fashion suddenly had to be conducted at the same time. The concept of "concurrency" pursued by the Western Development Division of the Air Research and Development Command called for the parallel execution of development, production and operational deployment, assuming the risks involved with the development of a completely new concept with unproved technology. Success was achieved when the first Atlas became operational. "Success" in this case refers not to the parallel development of the Atlas and Titan systems, but to the concurrent development of the missile, the support systems and the training of personnel. However, although no data is available that describes the interaction between these areas in detail, some pertinent comments on it are documented. Ritland states that "If concurrency is simple in concept, it is anything but simple in practice" (1960-61, 240). "Translating these ideas from design to hardware to installation, however, has proved to be a challenge to ingenuity and flexibility" (op. cit., p. 248).

A point of special interest is that, in order to achieve efficiency in planning and executing the training of the operation personnel itself, there was a newly perceived prerequisite. In the sequential approach, when this phase started, development was advanced and there was already a good definition of the product. As Briggs (1960-61, 190) points out: "at the beginning of training, it is necessary to use research data and to develop training aids that will depict principles and fundamentals. Often, there are no guides, no textbooks, no past to draw on". He argues that, in order to achieve the necessary flexibility, a better background in basic principles and fundamentals is required, in addition to specific equipment-oriented training. This is only one of the many changes concerning personnel qualifications incurred by the "concurrency" approach.

Similar to SE, organizations might notice the different training measures necessary to enable professionals to cope with issues like those in *Table 12* and to update their knowledge in their specific discipline(s). The industry is presumably still experimenting with the approach, and there is still a great deal to be learned. The potentially negative effects of "long-term exposure" are well known, such as exhausting (in terms of technical competence and health) the resources of professionals working in intensive processes in an effort to shorten the time-to-market.

Many of the variables that affect the performance of a team will be examined later in this book.

2.8 Concluding remarks

This chapter presented the historical perspectives relevant to the quality of the product development process. As such, it does not provide comprehensive coverage of the evolution of the pertinent organizational theories, the field of work psychology, nor the history of science and technology alone.

We examined the relevant social science theories (and organizational theories) related to product development. Concurrently, scientific concepts associated with engineering sciences as well as the necessity of a critical appraisal to improve quality have been presented. The myths surrounding agreements and the impersonal relationship to the subjects under study were discussed, followed by an introduction of inquiry systems and the effect of the style chosen on the results achieved. Serendipity was also described not only as a "valid" scientific approach but as a potential source of relevant information. Concrete cases of successes both in science and in product development were presented as well.

Quality and reliability systems were explored in terms of a historical perspective. Concepts of bureaucracy, both the basic utopian idea and the real-world implementations, were discussed. The contrast between mechanistic and organic organizations was examined, and the appropriateness of using standards to achieve quality of development was also discussed.

Project management was presented, with special emphasis on the tools used for scheduling, progress reporting and resource allocation. More on this topic will be presented in the next chapter (Section 3.2), especially concerning the implicit Cartesianism. The aspects of cost and time estimation were examined and, above all, the importance that their function not be to control but to protect workers to perform well.

The two key notions of Skunk Works and simultaneous engineering were presented. Skunk Works is a classic example that deserves reconsideration. The analysis of this concept revealed essential factors that can lead to quality improvements and hence that should be explored to shape the future of development organizations. Simultaneous engineering deserves consideration as an issue that primarily concerns human interaction, not information technology.

In an effort to preserve the unity of the text, some historical points were omitted from this section, but will be examined in the appropriate sections of the discussion. This is the case for such techniques as design review and the description of R&D as a *communication process* or when a technique appears only in the context of one of the supporting cases.

Chapter 3

SHAPING THE DEVELOPMENT ENVIRONMENT

3.1 Modeling the complexity of R&D processes

3.1.1 Phase models

There are many implementations of phase models applied to product development. This is perhaps the most frequent approach to organizing product design. It has many advantages: it provides a reasonable structure to arrive at a functioning product, it allows a systematic breakdown in complexity into smaller, manageable tasks, and it uses a stepwise model for progressively reducing uncertainties and alternative possibilities, leading to a direct convergence toward the intended product. *Figure 5* depicts just one of these phase models (VDI 2221).

Many project management authors and companies using phase models of this kind report having had success with them. They provide a framework to partition activities and hence make them manageable. However, there are certain points that have to be considered in these phase-model approaches. Roozenburg & Eekels (1995) mention that sharp divisions between the various phases cannot be drawn, and that the stages and phases do not necessarily have to follow one another rigidly as in the proposed models.

Takeuchi & Nonaka (1986) also proposed eliminating the interface between the phases with an overlapping phase system, as described above on page 76.

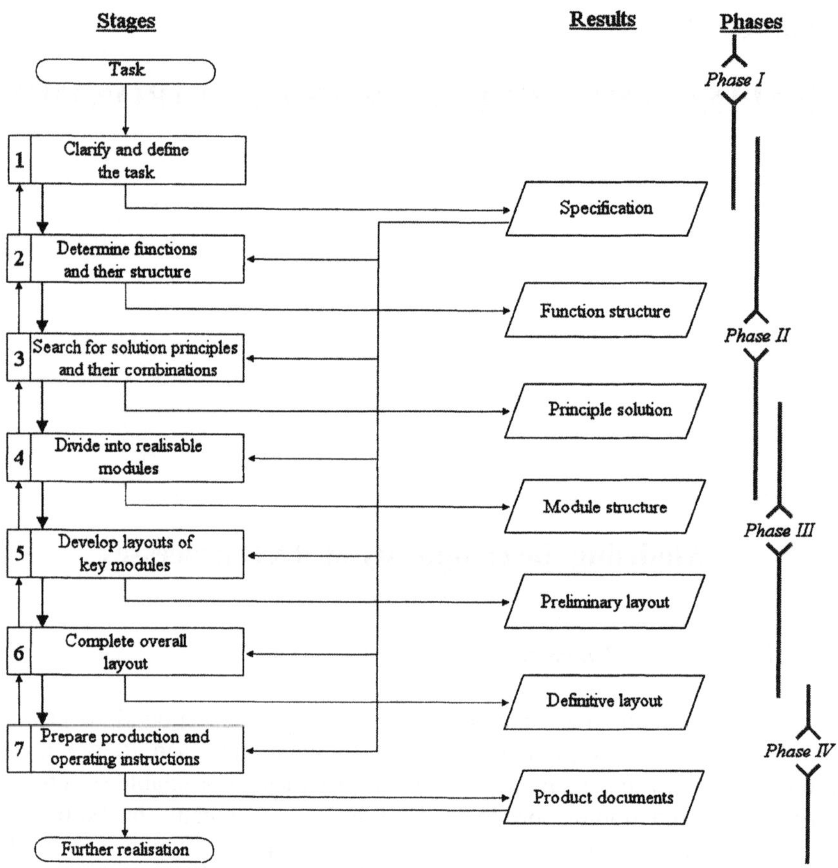

Figure 5. General approach to design according to the VDI 2221 guideline.[69]

The standard technique for project management is to delimit phases via so-called *milestones* that formally signal the end of a phase. Sometimes, reaching a milestone is associated with money transfers when a customer is

[69] From: VDI - Verein Deutscher Ingenieure (1987). Systematic Approach to the Design of Technical Systems and Products, Berlin: Beuth; Düsseldorf: VDI Verlag. Reproduced here by kind permission of the Verein Deutscher Ingenieuere e.V.

paying for the development. They are also a common measurement upon which professional evaluations are based. As a result, milestones act against the implicit fuzziness of design phases. Furthermore, they act against the need to repeat previous phases to clarify points disclosed after the phase transition.

The VDI 2221 model was chosen in this work because it represents phases (cf. *Figure 5*, right-hand side) as partially overlapping. Moreover, it leaves open the possibility of backward transitions on the left-hand side. This illustrates the ambiguities described above.

The greatest limitation of the phase-model approach is that does not show the problem-solving process behind it.

Phase models such as this are implicitly Cartesian. Consequently, it would not be surprising to find that the persons following this model would limit themselves to the Cartesian way of thinking, which of course is not the only way to cope with development issues. The model functions well for establishing a (Cartesian) product definition structure, which is usually tree-like. Moreover, it yields a very satisfactory understanding of the product, its structure and functioning *when it is ready*. It is not, however, necessarily good *during* product conception.

It functions well for problems with a clearly "correct" solution, but not for problems with multiple correct solutions, where different ways could be followed, where consensual or conflictual problem solving is to be used. Consensual or conflictual inquiry systems (cf. page 47) occur in practice, but are used in spite of the model, sometimes in contradiction to it. With such limitations, the model itself limits its usefulness for activity planning, even though it is implicitly task-oriented. This task orientation is the main advantage of maintaining this model.

Not surprisingly, it is correlated with the work breakdown structure and hence with task division. Structuring a task in terms of phases calls for dividing the task into individual activities to be assigned to different persons. This can potentially result in an interruption of the necessary interaction process from phase to phase and, much more importantly, among the persons working on the various modules, thus leading to implementation problems. This approach has similarities with the Tayloristic division of activities, which leads to the separation of "thinking" and "doing" as the following case shows:

Box 10. A case of Tayloristic division in development (case #9).

In a research institute in the early 1990s, an electrical engineer was working as a trainee to become familiar with the implementation details of surface acoustic wave component technology used for signal processing. The

task he was assigned was (partially) clear. In the first part of the work, he had to generate component data by:
1. calculating a series of Gaussian curves,
2. displacing them in time corresponding to response-shifted pulses,
3. adding or subtracting these curves and
4. using the resulting curve to FM-modulate the signal.

This was considered to be a very simple data generation task, so the origins of these steps were not sufficiently explained. It was assumed that this would not affect the results of such a straightforward mathematical computation.

When the data generation was completed, however, the curve was clearly different than expected. The engineer's first reaction was to search for errors in the data-generating program, which he did several times. Yet the more he rechecked the program, the more certain he became that it was correct, and the more it seemed that something was missing in the problem formulation. His concern was as follows: "If this Gaussian curve is the infinite impulse response of the pulse-shaping filter (an ideal condition to describe it), and as those shifted pulses are of finite duration, then the *convolution* of the Gaussian and the pulse should be used, not the unprocessed Gaussian." This is of course a very basic consideration of filter theory.

The engineer tried fruitlessly to discuss the point with his colleagues, who suggested that he should not complicate matters by "inventing" things. When he tried to explain the reasons for his misgivings to his supervisor, the latter replied that it was not the engineer's task to question the procedure he had been assigned. The assignment was to execute a sequence of steps. The engineer still tried to bring (basic) filter theory into the discussion and insisted that the results were not coherent. His concerns regarding the unsuitability of the procedure for the product were then countered by the argumentation that someone else had worked on the problem upstream and that the engineer had been given the next task in the succession. The previous task had already been completed, so it was not up to the engineer to second-guess it. There was probably an error in his program, his supervisor insisted. The engineer, knowing this was not the case, told his supervisor that, using filter theory, he had in fact tried the convolution operation, and the result turned out to be exactly what was expected. To which the surprised supervisor replied: "Oh, really?! Well, uh, in that case let's use it as you say."

The course of this conversation was predominantly non-technical. It could be even considered non-scientific. The technical arguments were completely overshadowed by the principles of task (labor) division. It was not "allowed" to think beyond the narrow limits of the problem. If the error, or better: its solution, had not been so clear, the engineers in this story might still be searching for the problem, or the project presumably would have failed and been canceled.

Another issue associated with phase models is the—implicit or explicit—tenant that the convergence of the entire process, which should result in an unambiguous product definition, consists solely of convergent subprocesses, the details of which are progressively refined and integrated at the end of the process. In fact, the process should converge macroscopically, but in the microscopic view, it is composed of a succession of convergent *and* divergent subprocesses. In the course of coping with development challenges, new problems arise, new possibilities emerge, and changes in the initial concepts become necessary. There are many occurrences of divergent and convergent processes in each phase. Problems encountered prompt a search for possible solutions and alternatives (divergence), followed by comparison, choice and implementation of the best solution for each problem or bundle of problems (convergence). As new problems arise, the divergent – convergent process begins again. Certain issues involving integration can arise, especially in complex, interacting systems that are not accommodated by the Cartesian method. Sometimes there is a need for a change of strategy (divergence) to arrive at the appropriate implementation.

The real flow does not resemble the phase model.

This convergent – divergent combination of subprocesses is depicted in *Figure 6*. Note that, although Goltz also uses the term "phases", the model describes a different approach.

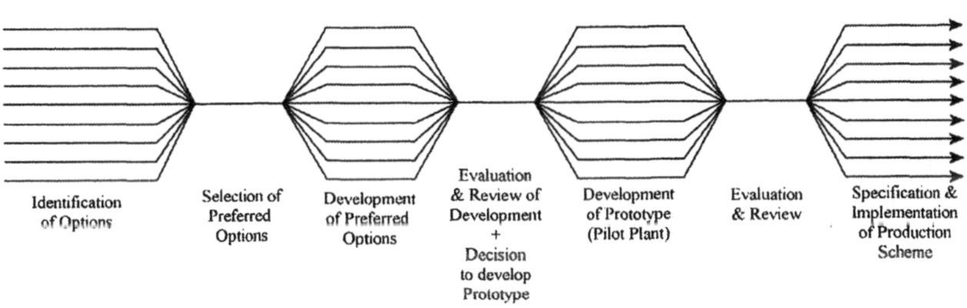

Figure 6. Convergent and divergent character of product development (Goltz, 1986, 245).[70]

Other phase models propose divisions such as: advanced studies, product concept, preliminary design, detail design, production design, product test

[70] From: Goltz, G. E. (1986). A Guide to Development, R&D Management, Vol. 16, No. 3, pp. 243-249. Reproduced here by kind permission of Blackwell Science Ltd..

and production phases. These models can *potentially* lead to an even less optimal division of labor, yielding poor product characteristics or too long a development time, which can of course adversely affect the persons participating in the process. The problem, however, does not lie in the models, and models such as the VDI 2221 might still be feasible and useful, provided the processes that are not considered in the models can nevertheless be taken into account and accommodated.

This need of considering the underlying processes calls for aspects of human interaction in the product development process to be examined. A similar conclusion was reached by Rubenstein, Chakrabarti & O'Keefe (1974) , who state that "the process of transforming scientific/ engineering knowledge into marketable items is often described by means of a sequential model. [...] The flow is not orderly, it does not appear to occur in the sequence implied by this model. Work may skip stages; it may move back to an earlier stage; an idea may start in engineering and turn out to involve fundamental development. A problem with a seemingly trivial item in production may require the talents of a basic research scientist to adequately solve it and may even span a new line of research"[71] (op. cit., p. 9). The converse is also true. Some advanced studies require the experience of a factory operator to contribute essential information about the execution of tests.

An alternative approach to a stage model of the technology transfer/ innovation/ adoption/ diffusion process is cited by the same authors (op. cit., p. 9) from Rogers (1962) containing the stages listed in *Box 11*.

Box 11. Stage model by Rogers (1962).

- Awareness
- Interest
- Evaluation
- Trial
- Adoption

Noteworthy in this model is the consideration of mental processes, thus reinforcing the need to consider the human aspects involved in the related processes. These five stages can occur at any point in the product cycle.

Similar findings were also obtained independently by Rauterberg, Spinas, Strohm, Ulich & Waeber (1994), who investigated the software development process. The traditional *waterfall model* of software development is an attempt to provide a way to progressively integrate generated code in a

[71] The authors mention the case of the efforts to prolong the life of tungsten light bulb filaments, which led to fundamental studies of crystal growth in metals.

cyclical manner up to the customer support level. However, as the authors describe. (i) "The attractiveness of the phase models stems from the fact that the single phases seem to be logically depicted directly over a linear time axis" (ii) "The milestones serve as a popular—even if frequently insufficient—controlling instrument". (iii) "For the appropriate organization of the course of activities,[xxii] there is the need of a model that is adjusted *in content* to the structure of the software development process that is independent of the time dimension". (iv) "In the traditional phase model, the work orders are passed on from a phase to the next. It is in this aspect assumed that all the essential conditions for processing the work orders in the following phase are available and that no unplanned or unforeseen influence exist". And finally, (v) "project leaders have a complete knowledge of all the planned activities following" (op. cit., pp. 81-83, italics added). The latter point deserves more attention and, as the authors point out, there are three prerequisites for this "complete knowledge":

Box 12. Prerequisites for "complete knowledge" (Rauterberg et al. 1994, 82-83).

- Exact knowledge of the reactions of all technical and social components,
- Exact knowledge of all factors that might impair the activities,
- Sufficient *previous* knowledge of the needed interventions in order to adequately face all adversities.

"These prerequisites, even in the most experienced project leaders, are practically never completely fulfilled" (op. cit., p. 83, cited from Schmienz 1979, 1024). The authors also mention the well-known weaknesses of the traditional phase model examined by Hesse, Merbeth & Fröhlich (1992, 75) and listed below.

Box 13. Well-known weaknesses of the traditional phase models (Hesse, Merbeth & Fröhlich 1992, 75).

- No systematic naming and division of activities, results and phases,
- No development cycles extending over two neighboring phases,
- No cross references between early and later development phases,
- Long development length between specification and implementation of the system and its components,
- Inexistent or insufficient consideration of modern tools possessing far-reaching possibilities that change the entire course of development,
- Inexistent or insufficient treatment of maintenance problems,
- No provisions for the cooperation between users and designers, the involvement of the user in the development process, nor of evolutionary system development.

3.1.2 Process models

In a series of studies involving more than 500 engineering professionals,[72] Thamhain & Wilemon (1987)[73] found that 90% of the respondents indicated the following measures of team performance as being the most important (and more than 60% of them put them in this order):
1. technical success,
2. on-time performance,
3. on-budget/within resource performance.

Engineering managers described a high-performing engineering team as one that not only scores high in terms of the three measures above but also possesses the task-related and people-related characteristics listed in *Table 15*.

Table 15. High-performance development team characteristics after Thamhain & Wilemon (1987).[74]

TASK-RELATED QUALITIES	PEOPLE-RELATED QUALITIES
Oriented toward technical success; Committed to the project, result-oriented attitude; Innovative and creative; Concern for quality; Willingness to change project plan if necessary; Ability to predict trends; On-time performance; On-budget performance.	High involvement, work interest, and energy; Capacity to solve conflicts; Good communication skills; Good team spirit; Mutual trust; Self-development of team members; Effective organizations interfacing; High need for achievement.

By quantitative analysis, the authors found an association between these characteristics and team performance with an average rank-order correlation of τ=0.37, thus a statistically significant association (p=0.95 or better). Furthermore, by investigating the effect of 60 influence factors mentioned by engineering managers as drivers or barriers toward high team performance, the authors could identify 12 of them as statistically significant (p=0.95 or better),[75] *Table 16*.

---

[72] It should be noted that there is no explicit indication that production, quality or marketing professionals were involved. This case probably involved development engineers.

[73] Excerpts from: Thamhain, H. J. & Wilemon, D. L. (1987). Building High Performing Engineering Project Teams, IEEE Transactions on Engineering Management, Vol. 34, No. 3, pp. 130-137. © 1987 IEEE. Reproduced here by kind permission of the IEEE.

[74] © 1987 IEEE. Reproduced here by kind permission of the IEEE.

[75] Team members were asked to describe their environment (drivers and barriers) on a 5-point Likert scale. These rankings were correlated against team performance ratings given by senior managers for the result-oriented characteristics and by project managers for the

Table 16. Strongest drivers of and barriers to development team performance after Thamhain & Wilemon (1987).[76]

DRIVERS	BARRIERS
Interesting and stimulating work	Unclear project objectives and directions
Recognition of accomplishment (of individual or team)	Insufficient resources
Experienced engineering management personnel	Power struggle and conflict
Proper technical direction and leadership	Uninvolved, disinterested senior management
Qualified project team personnel	Poor job security
Professional growth potential	Shifting goals and priorities

Their findings also show an association with the characteristics listed in *Table 15*. The authors then propose a model for the major drivers and barriers affecting the people-oriented and the result-oriented high-performance team characteristics that lead to project team performance (*Figure 7*).

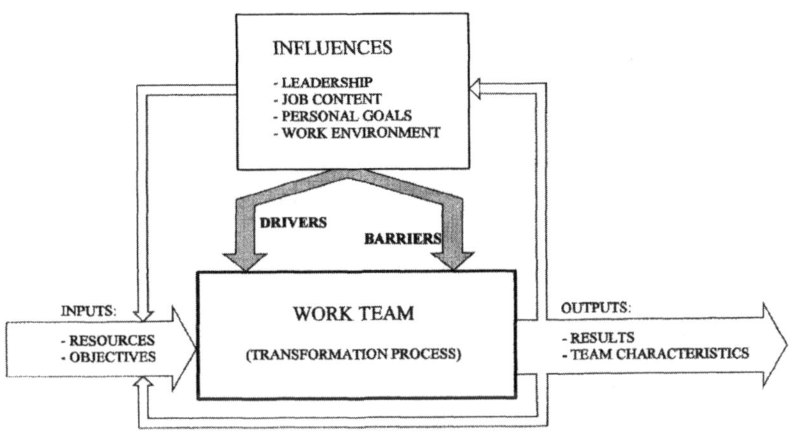

Figure 7. Process model for product development Thamhain & Wilemon (1987).[77]

However, there are some curious points not mentioned by the authors. Namely, among the initially identified drivers that were not found among the final six were:

people-oriented characteristics. The influence factors that correlated predominantly positively were labeled "drivers" and those that correlated predominantly negatively were labeled "barriers".

76 © 1987 IEEE. Reproduced here by kind permission of the IEEE.

77 © 1987 IEEE. Reproduced here by kind permission of the IEEE.

- proper planning,
- open communication,
- good interpersonal relationships among team members and with management,
- sufficient resources,
- few interpersonal conflicts.

Some of these discrepancies can be explained even without full access to the authors' data. For example, sufficient resources cannot be a driver of performance, but insufficient resources can certainly be a barrier. This resembles Herzberg's *two-factor theory* (see page 41) in the sense that removing a barrier does not necessarily result in the enhancement of a driver.

Likewise, it can be inferred from these findings that having few interpersonal conflicts is not a potential driver, nor are good interpersonal relationships, whereas their absence in the form of power struggle and conflict or uninvolved management can prove to be a barrier.

Furthermore, the initial candidates for barriers included:
- different interests and priorities among team members,
- lack of team definition and structure,
- wrong capabilities,
- poor selection of project personnel,
- low credibility of project leader, and
- poor communication,

but the converse of these factors was not found to be drivers of performance.

Also not discussed in the article, the most puzzling findings in the same research data were that:[78]
- *Unclear objectives* were negatively correlated with *involvement and energy* ($\tau = -0.45$) and *technical results* ($\tau = -0.40$) as would be expected, but was at the same time positively correlated to a *capacity to solve conflict* ($\tau=0.45$), *mutual trust* ($\tau=0.40$), *commitment and result orientation* ($\tau=0.55$) and *ability to predict trends* ($\tau=0.35$).
- *Insufficient resources* were positively correlated with *on-time performance* ($\tau=0.40$), *on-budget performance* ($\tau=0.55$) and *commitment and result orientation* ($\tau=0.40$), whereas they were negatively correlated with *technical success* ($\tau = -0.35$) and *involvement and energy* ($\tau = -0.30$, $p \leq 0.05$).
- *Poor job security* was positively correlated with *team spirit* ($\tau=0.40$) and *mutual trust* ($\tau=0.40$), but negatively correlated with *technical success* ($\tau = -0.30$, $p \leq 0.05$).

[78] All data from Kandell Tau correlations: 99% confidence level, $p \leq 0.01$, unless indicated otherwise.

- *Shifting goals and priorities*, which had a negative correlation with *involvement and energy* ($\tau = -0.30$, $p \leq 0.05$) and with *technical results* ($\tau = -0.20$, no data on confidence level) had a positive correlation with *on-time performance* ($\tau = 0.35$) and *concern for quality* ($\tau = 0.40$).
- *Power struggle and conflict* was positively correlated with *capacity to solve conflict* ($\tau = 0.60$), *team spirit* ($\tau = 0.40$) and *mutual trust* ($\tau = 0.45$), whereas it was negatively correlated with *technical success* ($\tau = -0.20$, $p \leq 0.05$) and *involvement and energy* ($\tau = -0.25$, $p \leq 0.05$).

All this seems to indicate a reaction of teams to the constraints of the perceived situations: they tended to compensate for the barriers imposed by the environment, although the likelihood of achieving the overall goals was low. This is very easy to see in the case of the correlation *between poor job security* and *team spirit and mutual trust*, but is less clear for the correlation between *shifting goals and priorities* and *concern for quality*. Of course, this might also be due to the study design or to particularities of the population, so conclusions should be drawn from these reports with care.

Nevertheless, these considerations provide a clue about the complexity of the interaction among variables. This interaction is not easy to assess without additional analysis instruments. In any case it should be clear that it is futile to offer simple "recipes" to enhance development performance.

3.1.3 Search for the "one best way" vs. interaction theory

A study by Rubenstein, Chakrabarti & O'Keefe (1974) involving field research into more than 1,000 organizations over a period of 20 years, an examination of empirical literature, and the insights of participants in and observers of the Research, Development & Innovation (RD/I) processes revealed about 240 pertinent variables.[79]

In a subsequent study, they analyzed data from six companies and 103 projects out of a population of ten companies and 140 projects having a typical project duration of four to five years. A total of 118 respondents provided data directly concerning the projects. The study's authors concluded that, although many studies revealed that a "product champion" or similarly designated person is a *necessary* condition for project success, it is not a *sufficient* condition. This statement is based on the many projects

[79] The number of different variables was lower because many of them were associated with each other and sometimes reflected the same aspects. Reduction of the number of variables was not easy to accomplish owing to the inexactness and ambiguities in their usage (op. cit., p. 13). Examples of these ambiguities are found in "top management *support*" and "top management *interest*", or in the roles played by "product champions", "technical innovators", and "technological gatekeepers", among others (op. cit., p. 148). Hence several parameters appear to be very similar, yet have essential differences.

studied that "were clearly unsuccessful, despite the presence of a product champion or technological innovator" (op. cit., p. 145).

They also pointed out that "several projects illustrate that an energetic and committed product champion or innovator, *plus* a clear market need can sweep away many of the usual barriers to successful RD/I (e.g., management apathy, insufficient information, less-than-encouraging state of the art)". On the other hand, other projects illustrate that "such [...] persons can succeed even in the absence of a clear market need if they have some other things going for them (e.g., lots of money; a good product development track record; lazy or ineffective competitors; a keen appreciation for the pitfalls in their particular market, segment, etc.)" (op. cit., p. 145, emphasis as in original).

Another significant contribution was the Hughes R&D Productivity Study (Ranftl 1978). Sixty organizations in different industries (not only aerospace) participated with 28 leading consultants in R&D and innovation participated in this study, which also included a comprehensive literature search. The study introduced "productivity profiles", which comprised lists of 173 characteristics. These characteristics were sorted according to (i) employees, (ii) managers, (iii) leaders (the latter two were explicitly separated), and (iv) organizational characteristics. Although not explicitly mentioned by the reporting author, many of these characteristics reflect the *form of interaction* practiced either by individuals or fostered within an organization. For example, a productive manager "gets involved when subordinates have problems—helps them and backs them up—and promptly corrects employee grievances" (op. cit., p. 100).

The study did not limit creativity and innovation to the R&D department and confirms that "the specific means for improving productivity are unique to each individual and organization; *there is no universal formula*" (Ranftl 1977, p. 25, italics added for emphasis). This confirms the inadequacy of the "one-best-way approach to the R&D environment.

This causes despair among engineers and economists who cannot accept that there is no "one" best way. They would prefer a list of characteristics that should prevail in the product development environment to ensure that productivity skyrockets, regardless of the situation or organizational context, the project of interest, the issues that arise, and, worst of all, the persons involved. These findings are clearly a source of frustration for those who subscribe to such thinking.

The situation is not hopeless, however. It has a parallel to leadership theories, where the search for leadership traits or characteristics (leadership traits theory)[xxiii] produced a list of 500 characteristics. Unfortunately, the studies are not only inconsistent with each other, there are some flagrant contradictions in their results. This led to the *interaction theory* of leadership, which maintains that interaction between the leader and the followers is the factor that determines good leadership (Ulich, Baitsch & Alioth 1987, 9). Similarly, studies of the characteristics that foster product development and innovation led to reports listing more than 200 characteristics, sometimes with the same confusing effect. The analysis of these interactions indeed provides many clues on how to improve the product development environment: interactions between individuals and the activities to be performed, interactions among individuals and, especially, with individuals having two functions: as leaders and as quality personnel. Two sections in this book[80] examine this aspect in detail.

In many cases, the interaction process constitutes the success factor for product development because it helps individuals adapt to the situation at hand.

3.2 Project management approach

3.2.1 Network resolutions and the Tayloristic tendency

Following the Cartesian principles, an issue is subdivided into as many parts as possible and necessary to facilitate its solution (Cartesian principle #2).

One of the elements of project management is to divide the product specification or statement of work (SOW) into smaller sub-specifications, hence creating a work breakdown structure (WBS). This structure is almost always tree-like, and the extremities of its ramifications are called work breakdown elements (WBE).

It is logical to structure a problem in order to solve it. Dividing it into subelements seems a reasonable way to gain an overview of the necessary steps toward achieving the established goals. However, it has been observed that not only does this not necessarily lead to product success, it sometimes can be misleading and can ultimately hamper the achievement of the desired product. Let us therefore re-examine the hypothesis that structuring a task by dividing it into smaller subelements can cause one to lose sight of the overall goals, thus and eliminating an essential motivational justification for this approach. It seems that by attempting to make a task manageable, one risks excluding the most important contributor to its accomplishment, that of the human activity behind WBEs.

---

[80] Cf. Chapter 6.

Investigating the first wave of the use of PERT, Schoderbek (1965) conducted a survey of 200 companies selected randomly from the Fortune Top 500 directory, the majority of which were not defense-related. Among Schoderbek's findings are such typical user statements as:

Box 14. Some problems encountered using PERT in 1963 (Schoderbek (1965, pp. 203-204).

- Too much detail causes loss of sight of the main problem
  PERT encourages excessive detail if not closely controlled.

Remarks provided by some respondents indicated that

- there were "attempts by government agencies to make this a one-way reporting system to and for government only".

Network resolution was limited during the first wave of PERT usage by the computational power available. With the microcomputer technologies of the 1980s, this computational power was no longer a limiting factor: neither drawing Gantt charts nor the resolution of the planning posed a challenge. One of the companies interviewed in 1996 for this study revealed the following:

Box 15. Problems associated with excessive division of activities (case #10).

A development manager of a specialized vehicle manufacturer—a mechanical engineer by training—had previous experience with the *finite-element method.* This consists of a fine network modeling of a given part to calculate, for example, the mechanical stresses on that part. The finer the resolution of the network, the more accurate the results. The development manager had used project management tools in the past with admirable success. Consequently, every development in the company was executed using this technique. When the development manager attempted to adapt the principles of the *finite-element method* to project networking, however, i.e. to increase the resolution of the PERT network to achieve better results, it no longer worked. Conflicts occurred as the resolution was increased, especially with regard to resource allocation. With a fine resolution in the subdividing of activities, the personnel working on each activity no longer had sufficient flexibility and to react to routine occurrences. If, for example, a change in the manufacture planning is made, then a change in the prototype execution is necessary because the resources have to be shared (prototypes in this case are manufactured in the production facilities). The tools for resource allocation actually prevented people from working on the diverse activities. Consequently, the resource allocation program was not used, and the people working on the various activities were left to solve the resource allocation problems themselves. The development manager ultimately recognized that one does not necessarily needs "fancy graphics" or powerful computers to achieve results.

The same practitioner confirmed that the network technique does not provide an overview of the interactions between the activities. Project managers have to ensure this by other means.

The network resolution or the fineness of the work breakdown structure recalls attention to the effect of subdividing a manufacturing process into its basic elements in an attempt to achieve control over the activities (cf. Effects of Taylorism or scientific management in Chapter 2, Historical Perspectives, page 33). If this error is repeated with product development, unsatisfactory results are to be expected. If developers lose sight of what they are doing, their ability to solve the numerous problems associated with the object under development can be impaired. Moreover, if developers are faced with an excessively stiff framework—including the attendant tasks of reporting and justification—they might lose not only efficiency, but also the motivation to perform.

Note furthermore that resource allocation is also associated with other non-technical aspects. Resource allocation issues are not necessarily only a question of the availability of resources, it also requires sharing and cooperative behavior amongst the persons involved.

When resources must be reallocated, a mere mathematical reallocation might not work. In his book "The Mythical Man-Month" , Brooks (1982) describes why adding people to a late software project may actually prolong the delay. If an activity is not additive[81] in the sense that activities can be performed separately and without mutual communication, then one person can do the same amount of work in ten months as ten persons in one month. If the ten persons have to communicate with each other, then the activity originally allocated to each person will take *more* than one month. If a great deal of information exchange or interaction is required, then the ten persons might take even more than the original ten months. If the activities are sequential in character, then the time needed for their completion is not significantly affected by the number of persons working on them.

3.2.2 Time and cost estimation

The PERT technique is also dependent on an estimate of the time to execute a given task. An experienced developer can do this fairly accurately. Some companies require developers to provide three estimates: one of the most probable time, one of the soonest completion date and one of the latest

[81] Task typologies are discussed in the section on group work, page 213ff.

completion date of each activity in the network. This three-fold estimate is then combined to obtain the mean value of task duration. A basic assumption of PERT is that the time needed to complete an activity possesses a beta distribution with a standard deviation of one-sixth of its range. This leads to the simplification expressed by the three-fold estimate rule, where the time to be entered in the PERT network is given by

$$t = (a + 4\,m + \mathrm{b})\,/\,6\,,$$

where m is the most probable time, and a and b are the soonest and latest completion times, respectively.

MacCrimmon & Ryavec (1962) investigated the effect of these assumptions and concluded that they can theoretically produce an error of 33% on the mean and one of 16% on the standard deviation. These errors may be compensated within the entire network. Assuming that all the tasks have the same range—an unlikely event—the total error is 17%. Apparently, this assumed beta distribution was never verified in real situations.

Realistic time estimates are very difficult to make. The time orientation of the individuals involved play a role not only in obtaining a match between estimated and achieved times-to-completion, but in the achieved results as well. A development manager once remarked that, paradoxically, the most highly innovative engineers often have difficulties adhering to a schedule they had proposed themselves, whereas less innovative engineers usually did not (case #11). It therefore makes sense to use a differentiated task allocation system for different kinds of people. This suggests a different association of individual time and result orientations.

Since the introduction of the project management technique, management has been faced with the following questions (Dooley, 1964, 167):

- Can personnel really be expected to consistently generate meaningful (and "unmanipulated") input data of the type required by these techniques?
- How can I help people develop their skills—and the motivations—this will require?

Weinberg (1971, 20) describes the estimation problem as follows, "The average programming manager would prefer [that the duration of the task] be estimated at twelve months and take twelve months than that the same project be estimated at six months and take nine." This problem poses a significant challenge to evaluating performance. Should one be evaluated primarily in terms of their adherence to a schedule, or should one be evaluated with regard to more relevant parameters such as innovativeness or the efficient use of resources?

The three-estimate rule for determining task duration should accommodate the effect of schedule slippage due to *expected* problems. However, performing the weighted average cancels out such an effect, especially if the parameters are symmetrical to the most probable time. As such, mere entering the average value in the PERT/CPM network will render the reflection useless in terms of technical effectiveness. This was confirmed by Müller (1989, 293), who states that the three-estimate method has long been used primarily for psychological reasons.

Whenever there are uncertainties involved in a technical program, there is the risk of encountering *un*expected problems. Even coping with *expected* difficulties that have not been faced before can lead to delays. This is not only a characteristic of military programs, but of any complex or innovative design.

Concerning the original case of project management (Polaris), it would be a mistake to believe that the Navy SPO's success meeting self-generated estimates was due to its *"managerial innovation strategy"*. According to Sapolsky (1972, 182), the same problems that caused substantial overruns in other programs seem to have been present in the Polaris as well, but its ability to meet the estimates stemmed from the overwhelming and dependable political support it enjoyed—a unique advantage. The only overly optimistic and hence inaccurate cost estimate was the initial one made in 1956. It was caused by the uncertainty of gaining approval owing to the Eisenhower administration's declared intention to reduce defense budgets. The launch of Sputnik in 1957 changed all this, leaving no doubt that Polaris would receive sufficient resources. As a consequence, the SPO and its contractors could afford to provide honest, accurate estimates.

Let us take for example the military equipment development of the tactical fighter experimental TFX[82] and C-5A transport aircraft. Developed in the 1960s when PERT was obligatory, they were plagued by cost and time overruns. There are many other examples of innovative programs that reported overruns even in spite of PERT usage.

Part of the reason is the perception of estimating time and costs for completion and progress reporting. There is a rule called Tyler's law that describes this phenomenon, formulated by Whyte (1978) as follows:

Box 16. Tyler's law for cost and time estimates Whyte (1978, 4).

- If you have done a similar development before, multiply the first estimate by $\pi/2$.
- If you have not done a similar development before but someone has, multiply the first estimate by π.
- If no one has done a similar development project, multiply your first estimate by 2π.

[82] TFX was the initial denomination for the F-111.

Tyler's law was firstly announced in relation to costs, but it is equally applicable to task duration.

Such a formulation might seem more amusing than useful. Whyte, however, cited historical cases that took place between 1780 and 1860 that actually *illustrate* its validity. He also cited contemporary examples, such as the case of the US Navy DSRV submarine, the Sydney Opera House, the Munich Olympic Stadium tent, the Concorde, the Rolls Royce RB211 jet engine, and others. All these cases *illustrate* the underestimation of time and cost with a high variation in the reported cases. Norris (1971) investigated the accuracy of cost and duration estimates for industrial R&D projects based on the records of 475 projects in four different organizations. The results are presented in terms of mean ratios of actual-to-estimated values ranging from 0.97 to 1.51 for costs and 1.39 to 3.04 for duration, and confirming similar British and American studies of the kind. Mainsfield, Rappaport, Schnee, Wagner & Hamburger (1981) investigated a pharmaceutical company and obtained the values of 2.11 and 2.95, respectively.[83]

Care should be taken in interpreting these results, however, because the processes behind them are yet to be investigated. The reasons for underestimating time and costs include (i) pessimism/optimism bias, (ii) prospects of gaining project approval (as Sapolsky described for Polaris), and (iii) the time horizon (Epton 1972).[84] Kanter (1991, 41) attributes them to the uncertainties inherent in product development, which require a high amount of flexibility. She lists the four main sources of "vulnerability of the mid-stages" as being: the difficulty of predicting problems, unexpected obstacles, criticism that surfaces only when the project is under way, loss of momentum (of thc working group). Bureaucratic delays were also a possible obstacle. Whyte mentions three factors that engineers face when estimating a novel project. (i) The degree of extrapolation factor (i.e. success depends on the accurate assessment of several unmeasurables, difficulty in predicting the future situation the product will have to face); (ii) the prevailing optimism/pessimism or self-deception factor; and (iii) the management/design competence factor. The applicability demonstration of Tyler's law is not relevant in itself, but one point must be emphasized: many managers tend to dislike it. First, they fear an explosion of estimates if everybody were to calculate costs and durations in this manner. Second, it suggests that slack is intentionally built into an estimate, and is therefore dismissed as waste or ballast. Unfortunately, it is seldom regarded as a way to plan for contingencies.

---

[83] Cited in Kanter (1991, 42).

[84] Associating the problem with the work of Fraisse (1964) and Shackle (1961).

Let us return to an example from the early days of the space age. After the launch of Sputnik in 1957,[85] Wernher von Braun asked to be given 60 days to launch an American satellite. In fact, his team had previously almost achieved this task in the course of testing reentry shapes, for which the same rocket had been used. Von Braun was promptly contradicted by his boss, Gen. J. B. Medaris, who corrected this estimate to 90 days. Explorer I was successfully launched in 1958,[86] 119 days later.

The difficulty of making accurate estimates persists, especially if innovations are to be made. The problem may even increase in response to current efforts to further reduce manpower costs or time to completion, where circumstances may force one to provide overly optimistic estimates.

3.2.3 Progress reporting

In addition to the difficulties associated with time estimations, it is significant whether a management technique is considered to be *driven by future actions* or regarded as a *justification of past events*. This provides a measure of progress of development activities.

Project managers have to rely on information provided by the people working on the actual tasks. The efficiency of this process depends on *how it is conducted* and *what purpose it serves*.

Critical path activities are known to determine the early completion of an entire project. Depending on the orientation of the manager, this can take different forms, as illustrated by the following cases.

There is a tendency to focus on critical path activities (and hence to put pressure on the individuals executing them) as being the factors most likely to delay a project. There have been cases where, due to the configuration of the activities, the critical path is constantly changing to accommodate the *declared* progress in the various activities. Something that is not on the critical path today might be on it tomorrow, thus deserving attention today. But this attention is not usually given if managers focus exclusively on the critical path. Being on the critical path ensures attention to an activity. This in turn usually means that someone is "buzzing around", thus violating the "protection" of a working group by means of persons or groups whose function is (or appears to be) to investigate the course of the activities. Such intervention is commonly perceived by the people trying to complete the tasks at hand as a disturbance. Let us examine the case below.

---

[85] 4 Oct 1957.

[86] 31 Jan 1958.

Box 17. Case of an "unprotected" performer (case #12).

In an aerospace company, there was a scheduled activity, the result of which had been declared a milestone with the customer. The activity was the planning of active measures to improve quality, which encompassed development, production and installation work. Achieving the milestone was an important event, and it was also tied to the financial transaction to the contractor. Owing to the crisis in the aerospace industry at that time, it was also a very important event for the company.

On one occasion, the company's president happened to see the engineer in charge of the project, and asked him casually whether the schedule could be met. The engineer assured him that, in spite of the number of tasks yet to be accomplished, the document would be ready on schedule. There was indeed considerable pressure on this engineer, and he was quite conscious of the multiple importance of the document.

Perhaps because the project was the subject of such intense scrutiny, and "everyone" was talking about it, the engineer's supervisor phoned him nearly every hour to ask whether the document was ready. At first, this did not really bother the engineer, although each phone call cost him about five minutes on average (i.e. a 10% decrease in time efficiency), including the time to reimmerse himself in the task at hand. After numerous interruptions of this kind, the engineer began to notice that it took longer each time to resume his work. Finally, he became irate at yet another interruption by his supervisor. This escalated to a lengthy confrontation, after which it took the engineer at least 30 minutes to recompose himself. And who would not be upset in such a situation? With the deadline approaching in just a few days and the internal approval of the final document still outstanding, every minute, every bit of concentration counted.

The activity was nevertheless completed on schedule, despite the problems encountered. The engineer in this case is an example of an "unprotected" performer.

This kind of action based on an activity in the critical path or on any problems occurring in the course of the such activities can be self-defeating in the hands of hapless managers. Of course, the critical path is a "fact of life" and if a schedule is to be met, heightened attention should be paid to the first sign of trouble. However, in trying to avoid further schedule slippages, a manager must take great care not to exacerbate the situation.

Depending on the configuration of the project activities, the critical path might actually divert attention from other, equally real problems. Consider the hypothetical case in which an activity not on the critical path were the very one bound to reveal problems that were not known beforehand. If left unattended, measures to anticipate the reduction of this inherent uncertainty would not be taken, and this might actually cause schedule slippage or even failure of the entire project.

Box 18. The umbrella

In the American moon program, many problems were encountered in connection with the rocket's second stage (S-2). The technical difficulties and the design constraints, together with major incidents (e.g. the stage catastrophically failed twice during ground tests) caused its schedule to slip. Technical uncertainties raised grave concerns, even among the public. In his account on behalf of North American Aviation, which developed and fabricated the S-2, Gray (1992, 159) describes the following insight of the participants: "The second stage was on the critical path from the very first (PERT) computer run. Among the other contractors—who were having plenty of problems of their own—*the S-2 became known as the 'umbrella' for the way it shielded them from NASA's wrath*" (parentheses and italics added).

This situation is not rare in cases were too much the attention is directed toward the critical path. Those working critical-path activities are put under disproportionate pressure, whereas those "fortunate" enough to have other duties are virtually free of scrutiny. Obviously, the latter have no interest in changing this situation, not only to avoid stress, but to be allowed to work undisturbed.

Another description of the psychological effects of using PERT stems from an incident during the Saturn V development. Examined from management's perspective, it illustrates some of the so-called "soft" difficulties in handling the critical path.

Box 19. Handling schedule slips on the critical path.

"A 'slip' is the admission by a contractor in the critical path that he will not complete his work by the time it is due. This has consequences for all of those who subsequently expect to find themselves to be in the critical path. If the slip were known to involve, say, three working days, the Control Center personnel would parcel the slip one day at a time because they didn't want others to relax. Consequently, the information was invalid and opportunities for others to improve quality and reliability were lost because they were held to unrealistic due dates." (Tompkins 1977, 17-18) (cf. Description of the Apollo/ Saturn Control Center on page 64)

As Sadow (1964, 146) states: "What can kill PERT while it is still in its early stages is irresponsible information mongering and evaluation by individuals who view the critical path on the PERT computer printout for one random reporting period, and from this form conclusions upon which they base management decisions or alarm others. This type of PERT misuse shakes the Government agency attempting to develop and operate the technique and frightens contractors who are endeavoring to report complete factual information. There must be some management maturity if we expect these advanced management techniques to work." In 1963, a panel of 57 members of the aerospace industry were asked to list thirteen "unethical" practices and to estimate "how common and how serious a breach of ethics they believed each one to be". The offense regarded by the panel to be the most serious was "lying or deliberately giving incomplete information to the customer about progress on a contract". Most of the panel members indicated they believed this was an "occasional practice in industry". One of the panel members, referring to incomplete information, said: "it depends on whether you're providing the information to an intelligent recipient who will not use it against you—common practice at the lower follow-up levels of Government. If the information is delivered directly to a competent person, then it would be best to keep the card on the table"[87] (anon., 1963, 13).

Another issue associated with time (and cost) considerations is the nonlinearity of the perception of progress. Wischnewski (1996, 368) called this the "90% syndrome" : the tendency to report after one-third of the total duration that 90% percent of technical results have been achieved. The remaining 10% of estimated work, comprising details that make essential differences in the quality of the product, takes the remaining two-thirds of the time. This nonlinearity of progress (90% of progress does not correspond to 90% of time) is usually not accommodated in detailed progress reporting.

3.2.4 Uncertainty and the nature of management

It is thc nature of management to (try to) eliminate uncertainty. The problem with this does not lie in reducing the uncertainty in a given development program, nor in guaranteeing that something useful will result from development "chaos". The problem lies in denying the uncertainty inherent in the creative process, in overemphasizing the need to make things manageable, in claiming to have everything under control.

If the success of the project manager were measured by how far he can reduce uncertainty, then he will strive to eliminate it, or at least act as if he does. He may truly believe he is doing so. This might place the project manager in an ambiguous situation, depending on his position and by whom he is evaluated.

A worldwide study on innovation management in large and small companies was conducted by Quinn (1985, 76-77). One of the initial

[87] Note: for non-English language native readers– *to lay one's cards on the table:* to divulge one's resources and intentions freely.

findings is what he calls *excessive rationalism,* which he considers one of "most notable and common constraints on innovation in larger companies". Quinn observed that "managers in big companies often seek orderly advance through early market research studies or PERT planning. Rather than managing the inevitable chaos of innovation productively, these managers soon drive out the very things that lead to innovation in order to prove their announced plans." Conversely, with regard to the factors crucial to the success of innovative small companies, his research suggests that "technology tends to advance through a series of random—often highly intuitive—insights frequently triggered by gratuitous interactions between the discoverer and the outside world. Only highly committed entrepreneurs can tolerate (and even enjoy) this chaos. *They adopt solutions wherever they can be found, unencumbered by formal plans or PERT charts that would limit the range of their imaginations.* When the odds of success are low, the participation and interaction of many motivated players increase the chance that one will succeed" (op. cit., p. 77, italics added).

Another of his findings is that "reward and control systems in most big companies are designed to minimize surprises. Yet innovation, by definition, is full of surprises. It often disrupts well-laid plans, accepted power patterns, and entrenched organizational behavior at high costs to many. Few large companies make millionaires of those who create such disruptions, however profitable the innovations may turn out to be. When control systems neither penalize opportunities missed nor reward risks taken, the results are predictable" (op. cit., p. 76).

Kanter also emphasizes that "the innovation process is uncertain, with unforeseen delays and setbacks, with possible occurrences of anticipated cost overruns, and uncertainty in the ultimate results" (Kanter 1991, 17). Furthermore, "Cost overruns and missed deadlines are common because of the inherent high uncertainty of the development process" (op. cit., p. 41).

In coping with these uncertainties, an ambiguous situation can arise for the people managing this process, and this situation be tied in turn to the manager evaluation process.

In order to cope with uncertainty, "flexibility is the most important issue: to allocate new resources, to mount secondary projects to remove the unexpected roadblocks to renew the spirit of the working team, and to rethink the approach in light of the learning that emerges" (Kanter 1991, 43).

3.3 Quality systems

Quality systems can contribute the following elements to the development environment: (i) a mature set of procedures, (ii) homogeneity of solutions, (iii) a clear definition of responsibilities, (iv) the institutionalization of solutions, (v) avoidance of duplication of tasks, and (vi) a certain transparency of methods to customers and suppliers. However, in spite of these elements, which are considered necessary to the management of quality, certain issues can arise in the real implementations in the R&D context that sometimes overwhelm these advantages.

Referring to the product development process, Seghezzi (1988, 332) states that "from the experience with the ISO 9001/9004 implementations, it follows that only a formal installation and description of the quality assurance in the development is insufficient. It was indeed perceived that different companies with the same formal fulfillment of prerequisites achieved very different results in optimizing and mastering the quality of new products." (cf. Contrasting cases of quality improvement with ISO 9001, page 58)

Defenders of these systems claim that they enforce better "discipline" in R&D, which in some cases might admittedly be necessary. But misplaced discipline also bears many potential disadvantages in the development environment.

Let us study a few examples that illustrate the problems of formalized bureaucratic structures associated with technical problem-solving.

Box 20. Problems associated with procedural systems.

The first version of the DC-10 aircraft exhibited a problem concerning the locking of the cargo door. In many situations, this induced the explosive decompression of the cargo bay, causing the floor to collapse and disrupting the (redundant) hydraulic control lines passing through it. Loss of control of the aircraft was the final result.

The problem was known among the technical people involved, but it could not be adequately solved for political and financial reasons. There were at least nine possible failure sequences of the latching mechanism of the cargo doors that could result in a life-endangering situation. They were all known *a priori* because they had been revealed by the failure mode and effect analysis (FMEA) (French, 1992, 185). After some real occurrences of the failure (fortunately with no casualties) starting in 1970 in a ground testing incident with Ship 01, it was decided to install a mechanism to prevent the pressurization of the aircraft if the locking mechanism failed to engage. This mechanism was itself failure-prone and caused a major accident in 1972 when an aircraft in Windsor was forced to execute a controlled crash. Further modifications were needed and, although the technical problem and its solution were completely characterized, the organizations involved failed to avert the 1974 crash of Ship 29 in Paris, in which all the occupants were killed (Fielder, 1992).

The system used by the organization that had manufactured the aircraft had unsuccessfully called for a modification in existing aircraft to prevent a recurrence of the 1972 Windsor accident. "In July, 1972, Ship 29 was in the 'Rework for Delivery' area of Long Beach plant of McDonnell Douglas. Included in the work to be done on Ship 29 were modifications to the cargo door latching mechanism. The plant records for July, 1972, indicate that *three inspectors stamped the work records for Ship 29* to indicate the modifications had been completed and that the plane was in compliance with Federal Aviation Administration (FAA) guidelines. None of the work on the cargo doors had actually been done" (French, 1992, 187, italics added). An inherent weakness of sole reliance on a "system" to accomplish tasks was subsequently reported by Eddy, Potter & Page (1992, 113): "Douglas employees later testified that they were simply unaware of the significance of the various things that were supposed to be done to DC-10 doors."

Another case, reported by Kutzer (1989, 371), occurred during the integration of the European reusable space laboratory *Spacelab.* It serves to further illustrate the problems that lurk in bureaucratic principle. "One morning, the clean room in the integration hall was flooded because an integration team had not closed a water tap as prescribed. In an attempt to avoid a repetition of the incident, a procedure was introduced that required six signatures to confirm that the water tap was appropriately closed. Two weeks later, the hall was flooded again, despite—or maybe because of—the six signatures."[88]

Yet another case illustrates a concept associated with the constraints imposed by formal, bureaucratic organizations. It was reported by Feynman (1988, 171-173) with reference to the Rogers presidential commission to investigate the Challenger explosion. In order to check the circularity of the sections of the solid rocket booster prior to field assembly, the approved procedure was to introduce a rod in two diametrically opposed holes out of a total of 180 holes and repeating this three times for each joint interface. Each of the two boosters has three field joints (twelve interfaces). The task required the assembly personnel to "climb up into an awkward place to count the [90] holes [around the external circular perimeter]. It's very slow and very difficult". Hence, an alternative to the official procedure was devised. It entailed painting four tiny marks on each interface, 90 degrees apart, such that in the worst case one would not have to count more than 22 holes from one of these marks and repeating this process from the diametrically opposed mark.

Management rejected the proposed solution because it was too expensive to implement: *not to paint the marks, but to change all the manuals.* The

[88] Kutzer's original wording has been paraphrased for clarity.

assembly personnel also brought to management's attention the fact that measuring three diameters was not sufficient to guarantee the roundness of the booster, but this was equally futile.

It was imperative that the notorious o-rings be located on the booster sections of 3.71 m (146 inches) diameter and be compressed by their flanges by 0.5 mm (0.0000013%) to provide adequate hot-gas sealing. Known failures to fulfill a given specification were overlooked as long as an established procedure was followed properly. This situation did not contribute to the Challenger tragedy but was strongly related with the hot gas sealing function of the o-rings.

These three cases were selected because they reinforce the professionalism of the workers involved. Moreover, they occurred in work environments that place great emphasis on quality. Yet they occurred because quality consciousness was overshadowed by a system that does not help people solve such issues. Quite the contrary.

Nevertheless, there are valid reasons for having approved procedures. The following example describes such a case.

Box 21. The case of the mysterious sulfur (case #13).

In a leading telecommunications company, back when relays were the basic building block for telephone switching,[89] corrosion problems arose. Chemical analyses traced the cause to the presence of sulfur. Company management knew there was no sulfur in their production processes, so they began to search for the possible entry point of the offending substance. They checked the water and the air, but found no traces of sulfur.[90] A quality engineer continued to ponder the problem, when someone from the production staff asked him quite euphorically: "Did you see the quality of the solder since we've been using *that*?" The substance in question was a "wonderful preparation" the production employee had found in a small shop in town. It rendered the solder shiny (shiny solder is the visual indication of a good metallurgical bond). Sure enough, an analysis of the substance revealed that it contained sulfur. The employee had *smuggled* it into the production line without previous analysis or consultation. Despite his undoubtedly well-meant intention to improve the performance of the solder line, he had in fact created a much greater problem than the one he had endeavored to solve.

89 The technology transition to full-electronic switching took place from the mid-1970s to the early 1980s.

90 One fact not mentioned that might or might not be related to this case is that approximately at that time, an auto manufacturer operating in the same region experienced many corrosion problems incurred by a nearby chemical factory, causing airborne contamination of automobile bodies.

Similar things are frequently done with numerical values: they are "improved" in one respect, but without comprehensive regard to their full implications.

Standard procedures (in this case, regarding soldering techniques) are usually mature, and they reflect valid reasons for following or eschewing a given approach. Working "by the book" was generally a good way to avoid unexpected problems, of ensuring that everything is under control. But let us examine the underlying causes that can prompt a conscientious employee to deviate from the established procedure. The problem does not lie in the fact that an individual shows initiative, but whether this initiative is properly channeled. Initiative is a potential source of a great solution. As illustrated by the case of quality circles, properly channeled initiative can be a considerable motivator. In fact, quality circles in Germany never quite replicated the quality improvements that emerged from Japan, but the motivation-enhancing effect was clearly present (cf. Bednarek 1985). Organizations would do well to channel the initiative and involvement of its staff in a constructive way.

Many companies have faced the inherent paradox of encouraging and yet channeling the innovations devised by their staff. In early cases, the relatively slow dynamics of technical development allowed sufficient time for the formal structure to analyze and approve a solution before adopting it, but the dynamics of technology evolution have changed considerably since the time of electromechanical relays.

As a result of technological advances, the electronics industry is tearing down its own dogmas regarding production processes. Procedures that once were abhorred are now used freely. For instance, boards are now cleaned with water-based compounds. In some cases, the solder flux of the finished assemblies is not removed, although the complete characterization of long-term effects on reliability are still under way. Such procedures would have been unthinkable in the relatively recent past. Moreover, companies are increasingly under pressure to shorten their product cycle times and to considerably accelerate their internal dynamics to bring a product to maturity.

For production groups that operate machinery it is relatively straightforward to establish a set of detailed "standard" procedures, even though the procedures may not be complete and must sometimes be deviated from. For problem-solving groups, establishing an effective set of detailed procedures is a much more difficult task.

Let us turn our attention back to the Apollo moon program. Alexander (1969, 156) stated that "everything—every technician's task, every test, every lot of every material, every component—has a formal document associated with it that can be used to feed information to a higher level, to fix responsibility, or to trace a component or material failure and prevent its occurrence". However, he continues, "things like configuration management smack of what is otherwise known as bureaucracy and all the documentation seems difficult to distinguish from red tape". Moreover "many of the Apollo's own engineers chafe under such requirements, which they sometimes feel to be peripheral and time wasting."

It should be noted here for non-aerospace readers that NASA is an extreme case in this respect because of the extensive requirements placed not only on the documentation but also on the production process. A failure could usually be traced based on the meticulous records. After all, a failed prototype is generally no longer existent, and a functioning prototype would be "in orbit" or lost in space. The authorities require thorough documentation of the exact configuration of every item. However, the same kind of approach can also be found in non-aerospace companies under ISO 9000 systems because the same elements present in the extreme case tend also to be present in their implementations.

Alexander continues: "In any case, most of the managers in Apollo seem to agree that the *real key to success lies not in the specific formal procedures but in the operational style*—the dedication, excruciating attention to detail, and mutual cooperation among a variety of talented and strong-minded individuals" (op. cit., p. 156, italics added). The success of the *system* during the moon program was based on the persons using it, not on a set of procedures that would function independently.

According to Seghezzi (1988, 339), the essential success factors of quality management during the development process are that

- the development process must be clearly structured and organized,
- the quality processes are to be defined and made transparent,
- the quality assurance activities are to be united in order of events,
- the involved parties (and there are many of them in the development) have to have a clear assignment of functions, competencies and responsibilities.

Much more importantly, however, he adds, "The resulting organization must take special care of the numerous interfaces, where the *harmonic interplay of the participants is more important than the precise demarcation of their scope of duties*" (italics added).

Lessons unlearned

The National Aeronautics and Space Administration was once cited as a model for the public services in America. The prevailing belief in the post-Apollo era was that many other problems could be solved if similar approaches were taken to other public services: education, transportation, food, industrial and urban development, etc. Lessons learned from the space industry should be applied to other institutions, people thought.

NASA of the post-Apollo era was investigated by Fries (1992), who interviewed engineers at NASA centers that were representative of the Apollo era. She found that NASA was moving toward almost total reliance on contractors and an expansion of bureaucracy (op. cit., p. 181). Her study is becoming increasingly relevant with the current trend of the space industry to "go commercial."

"NASA's engineers *complain bitterly of bureaucracy*, its frequent absurdities, its incessant drain on one's time and energies, as do most employees struggling against paper barriers, hierarchical protocols, and the shackles of central administration everywhere. On this subject they are merely sections of a larger chorus and have little to add that is peculiar to NASA. But they are not wholly devoid of observations that suggest forces exacerbating the tendency toward bureaucratization on the nation's civilian space program.

One of those forces is procurement—contracting—which contributes its own special amount of paperwork and procedures to comply with the latest federal acquisitions regulations. At Langley Ed Beckwith ventilates vexation: 'I'm right now in the throes of trying to get a purchase request through so I can get two contracts without going through a full and open competition.' In Langley's procurement organization 'they start talking to you about a JOFOC [Justification for Other Than Full and Open Competition]. I didn't know what that was. My memos should not have to go into detail to tell them how I ought to do this to get this contract out. My memo should say this is the reason for that. ... Oh Lord! You can see the frustration!' The reason for sluggishness in the procurement process, explains Henry Beacham at Goddard Space Flight Center, 'is fear. Fear of getting a protest on a contract award. After you deal with a couple of them the system tells itself, We'll never let that happen again.' And the way to prevent protests on a contract award is to cross every 't' and dot every 'i' in a complex procedure designed to ensure that every eligible individual or firm has been given a fair shake in the scramble for government funds." (op. cit., pp. 181-182).

"Derek Roebling and Bill Cassirer, at places as different as Kennedy Space Center and Langley Research Center, agree that the way NASA has adapted to its political circumstances and environment has much to do with the degree to which it is afflicted by the worst handicaps of a bureaucracy. Roebling sees, most of all, a 'cultural change' as the 'major' change to have occurred within the agency. With the massive organizational mobilization required to carry out the Apollo program, NASA 'became very institutionalized,' he asserts, and was soon transformed into a 'corporate bureaucracy'. More important, 'the agency has matured from a small group without an agency culture into another federal agency. ... We're no longer in the laboratory; we're now the administrative kind of thing... Bureaucratization was carried to extremes in many cases...[with] increased complexity, less personal responsibility, and more organizational responsibilities achieved through division, multiple signatures, checks, and balances, more reviews, more meetings, more formal systems to keep track of different items,' and the replacement of 'personal responsibility' by 'organizational hierarchy.' " (op. cit., p. 182).

This loss is in great part because of the changes introduced following the Apollo program. The system was kept, but it seems that the consideration of the operational styles and the interplay of the participants, both internal and external, were relaxed. The effects on the development environment are clear.

3.4 Control of development

Associated with systems, especially quality systems, is the element of control. The first edition of the ISO 9001 standard called the activity related to the quality of development *design control.* Its focus on the controlling action is clear from its transcription:

Box 22. Design control as required by the ISO 9001 standard.

"The supplier shall establish and maintain procedures to control and verify the design of the product in order to ensure that the specified requirements are met." Even the explicit mention of the term *verification* is specified by the ISO in terms of control: "*Design verification shall establish* that design output meets the design input requirements..." (ISO, 1987a, 3, italics added).

Problems are encountered with every new product. They are part of the maturation of a product, and they are expected to show up during development.

The isolation of the development and manufacturing teams was one cause of the numerous modifications needed to achieve product maturity.

In an extrapolation of production control, people from the quality control staff tried to control (cf. Discussion on semantics of *control*, p. 28) development progress.

Development personnel reacted with a very strong resistance to such control measures, thus responding to requests for information, even in preliminary form, by saying "it is not ready yet". This meant that no product or process could be quality-controlled until after the work, or at least major parts of it, had been completed.

Box 23. A suggestion from quality control (case #14).

At a company in the avionics industry, an engineer on the quality control staff proposed to determine the quality of product development by taking the product definition documentation and checking whether the designer had complied with the component datasheet. This proposal was a response to a case in which a designer had in fact not used an electronic component as directed. The lab prototype of the product had worked fairly well, but the missing protection circuitry recommended by the datasheet caused problems to occur when the product was manufactured and installed. The cause was found and corrected, but the incident sparked the proposal that this kind of issue should be controlled (in the narrow sense). The quality engineer in fact proposed it as *the* method to ensure quality in development, apparently completely overlooking the problems that this would cause among developers. Moreover, this approach would have been entirely insufficient because many more factors affect product quality than merely following a component manufacturer's instructions.

The intention of quality systems is to systematically approach the quality of the product development process, but when systematization is performed in a controlling manner, it can have negative consequences on the product development environment.

This explains Seghezzi's observation (1988, 358) that the culture among development personnel was originally so adverse to the ISO 9001/9004 requirements, until their necessity became less controversial. The author describes the culture of the development department as follows: "In development, the activities of planning, executing and verifying are very often in the hands of one entity, whereas in production, tayloristic methods of work division prevail. Development staff usually feel completely responsible for the quality of their product and, with correspondingly high motivation, they exercise self-control in the best sense of the word. An additional quality evaluation by an independent organ, even external test guidelines,[xxiv] is perceived not as a measure to increase the margin of certainty, but as interference in one's freedom. Generally, development staff are technically well equipped and competent to perform all the necessary tests for a quality evaluation.[91] The participation of other organs is perceived as unnecessary interference. The execution of complementary quality tests by others is interpreted as an expression of distrust." (op. cit., pp. 357-378).

[91] The same is not generally applied for reliability evaluations because considerable additional knowledge and equipment are required to perform these tasks effectively.

Some companies claim to have achieved product development quality or production quality thanks to this control (in the narrow sense) approach. However, it is important to recall their two basic differences: (i) the type of activity performed and (ii) the qualification of the persons involved are significantly different. The comparison between the QC/QA approaches is not based on which of them provides the best production quality, but on which is more efficient under what conditions, within what limitations, and especially, what side effects do they introduce into the processes (cf. Discussion in Chapter 2).

To answer these questions it is crucial to consider the processes being "controlled" in the development activities.

3.4.1 Technical issues

In the production hall, one finds many various conditions that influence the effort to achieve product quality. Put in trivial terms, the task in production is to "clone" a design. The experienced practitioner knows that the cloning process is not entirely devoid of intelligent challenges. To *re*produce a given article, especially if it was not designed to be easily reproduced, requires creativity. As manufacturing evolved into industrial engineering a long time ago, the separation between those who think (in the so-called engineering department)[92] and those who do the job (in manufacturing) was no longer valid. Manufacturing activities gained a "brain," which frequently worked independently of the design engineering group. However, the tayloristic division of those who "think" and those who "do the job" was frequently maintained internally.

The conception of a manufacturing system is as challenging as the conception of the product itself. There are many cases in which the actual reason for an enterprise's success is its well-conceived production facility.

Establishing a reference for quality production work is straightforward. The product to be *re*produced can easily be decomposed into controllable variables such as: the dimension of a part must be within certain numerical tolerances, the component leads should be bent according to pre-established criteria, the amount of solder should be within a certain range, welds should comply with workmanship standards, and many more. Consequently, it is not only easy to draw up suitable checklists, but, more importantly, the criteria were very well defined before the manufacturing effort took place.

Frequently, a different person is allocated to control the work of a group of persons, although at some point in history the workers performing the

[92] In some companies, the development department is still called the "engineering" department.

production job *re*gained the responsibility for ensuring for quality of the parts they send down the production flow. In other words, they controlled their work themselves. Not surprisingly, however, the self-control approach met with the resistance of quality control as well as manufacturing managers.

For development groups, it is not so straightforward to achieve this built-in quality. Natural to many established disciplines, one first tries to extrapolate a known technique to a new field. This was the case for development quality. Many techniques to control the quality of product development originated from manufacture control techniques, most of them from quality control. Some people even tried to force these control techniques upon "undisciplined [development] engineers".[93]

Box 24. The case of the "independent development control department" (case #15).

At a digital equipment manufacturer of hardware and software for bank automation in the early 1980s, many problems were found in the product definitions that the development department had sent to the production department. At that time, there was a strong trend to compartmentalize activities. This meant that there was very little involvement of production people in the product design phase, although initial efforts were being made to enable this involvement. Not surprisingly, many design problems emerged, most of which were related to product engineering. There was a broad spectrum of typical problems in the digital technology industry, ranging from

- non-compliance with drafting standards,
- inadequacies of constructive aspects such as
  - (i) printed circuit board layouts and
  - (ii) accessibility of essential parts,
- difficulty in testing and maintenance such as
  - (i) the absence of a test concept,
  - (ii) too few test points,
  - (iii) excessive use of monostables,
  - (iv) restrictions in the digital sequential circuitry to be fully controllable,
- insufficient considerations on digital timing,
- inadequate program documentation,
- reliability weak points due to component overload and
- inappropriate component technology choices, among others.[94]

---

93 Heard during an ISO 9000 seminar.

94 During the 1980s, the awareness of this kind of problems increased among electronics companies worldwide as a result of the digital revolution, which was accelerated in turn by the advent of the microprocessor.

In order to improve the quality of the development process, this company created a independent "control instance" for evaluating and approving the submitted product definitions. This control took place at the (then) usual hand-over from development to production.

Their independent status was meant to free them of pressure to release the definition documents. If something "wrong" were found, they would return it to the product development department for revision.

What happened with this independent department? To justify its existence in the company, it rejected *every* development, accompanied by a long list of development errors. It even rejected developments with tolerable errors that could easily have been coped with by the manufacturing department. It rejected developments in cases where it would have been possible to accept the definition temporarily pending the revision of minor changes.

Such behavior, according to a high-level official in that company, turned the group "into a department that could *only* give written proofs of the 'incompetence' of others and at the same time blocking the entire development process". This control activity was ultimately deemed inappropriate and was discontinued.

Attempts to extrapolate the quality control approach to the development activities face some very difficult issues:

How to control something: In order to follow the control approach, it is necessary to agree upon the form of controlling the job to be done. Lack of control criteria *prior to* development leads to the perception by developers of unfair treatment when such criteria appear only after the work has been completed.

Which points are to be controlled: A good reference for comparing a finished work with what should have been accomplished is needed to differentiate between a "good" and a "bad" conception. This is one of the principles of controlling the quality of manufacturing. In the case of quality of development, there is no absolute standard for comparison as there is in metrology, for example. Furthermore, reference criteria similar to *workmanship standards* used in production require a more flexible application, have a more dynamic character and require a critical appraisal when applied to the quality of development, because in many situations, even widely accepted rules may not be applicable.

Who will do the controlling: Whether an independent instance is to perform the task of controlling or self-control is to be exercised is merely the beginning of the problem. In controlling the quality of production, controllers must be familiar with the criteria to be controlled. Extrapolated to development, the familiarity of the controller with development activities, in addition to the control criteria, make an essential difference in achieving better design quality because the meaning and relevance of each criterion are much more important than the criterion itself.

These issues are frequently insufficiently addressed, leading to non-repeatable results.

3.4.2 Product specifications

The next step for implementing control is to break the general product specifications into specifications for each part, which would allow intermediate checks. This is done in practice in product design and provides a stable basis for technical discussions during the course of development. However, as it is not possible to specify every detail (recall Chapter 1), many important intermediate characteristics remain implicit.

An item that is required by specification has a much higher priority than one that is not listed but would improve the product. In many situations, the simple mention that an item is on the specification resolves the discussion. Consequently, there is a tendency to specify as many details as possible in an attempt to make quantitatively clear what is supposed to be achieved.

Within the scope of the present chapter, the control approach fosters the trend to pursue an excessive level of detail in setting the specifications so that they may serve as comparison criteria.

Although there might be detailed characteristics explicitly described in the specifications, not all the characteristics that lead to a quality product are necessarily listed. Hence only a simple (though very decisive) form of verification is possible. Again, the trivial comparison of intended versus achieved specifications is—at best—too late. Comparing the results only at the end of the development process would mean returning to the obsolete quality concept of discarding or *scrapping* the unsatisfactory work done.

Specifications can be used for intermediate controls, but with the complexity of modern products, the interaction in the integrated product of the individual specs makes this form of control a hindrance to achieving success. Moreover, the control performed by the controlling instance is not always task-centered, and, not infrequently, this is acutely perceived by the persons affected. In both cases, it is sometimes purported to be task-centered, but "lip service" of this kind is rarely effective and in fact creates new barriers to success.

3.4.3 Checklists, references, workmanship standards

Another way to implement this control is to draw up checklists of standard practices as guidelines for implementing product features or characteristics. The basic idea is to establish a set of design rules that summarizes the result of past experience with similar products.

Box 25. Design rules as quality criteria.

A trivial example of a design rule is a control knob that should be devised such that *rotating it clockwise increases the parameter controlled by it.* As linear control devices appeared, the rule had to be extended to *movement to the right or upwards increases the controlled variable.*

Checklists can be used to list "good" engineering practices not defined by the product specification. Birolini (1994) gives some examples of checklists used for design reviews. These lists help avoid known problems and are part of the expert knowledge necessary to achieve quality. The main advantage of these lists is that they inform designers of known issues *a priori*, and designers can in turn apply these rules to define the desired product. Checklists thus constitute one of the conditions for executing control, namely that issues known in advance can be controlled *a posteriori*. This is the constructive way to use them, provided that the lists are made available prior to the design activities. Nevertheless, especially in innovative designs, many of the problems faced in the development are recognized only during or afterwards, and new problems arise as well. Checklists can clearly only help solve the *old* problems.

Companies that used such lists achieved various degrees of success in developing quality products. There are documented cases from the 1960s, although it is believed that this technique was applied before that.[95]

Box 26. Example of design rules applied to aircraft development.

An example of the application of expert knowledge occurred at Boeing during the development of the 747 and was described by Irving (1993, 253 ff.). A document called "Model 747 Design Directives and Criteria" played an important role during the development of this aircraft by "accumulating wisdom of the world's best team of aeromen, structures and systems men, anyone who had ever contributed to the design of anything from a cargo hold door to the new-technology inertial navigation system." The 457-page compilation contained such items as "Wing fuel tanks shall be baffled to limit effects of airplane attitude and wing flexibility on fuel center of gravity movement". This document was revised and amended continuously "because the body of empirical knowledge, the static content, was being updated and supplemented by the ambitions of the moment, the dynamic content."

Owing to the dynamic nature of technological progress and to the natural evolution of the product, such lists are indeed subject to change. The rules

95 The verification of historical evidence of such lists is beyond the scope of this work.

for using components and materials change, and considerable efforts must be devoted to updating these lists, especially if a company expects to keep pace with the evolution on its field.

The following examples illustrate the difficulties of managing these rules.

Box 27. Difficulty of transferring existing knowledge to present problems in electronic equipment development.

Wrong transfer (case #16): One of the rules for applying TTL (transistor-transistor logic) integrated circuits that was valid in 1981 was to provide a pull-up resistor when an input line was to be fixed at a logic "1" instead of tying this line directly to the supply voltage. The noisy supply voltage would cause voltage spikes to be applied directly to the input line, risking a component failure. *One* pull-up resistor would protect these inputs and could be tied to *many* input lines. For the compatible component technology LS-TTL (transistor-transistor logic, low-power, Schottky diode) this rule no longer applied due to the inherent protection of the Schottky diodes that had been introduced to enhance switching performance. In one case in 1985, an engineer believing to have found a weak point in a design noticed by consulting the manufacturer's datasheet that this rule was not mentioned for LS components, although the rule was still applied generally to TTL-family IC's. The rule was subsequently omitted. At the same time, because of the need to make electronic boards testable, another rule was being formulated, namely that *every input* line should have a *separate* pull-up resistor in order to render the IC completely controllable by the test equipment. Once again the pull-up resistor was added. Note that the rules are essentially different and that merely transferring the old rule would have been insufficient to address the new issue.

Restrictive transfer (case #17): The application of C-MOS technology was of limited use in high-reliability equipment because of the inherent problem in this technology known as "latch-up". This led the estimation of failure rate of C-MOS components to be a factor of 10 higher than those of comparable other established technologies (learning factor MIL-HDBK-217).[96] As the mechanisms of this failure were investigated, the component manufacturers devised methods to avoid its occurrence, so the restriction was no longer technically valid. By the end of the 1980s, the semiconductor

[96] MIL-HDBK-217 was the U.S. DoD military handbook for Reliability Prediction of Electronic Equipment. Its main advantage still lay in providing models for component reliability. It was not the sole source of reliability data, but it was the high costs of determining empirical failure parameters for models for failure rates between 10^{-6} and 10^{-9} failures per hour. It is believed that the interruption was caused partly by the misuse of these values for reporting purposes but not for actively improving reliability.

industry had already the (declared) tendency to follow predominantly this technology. Yet the reliability organizations in some companies were still reluctant to apply this technology for high-reliability applications.

Invalid transfer (case #18): For a certain high-reliability product, it was explicitly prohibited to use electroluminescent displays as stated in the program-specific document "Product Standards". This rule originated from problems with cross-grid (*powder*) electroluminescent (EL) display devices with AC excitation, which exhibited limited lifetime and luminance. System and design engineers erroneously transferred the ban to electroluminescent light sources that were intended for panel *backlighting* but that did not exhibit the same problems that originally resulted in the exclusion. The devices were actually used (indeed quite extensively) in the intended environment. This, however, unnecessarily worried designers. One engineer cited the fact that EL devices were indeed used in related products that were subject to the same standard, and questioned the validity of the document. Finally, a reliability engineer resolved the issue. He also established that, from a reliability point of view, they could even use *film-layer* EL *displays,*[97] even though the control document made no distinction between the two display types because such a differentiation would be expected to be nonexistent at the moment the rule was *generated.*[98] The idea of static + dynamic contents was not considered in this case.[99]

Another example of invalid transfer (case #19) is the application of the rule stated in MIL-HDBK-217 when the component count of an electronic device is not known. The rule stipulates assuming the number of components in the equivalent electric circuit because it is an indication of the component complexity, which is at least weakly correlated to the physical implementation of the device. A quality engineer suggested applying this rule for estimating the reliability of a *mechanical* filter used for telecommunication equipment whose equivalent electrical circuit did not have any relation to its physical constitution. This rule was inappropriate for this case. The engineer stubbornly maintained that that was what the "standard" stated. Incidentally, MIL-HDBK-217 was never intended as a

---

97 At least from the point of view of the characteristics mentioned here.

98 Although this rule had become non-applicable many years before, the situation described here occurred as the program to which these rules belonged was being started. The reason for keeping the rule in the program is not clear. There is some indication that this rule, like several others, was "inherited" from the experience of a previous program that also called for the development of a series of other equipment.

99 The document should have been written: "EL displays should be avoided because of variation in luminance with aging, resulting in display illegibility". Instead, it read "EL-displays are prohibited <period>".

standard, but as a handbook (this is a common mistake in connection with reliability considerations).

Potentially problematic transfer (case #20): Surface mount technology (SMT) had been used with hybrid circuits mounted on alumina substrate possessing a compatible thermal expansion coefficient for a long time. At some point, SMT was used ubiquitously to reduce size and weight. For mounting circuits on fiberglass substrates, application rules for these components were simply transferred from those of alumina substrates. However, because of the different thermal expansion coefficients of the printed circuit board (FR4: 15×10^{-6} K^{-1}) and the component (chip capacitor: 7×10^{-6} K^{-1}) due to the relative movement of component lead and the printed circuit board pad, a potential problem was suspected that might lead to an unreliable solder joint. The characterization of this problem was incomplete, but serious doubts were raised concerning the amount of solder that would be necessary. A metallurgical characterization and analysis of this problem took from 1991 to 1994. In the end, no noticeable reliability problem could be inferred from the data. In formal scientific terms, no acceptance or rejection of the hypothesis was made. However, it provided important information to practitioners even if no effects were found in the conducted study. The technology continued to be used without concerns about this specific issue.

These rules change, and the organizations using them should know this and respond quickly to these changes. Otherwise, rules might exist formally that may not do any harm, but will lose credibility and consequently be ignored, potentially resulting in ineffective activities. It is also necessary to know when it is possible to deviate from stated rules.

Box 28. Examples of the difficulty of judging the applicability of rules.

Conflicting goals in the rules: One rule universally followed in high-reliability applications is to *derate* a component. This means using it systematically in conditions less critical than its maximum ratings in order to improve its reliability. One of the derating rules is to limit the *fan-out*[100] set up previously for a development program. In trying to uphold this rule without questioning it, an engineer might have to introduce another component that was not there before. The inputs of the new component should be tied to the outputs of upstream components. This would sometimes cause the output-derating rule to be violated by the component, leading to the need for another component there and so on. The ultimate result is a local "demographic explosion" of the components. Increased

[100] Number of loads on a digital IC output.

component count associated with the (failure-susceptible) electrical interconnections may lead to a solution that would be more unreliable than if one merely broke the derating rule for that case using the full fan-out for that output. The decision to break the rule is not easy because it requires knowledge of the effect of the possible failure this might introduce, and it depends on the criticality of the function, the cost of the devices involved as well as non-modeled parameters such as soldered connections. This case alone is potentially controversial. It serves to illustrate the difficulty of finding a solution to a technical problem that is not dependent on simple rules. In the typical cases, the design engineer does not have the required knowledge to make such a decision.

Controversial prohibition rules: The computer language statement GOTO or equivalent was banned from the software development scene for being an unreliable practice. The main reason for this was that the instruction allowed a loss of control of programming loops. Even for computer languages that were implicitly structured to ensure better control of the loops, the instruction was maintained. The *Pascal* language, for example, was criticized for retaining it, although with restricted use. It was retained because of its advantages when an error occurs or in exception handling (cf. Perrott & Allison, 1984, 281). In the *C* language and its associated variants, it was analogously retained. Years later, in the early 1990s, a development director (who had been a pioneer in the field of computers and apparently had experience with the *FORTRAN* language) recommended to an electronics company that prohibiting the use of GOTO was a well-known way to improve software development quality (case #21). Maguire (1994) give an account of a situation where an implementation with GOTO actually enhanced the clarity and simplicity of a program, thus improving its reliability. This was promptly recognized by the programmer, yet he preferred not to use it, believing it to be the mark of an "incompetent programmer".[101] According to McConnell (1993), the "debate (regarding GOTO) reappears from time to time in the working places, in the reference works, in the magazines, but nothing is said that wasn't already treated in detail twenty years earlier".[102]

These cases show how difficult it is to apply these rules: there must be a critical appraisal of the sense of its application otherwise a false clue might be followed blindly.

Similar results regarding the use of checklists were observed independently by Rauterberg, Spinas, Strohm, Ulich & Waeber (1994, 130-

[101] For the transcript of the entire conversation, see Maguire (1994, 35-36).
[102] Cited in Maguire (1994, 36, parentheses retained).

131) in their investigation of software development processes. The authors found that checklists were used to evaluate user-friendliness. As such, they possess the following strengths and weaknesses:

Box 29. Strengths and weaknesses of checklists for assessing user-friendliness of software (Rauterberg et al. 1994).

Strengths:

- the efforts to perform the evaluation are considerably less than with *user tests*;
- help avoid forgetting important aspects;
- help process routine cases;
- contain in concise form the aspects considered relevant.

Weaknesses:

- cannot substitute for expertise; working with checklists requires sufficient knowledge of tasks, problems, context and approaches;
- reliability is lower than other procedures, as conscious or unconscious misinterpretation of the questions is possible;
- limited opportunity to obtain a detailed explanation of the individual questions in the paper versions;
- special aspects not considered within the framework of the checklist cannot be considered.

3.4.4 Non-technical issues

The following situation describes the consequences of a quality assurance activity that was perceived as control.

Box 30. Believe me, I do not want to control your job (case #22).

In an electronics company, a quality engineer was assigned to implement quality assurance activities in the development phase. The company's technical competence covered a wide range of electronics technologies (analog, digital, wide range of frequencies, telecommunications, control systems). This meant a diversity of specific knowledge in each of the partial domains. The company was predominantly technically oriented, and most of its employees were engineers. In its 30 years in the business, the company had always placed special emphasis on the technical knowledge to be gained. When receiving a technology transfer, for example, it avoided accepting the transfer of mindless tasks that did not allow employees to know what they were building. It refused to be a mere parts assembler, to do what any "monkey" could do. This philosophy led to exceptional performance records with high-reliability equipment, and the company became well established in its market niche.

The quality engineer had a background in product development processes and some familiarity with production problems caused by the product conception. He also knew cases of unsuccessful implementations of quality assurance in similar companies. It was clear to him that it would not be easy to implement quality assurance because of the extent of detailed knowledge required in a broad range of areas and the interconnection with non-technical problems. His technical knowledge would cover a substantial part of the products, but he lacked experience with many of them. He needed additional practical knowledge of reliability techniques, recognizing as he did that theoretical knowledge alone is insufficient.

Based on all these considerations, the main points of his quality assurance guidelines were as follows:

1. It would be assumed that the conceived products are correct until the contrary was proved. Most activities would be conducted in this sense rather than trying to prove the contrary.
2. No one-to-one follow-up of the activities of the development personnel would be performed.
3. Responsibility for the development lay clearly with the development personnel.
4. Quality assurance should be conducted as an evaluation of product characteristics, not as an approval or rejection of the design.
5. The quality assurance personnel involved in the activities should possess experience in product development.
6. Owing to the different completion speeds and staggered completion of parts, the quality assurance evaluation would be conducted in a puzzle-like manner. Missing points would be progressively being cleared, using the basic product structure as a framework.
7. Results would be presented in a formal way.

The quality engineer presented these points to the head of the quality department, himself a good engineer with an excellent scientific background. The department head's comment on the first point was to question whether it would not be better to try to prove the contrary, to falsify the design. The engineer responded that this would be correct from a scientific perspective, but explained his intention to avoid the negative psychological effects associated with trying to prove that something does not work. He mentioned the case described in Box 24. The case of the "independent development control department") and explained the negative effects that can result from an attempt to "scientifically" falsify a design. The assumption that a design is correct is a better way to overcome this barrier, he said. The department head was persuaded.

Of the three new development programs in progress at that time, he chose to start with two of them: A very complex system but one that used technologies with which he was the most familiar, and a simpler piece of equipment that was already at an advanced stage. The engineer reported encountering the following problems:

- A development engineer considered by his peers and superiors to be a very competent designer was deep in thought as the quality assurance engineer entered the room. The latter immediately perceived the designer's unease upon realizing that he was being approached. As the quality engineer proceeded to pose some questions about the product the designer was working on, the latter was exceedingly reluctant, weighing his answers with great care and responding only after considerable hesitation. At the end of the discussion, the designer asked apprehensively, "Did you find anything?" This embarrassed the quality engineer greatly, who had had no intention of hunting for errors, but had hoped to discuss freely as an outsider with developers about their work. A former development engineer himself, he was disconcerted by the designer's reaction.
- A development engineer working on the complex system (the first of the two projects mentioned above) told the quality engineer proudly and, it seemed, warningly: "We were visited by the president of the company that was our partner in a former, similar program. He saw [the system] and said, 'Good job!'." The designer apparently assumed that the visitor had been able to appreciate the details of such a complex implementation in a couple of minutes. However, one would expect to find any problems of the product in the details, not in something obvious. What the designer was overlooking is that the quality assurance engineer had every intention of doing exactly the same as the first visitor (to say, "Good job!"), but after having fully appreciated the details.
- In anticipation of the quality assurance engineer's visit, another engineer, an expert in his field but known for his quick temper, was heard to say, "If he dares to come in here and try to teach me how to design my circuits, if he dares..."
- The material provided to the quality engineer to assemble the evaluation "puzzle" was incomplete information about trivial assemblies at the margins of the system, not those associated with technical difficulties. The main problems were expected to occur elsewhere, but no information about the core of the system was ever supplied. When asked, the reply was always: "It isn't ready yet".
- The incomplete information lacked, for example, a circuit description. Even though the concept was clear in the schematic diagrams, the central issue remained: something might function on paper, but how does it

perform in reality? The quality assurance engineer discussed his predicament with a colleague, who was also experienced in product development. "Look at this!" the first engineer said. "No information about component biasing, values or signals expected. They claim they have no documentation yet. But I refuse to recalculate it, for they have surely done so before." His colleague replied with a shrug, "Well, I don't think you have an alternative, do you?"

To the quality engineer, it was clear that the activity was off track and need to be reformulated. He suspended the search for information concerning the development programs until the situation could be rectified.

Another quality engineer who had been in the company longer proposed that the activity be conducted as a pure control instance, a suggestion presumably rooted in his experience in quality control. This engineer stated to the development personnel that nothing would escape "The Sieve" his colleague was preparing. This was utterly contrary to what the quality assurance engineer was trying to implement. He was actually trying to avoid this controlling (in the narrow sense) mentality.

In the second round, development personnel were still not willing to provide information that might have revealed deficiencies in their design until the development activities had been completed, when the design work had achieved a certain maturity. Usually, this is too late to be of use.

When quality personnel tried to engage developers in detailed discussions of product characteristics, they were successful only at very distinct intermediate phases or upon completion of a design. This happened in many cases, even when the people involved were quality assurance-oriented and were not trying to control and stamp other people's work. It was very difficult to collect preliminary information, and only limited information flowed. Quality assurance activities require a certain transparency to prevail, but any attempt to ensure this was interpreted as an invasion. Eastman (1980, 39) also noticed this phenomenon and lists two reasons for the resistance to release advanced engineering information:

- The designer's training encourages him to ensure that every feature of the design is acceptable before its final release.
- If the engineer's work has to be changed or if errors are discovered, he may be criticized and his career adversely affected.

Eastman states further that "under many engineering managers, [one] can avoid issuing advance information by simply saying 'no'. A certain number of extra design and drawing changes must be accepted if advance information is to be used."

Moreover, many design engineers resist releasing advanced information because they fear that wrong information could continue to circulate long after changes to the design have been made. They are afraid that conflicting information may cause more problems than the gain in time intended. Eastman mentions that advanced information is usually obtained informally through personal relationships with designers because the formality of the management and planning system does not cultivate this exchange. The reluctance of design engineers to submit to quality assurance activities is tied to the impossibility of controlling the destination of the information provided and is more suitably fostered by the cordial relationships between the parties involved (op. cit., p. 40).

A good level of transparence of the design process is an essential condition for quality personnel to be able to act "before it's too late". In their efforts to gain access to developmental data, independent of their orientation—whether quality control or quality assurance—quality personnel were frequently accused throughout the 1980s of "sticking their noses into our affairs." Even in production, the presence of quality control is not readily accepted. Its agents may be tolerated in production, but in development, engineers were far less willing to allow such "intruders". This perception clearly has very negative consequences. Not a few development engineers protested vociferously with questions such as: "Does he have the right to do this?", "Is it his place to ask these questions?", "Why does he drop in to question our work?"

Exacerbating the issue, some managers endeavored to exercise a form of control over "undisciplined" engineers, with the direct consequence that development personnel came to feel constrained by any and all quality assurance activities.

There is yet another issue connected with the control approach. Once a design has been approved, who carries the responsibility for its quality? The person or entity that stamped its approval, of course! What if a problem emerges only later (e.g. through use). Who is to blame?

Box 31. Quality assurance would the first to be crucified (case #23).

A development director at a high-tech company had experience in the quality control department. He was familiar with the quality approaches in use in the 1960s and 1970s. Possibly for that reason, he was known as a good "controller", especially concerning the work of other people. This knowledge of quality problems helped him thoroughly understand the problems encountered in production that had been caused by misconceived development. He tried—with limited success—to convince his personnel of this.

One day in the early 1990s, he was having a meeting with one of his project managers, when the engineer responsible for product assurance entered the room. The latter could not help but notice that his arrival caused an abrupt end of the conversation. Indeed, the director had just been telling the project manager that the product assurance engineer had every right to pose "all those questions" about the developer's tasks, because "if something goes wrong, quality assurance would be the first to be crucified".

Later, when the project manager and the product assurance engineer were alone and the atmosphere was collegial, the latter asked the project manager what had brought about the discussion of the role of quality assurance. The project manager, however, refused to discuss the point further, despite his colleague's sincere attempts to make it clear to him that such information could only help him improve the quality assurance efforts.

Wrongly attributing fault to the quality department is still prevalent in various disguises. It arose from the quality control approach in production, where the main responsibility for product quality was indeed in the hands of the quality control personnel (it was exclusively there until the early 1980s). The same tended to be true of the development process, and there is a trend to implicitly shift the responsibility to whoever stamped a QC OK on the product definition, even though shared responsibility is implicit.

Ineffective formality

The formality connected to the control approach has other implicit disadvantages. First of all, it can be performed only when substantial parts of the work are completed or in an advanced state. When this is done, its control is slow because of the high number of items to be controlled in order to be effective. In addition, there is the overhead of having to generate formal documents to be presented to the controlling instance. With the case of the DOD-STD-2167 standard for software development (page 59), things went too far in the case of *Visibility & Control*. The formality of the control mechanism also led to the situation where there was the clear perception that "too much time and money must be spent on the contractually required documentation and the multitude of minute details called for by 2167" (Coad, 1988, 48).

The formal approach slows down the development process because of the time it takes to learn about the real product characteristics. The problem of controlling speed requires a vast controlling organization, possibly the same size as the development organization (manufacture development included) or even bigger, including personnel of equivalent competences and capabilities. Thus it is a fairly inefficient approach, at least with regard to resources and time.

The control approach slows down the evolution of the development methods or of the technology used. New methods or technologies lead to new or slightly different problems, rendering the control criteria invalid and hampering innovation. Again the example of the DOD-STD-2167 serves as a good example of the necessary flexibility that prompted revision A.

Although it is theoretically possible to control ongoing work or preliminary results, high inefficiency is expected when doing so in development.

3.4.5 Semantic aspect of control

Both the ISO-9001 and the AQAP-1 quality systems used the term "Project Control" to denote the activity related to product development. By virtue only of this title, persons in charge of this activity were inspired to impose a control mechanism very similar to the old style of controlling the quality of production. Asked about the way to approach the issue of the quality of development, they replied: "the standard says 'control'. We have to do it because it is in the ISO requirements" (case #24). Although ensuring that goals are achieved should imply control in the broad sense, recalling the definition stated by ISO 9001 (cf. Box 22 on page 118), there was no mention of the "regulating" sense of the word "control". Hence it is not surprising that many quality assurance activities became an exercise in the inappropriate form of control: in the narrow sense.

In the German version of ISO 9001, the term "Lenkung" (steering, guidance) was used, hence: control in the broad sense. The translators presumably wanted to avoid perpetuating the poor translation of the term "quality control charts" used in production (cf. footnote 18, page 29). This alone does not imply that better results were achieved with that version. This semantic issue constituted one of the essential differences between a good and a bad *description* in a company's quality manual of the processes to develop a quality product. Semantics refer here to the meaning, not to the labelling. Thus, simply relabelling a control (in the narrow sense) activity will not solve the problem.

Generally speaking, according to Mary Parker Follett (1949b), a trend was observed (probably in the 1920s) that in the best managed industries,[103] control was "coming more and more to mean fact-control rather than man-control" and task-based instead of person-related. In addition, "central control" was "coming more and more to mean the co-relation of many controls rather than a super-imposed control" because of the too widespread ramifications of "modern" industry, the complexity of its organizations and the intricacy of the problems it faced. It follows that it would be impossible for an industry to be "managed from commands from the top alone." Follett concludes that the aspect of central control, "does not mean a point of

[103] This probably refers to American industries, but can be generalized to *western* industries.

radiation, but the gathering of many controls existing throughout the enterprise". She stated this in an era of strongly hierarchical industrial organizations. Its application to (or confirmation by) partially autonomous, "flat" company structures in studies, evaluations, and use today is particularly surprising. It can be interpreted as a decentralized structure for actions and control (in the broad sense) and a centralized convergence of these actions and control (also in the broad sense).

Hence Follett emphasizes this aspect of control (in the broad sense), especially the need to *co-relate controls* in the organization, and lists the fundamental principles of organization as follows:

Box 32. Four fundamental principles of organization according to Mary Parker Follett (1949b).

Co-ordination as the reciprocal relating of all the factors in a situation where the involved parties have to adjust themselves to one another and at the same time to the situation, not as one's own sacrifice of his/her discipline and expertise, but as one's contribution to the whole. This is where all factors are expressed not as an *additional total,* but as a *relational total*—a total where each part has been permeated by every other part.

Co-ordination by direct contact of the responsible people concerned: the people closest to the matter at hand.

Co-ordination in the early stages: if the heads of departments confront each other with finished policies, agreement will be difficult. Of course, the tendency is that they then begin to 'play politics'.

Co-ordination as a continuing process: the various people concerned should not try to unite only when difficulties arise.

Once again, it is amazing to discover how applicable Follett's concepts from the 1920s still are today, and how they correspond to such contemporary concepts as *participative leadership*, *multi-functional product development*, and—needless to say—total quality.

3.5 Autonomy

There seems to be disagreement regarding the autonomy necessary in product development. Some companies have tried to liberate their development teams by creating a *lab-in-the-woods* environment in which, away from mundane pressures, their R&D teams could work "stress-free" to develop the new products that would "shape the future". Many such experiments failed. One main issue was that this isolation removed the R&D teams from the company reality. For example, their proposals did not take into consideration the production environment or the problems encountered by other departments.[104] They also became unacquainted with company strategy and were accused of being far "too scientific". They might be publishing good papers, but they were not solving the immediate or near-future problems of the enterprises they were supposed to be serving, it was said. Another issue was the loss of focus on approaching technical problems on a short time horizon. Some early attempts to provide a place where R&D personnel could exchange ideas, such as a separate cafeteria, were not only unsuccessful, they were even counterproductive. Many development managers cited such examples to justify why their development teams are kept in a narrow work frame with few degrees of freedom.

The *lab-in-the-woods* approach might have failed, but there are many successful examples of autonomous development groups that were not only able to react quickly to deliver successful products, they also developed essential innovations in their fields. Such groups succeeded in leveraging the inherent advantage of being able to overcome the resistance to change that characterize established practices.

One example cited in the literature is the first IBM PC™, which was developed by an "Independent Business Unit" within IBM. It provided complete freedom with regard to product planning, and ensured minimal intervention by the "central" corporation. This allowed simpler and more informal leadership methods than were applied for the development of mainframe computers, because the product development was deemed low risk (Krubasik, 1989, 27). The development of the IBM PC could be called a "zero-procedure" approach.

Other example of an autonomous structure is the development of McDonalds' Chicken McNuggets™, for which the development group was granted a significant amount of insulation within the large corporate structure. Its work environment was composed of three primary players, a minimal number of constraints imposed by the existing structure, and an agreed upon decision-making process that forced communication and speedy resolutions. Certain members of team could be added or subtracted as necessary. Bud Sweeney, the project leader, justified the choice of the structure by commenting: "Over the years, I've seen so many projects go dry at the corporate level because of layers of insulated bureaucracy. The group was flattered by the fact that the innovation they introduced was firstly copied by Kentucky Fried Chicken." (condensed from Larson & LaFasto 1989, 50-53).

---

[104] There is no indication whether these companies tried to include representatives from other functions in these "labs in the woods", even on a consulting basis. It is believed they did not.

Other examples of innovative, autonomous development groups are the Lockheed Skunk Works and the Navy Special Projects Office, which developed the Polaris as described above. Another was the K-3, the first Russian atomic propulsion submarine developed by the Malakite Development Office in Leningrad in 1952. The Aeronautics Institute in Moscow was involved in this effort, which led to the innovative tear-drop form of the hull, which is universally used today. Until then, the conventional hull form of submarines had—not surprisingly—borne a strong resemblance to that of surface vessels. As mentioned above, however, this independence is not immune to failure, because it can isolate the group from essential information. In the case of the K-3, decisions made regarding the reactor implementation yielded problems of radiation leakage from the reactor. The American Nautilus design used the "classic" submarine hull, but their compliance with stringent regulations governing radioactive leakage produced an inherently safer design, with very good records in use.

The question of autonomy is not simply how much of it should be given to R&D professionals, but under what conditions autonomy can be expected to yield better results. As cited in Ulich (2001, 540-541), Pelz (1956) concluded that "in research institutions, a high degree of autonomy is only accompanied by higher performance if there is also a high degree of contact[xxv] between the workers and the project leader". Ulich cites another study performed by Fisch & Daniel (1983), who analyzed more than one thousand research projects. Their findings support these results as follows, "autonomy ranging from middle to high levels, when associated with regular (frequent) joint conversations[xxvi] exerts a demonstrable positive effect on the research productivity." (parentheses added).

Table 17. Heterogeneity in the population of the Pelz & Andrews study.

N	ORGANIZATION	FIELD
641	5 industrial laboratories	Pharmaceutical, glass & ceramics, electronics, electrical equipment
144	Professors of 7 departments of a Midwestern university	Biological, physical & social sciences
526	5 government laboratories	Weapons guidance, animal diseases, commercial uses of agricultural products, basic research in several physical sciences

Other data supports these findings as well. A study by Pelz & Andrews (1976) , initially published in 1966, evaluated a heterogeneous population of 1,311 R&D professionals in eleven different laboratories (*Table 17*). Their findings showed the following.

Performance[105] generally improved with increasing autonomy[106] for lower levels of autonomy and decreased for higher levels. *Figure 8* depicts the relationship for *scientific contribution.* Similar results were found for other performance variables as well (op. cit., p. 27).

With the same population, Pelz & Andrews defined four categories[107] of persons and their ability to determine goals of the activities under their own responsibility. By extracting from the data how many of these four decision-making sources exerted a minimal influence of 10% on decisions affecting technical goals, they obtained the general result that performance increased with an increasing number of decision-making sources involved. *Figure 9*[108] shows the results for PhDs and engineers working in development. With the exception of the higher number of patents of single decision-source engineers, performance increased monotonically with the influence of external members. The findings regarding research activities differed; for PhD's in research labs, an increase in performance[109] was found for single decision sources, which in this case meant the individual researcher. Curiously, the number of published papers or unpublished reports did not increase for the same situation; in other words: these individuals contributed more, but published less (op. cit., p. 12-13).

Differences were found in the effect of autonomy on performance for a variation in the breadth of the area of action as illustrated by *Figure 10*[110]. Of PhD's in development labs with a strong interest in breadth, their scientific contribution was found to be stronger for a higher level of autonomy than of those with a weak interest in breadth in their field of work. Similar results were revealed in the PhD population working in research labs.

---

[105] Measured separately in terms of *overall usefulness*, *scientific contribution*, *published papers* and unpublished reports. *Overall usefulness* is a measure of how much the individual helped the organization carry out its responsibilities. The other measure is related to the *Contribution to general technical or scientific knowledge* in the field. The two measures were based on judgments by senior scientists from both nonsupervisory and supervisory levels. They are expressed as percentiles within each laboratory. The mean is necessarily 50%.

[106] Autonomy was measured by how much influence an individual has on setting his own work goals.

[107] Their data distinguished among: the person itself, subordinates, colleagues without higher supervisory authority, immediate chief, higher-level technical supervisors, non-technical executives, clients or sponsors. Subordinates were grouped together with colleagues (i). Higher-level technical supervisors, non-technical executives and clients or sponsors constituted the second group (ii). Finally, the scientist himself (iii) and his immediate supervisors (iv) constituted the two other possibilities of communication.

[108] Pelz & Andrews (1976, 12, 14).

[109] Measured by overall usefulness and scientific contribution.

[110] Pelz & Andrews (1976, 29, 31).

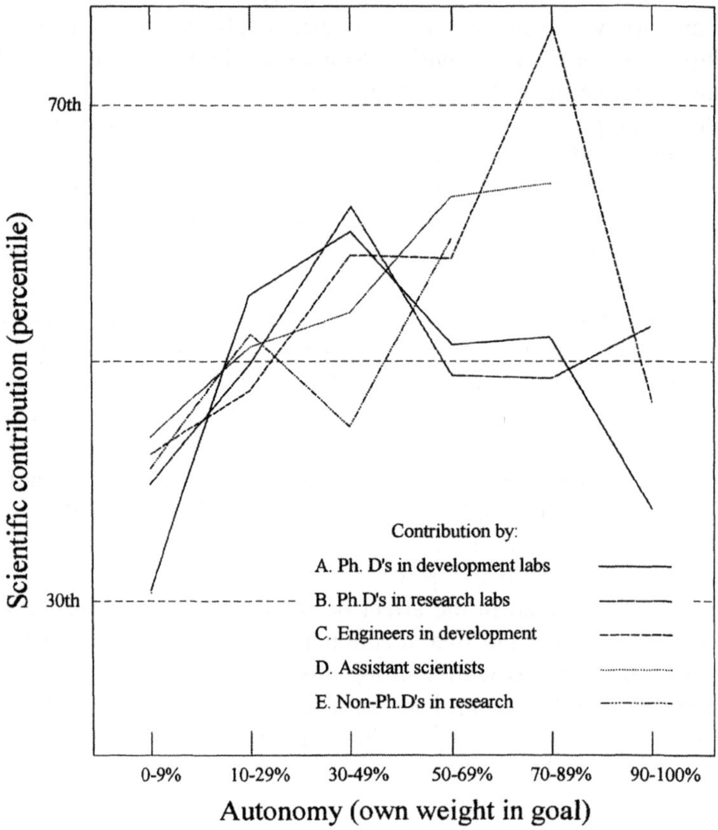

Number

	13	34	17	19	7	5
	2	5	9	14	13	12
	38	95	42	25	10	3
	34	31	11	6	5	0
	7	12	9	13	1	1

Figure 8. **Effect of autonomy on the scientific Contribution of R&D professionals by Pelz & Andrews (1976, 28).[111]**

[111] From: Pelz, D. C. & Andrews, F. M. (1976). Scientists in Organizations - Productive Climates for Research and Development, revised edition, Ann Arbor, MI: Institute for Social Research. Reproduced here by kind permission of the Institute for Social Research.

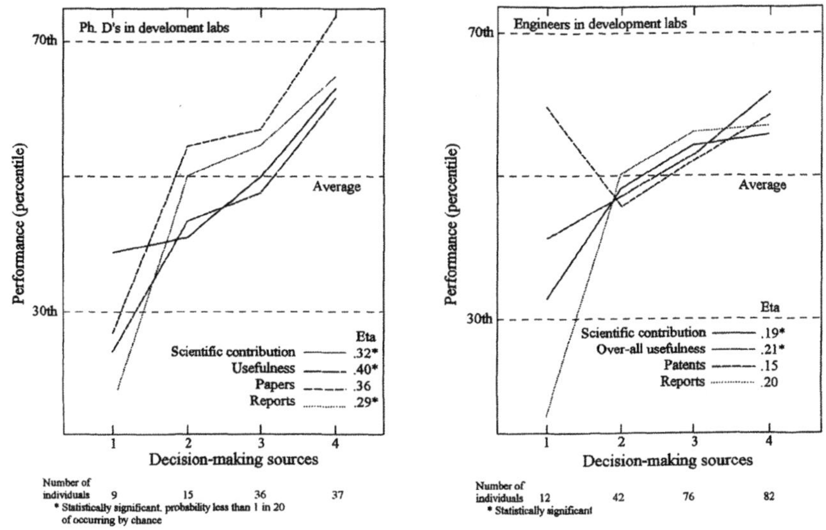

Figure 9. Effect of number of decision sources influencing a developer's job.[112]

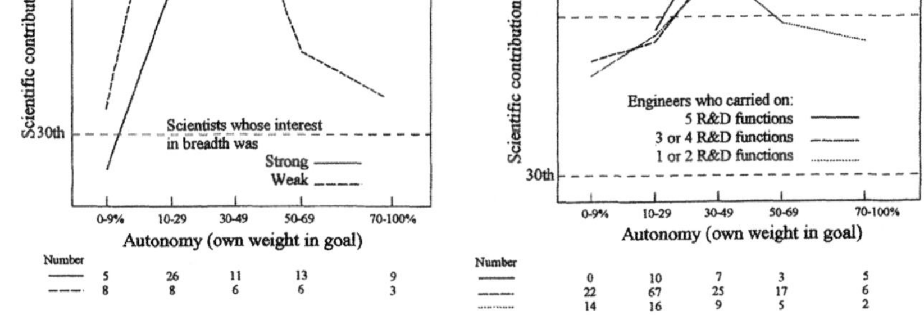

Figure 10. Increased effect of autonomy for higher diversity in range of duties.[113]

[112] From: Pelz, D. C. & Andrews, F. M. (1976). Scientists in Organizations - Productive Climates for Research and Development, revised edition, Ann Arbor, MI: Institute for Social Research. Reproduced here by kind permission of the Institute for Social Research.

This variation in the study population explains the observed decrease in performance for higher levels of autonomy (*Figure 8*). Of engineers working in development, the same increase in performance for higher autonomy was found for those whose activities included a high degree of diversity.[114]

These studies confirm the successes and failures of autonomous development projects. They also confirm the existence of moderating factors under which autonomy can influence the results of research and development activities.

Amabile & Gryskiewicz (1987) conducted interviews with 120 R&D scientists from 20 different corporations. Analyses of their interviews revealed qualities of environments that served to promote or inhibit creativity in organizations. They ranked the percentage of scientists who mentioned certain criteria during the interview. Their observations are in agreement with the findings of Pelz & Andrews. The results are shown in *Table 18* and *Table 19*.

Table 18. Ranking of qualities of the environment that promote creativity in R&D (Amabile 1988, 147).[115]

QUALITIES OF ENVIRONMENTS THAT PROMOTE CREATIVITY		
Freedom	74%	Freedom in deciding what to do or how to accomplish the task; a sense of control over one's own work and ideas. The most important type of freedom, as described by these interviewees, is operational autonomy (cf. Bailyn, 1985)—freedom in the day-to-day conduct of one's work, freedom in deciding how to achieve the overall goal or mission of a project.
Good project management	65%	A manager who serves as a good role model, is enthusiastic, has good communication skills, protects the project team from outside distractions and interference, matches tasks to workers' skills and interests, and sets a clear direction without managing too tightly.
Sufficient resources	52%	Access to necessary resources, including facilities, equipment, information, funds and people.
Encouragement	47%	Management enthusiasm for new ideas; creating an atmosphere free of threatening evaluation.
Various organizational characteristics	42%	A mechanism for considering new ideas; a corporate climate marked by cooperation and collaboration across levels and divisions; an atmosphere where innovation is

[113] From: Pelz, D. C. & Andrews, F. M. (1976). Scientists in Organizations - Productive Climates for Research and Development, revised edition, Ann Arbor, MI: Institute for Social Research. Reproduced here by kind permission of the Institute for Social Research.

[114] Defined by the number of R&D functions the person was in charge of.

[115] Reproduced from Research in Organizational Behavior, Vol. 10, T. M. Amabile, A Model of Creativity and Innovation in Organizations, pp. 123-167, Greenwich, CT: Jai Press., ©1988, with permission from Elsevier Science.

QUALITIES OF ENVIRONMENTS THAT PROMOTE CREATIVITY		
		prized and failure is not fatal.
Recognition	35%	A general sense that creative work will receive appropriate feedback, recognition, and reward.
Sufficient Time	33%	Time to think creatively about a problem; to explore different perspectives rather than having an already determined approach imposed.
Challenge	22%	A sense of challenge arising from the intriguing nature of the problem itself or its importance to the organization (internalized by the individual as a personal sense of challenge).
Pressure	12%	A sense of urgency that is internally generated from competition with outside organizations, or from a general desire to accomplish something important.

Table 19. Ranking of qualities of the environment that inhibit creativity in R&D (Amabile 1988, 147-148).[116]

QUALITIES OF ENVIRONMENTS THAT INHIBIT CREATIVITY		
Various organizational characteristics	62%	Inappropriate reward systems in the organization; excessive red tape; a corporate climate marked by a lack of cooperation across divisions and levels, little regard for innovation in general.
Constraint	48%	Lack of freedom in deciding what to do or how to accomplish the task; lack of sense of control over one's own work and ideas.
Organizational disinterest	39%	Lack of organizational support, interest, or faith in a project; a perceived apathy toward any accomplishments coming from the project.
Poor project management	37%	Manager unable to set clear direction; manager with poor technical or communication skills; manager who controls too tightly or allows distractions and fragmentation of the team's efforts.
Evaluation (tied rank)	33%	Inappropriate or inequitable evaluation and feedback systems; unrealistic expectations; an environment focused on criticism and external evaluation.
Insufficient resources (tied rank)	33%	Lack of appropriate facilities, equipment, materials, funds, or people.
Time pressure (tied rank)	33%	Insufficient time to think creatively about the problem; too great a workload within an unrealistic time frame; high frequency of "fire-fighting."
Overemphasis on the status quo	26%	Reluctance of managers or co-workers to change their way of doing things; an unwillingness to take risks.
Competition	14%	Interpersonal or intergroup activity within the organization; fostering a self-defensive attitude.

[116] Reproduced from Research in Organizational Behavior, Vol. 10, T. M. Amabile, A Model of Creativity and Innovation in Organizations, pp. 123-167, Greenwich, CT: Jai Press., ©1988, with permission from Elsevier Science.

Note the importance attributed to the aspect of freedom (first rank in occurrence in the interviews) and the relevance of its counterpart "constraint" (second rank in occurrence) associated with R&D autonomy and influence of good and poor management performance (rank 4 in occurrence). The results of this have additional implications in another chapter of the current work.

3.6 Design review

3.6.1 Basic idea

Design review is originally a technique from the aerospace sector that consists of formal meetings held at certain intervals during product development. They were basically intended as a critical examination of the product design, its configuration, the quality of design documentation, test program, test data and product characteristics to ensure that all necessary characteristics are incorporated into the product or that they are being taken into sufficient consideration. The scope of design reviews spans a wide range of disciplines, not only product development, but also manufacturing engineering, reliability and quality, program management, costs and customer utilization, among others.

The diversity of the related disciplines involved and the fact that design reviews took place in relatively long intervals had the consequence that they took one or more days to conduct.

The first sectors to adopt the technique were space and defense industries, followed by civil aeronautics. As project management literature referred increasingly to the technique, other technical sectors started to adapt it to their needs.

Its principal functions are, according to Samaras & Czerwinski (1971, 187), to "provide customer and company management with data for determining the design status, identifying problem areas, approving production of hardware, conducting trade-off studies, requesting changes to specifications, and establishing test programs involving the different types of equipment that are combined to form a system."

Additional contributions of design reviews were listed by Juran (1992, 201): They provide an early warning of upcoming problems, data to aid in finding the optimum and the challenge of theories (reducing the risk that unproved theories or unwarranted beliefs will prevail).

Frühauf (1990, 33) cites some secondary effects of design reviews as being:

- better spreading of information about the product,
- "informal" standardization of product quality,
- change of the work climate to teamwork and
- a better predictability of the project.

Other goals of design reviews are listed by Birolini (1994, 329):

- a critical review of the system specifications on the basis of the result from market research,
- comparative studies,
- patent situation;
- a critical review of all documents belonging to the assembly under consideration;
- comparison of the target achieved with the system specification requirements;
- checking of interfaces to other assemblies;
- critical comparison of prototype qualification test results with system requirements;
- formal review of the correspondence between technical documentation and prototype;
- verification of producibility, testability, and reproducibility.

The idea behind design reviews is not only to compare the achieved with the intended specifications, but also to discuss the requirements of the product specification and refine them under consideration of the new information stand. Its purpose is to identify in a timely manner the possibilities of improvements and sources of errors or problems. The advantages of successfully using this technique are (Seghezzi, 1988, 352-354):

- fewer problems during manufacture and use,
- fewer product changes in preparation for production and during product launch on the market,
- reduced costs for tooling and production, and
- avoidance of unnecessary delays in the interfaces.

In the defense industry, design reviews also require a formal response of the procuring agency (customer) in the form: approval/contingent approval/disapproval (cf. DoD, 1981b, item 4.2.4). Conducting a design review gives the "green light" to proceed to the next phase of development work, which is why it is usually conducted after a given number of predetermined phases. In many cases it is a necessary, contractually defined event before proceeding to the next phase of product design.

The conventional form of design reviews is a meeting at which a person with detailed knowledge of the items under discussion, usually the design engineer, makes a presentation to experts from other disciplines or departments. These experts will have received data about the items to be reviewed sufficiently prior to the meeting. They are asked to review the data within their area of expertise and are required to provide an oral critique and presentation of their findings (ESRO[117], 1973). The participants usually act as "devil's advocates" when discussing the product (O'Connor 1991, 160).[118] The point is that information about one's work is to be criticized by others.

A list of the points discussed and appropriate actions to be taken is usually provided after the review.

3.6.2 Implementation issues

Some of the problems experienced in implementing design reviews were:
- non-constructive criticism,
- loss of control,
- formalism,
- objectives other than improving the design.

Box 33. Example of a catastrophic design review (case #25).

In a telecommunications company that did not require design reviews by contract, development engineers had done their job. At that time, they used the concept of sequential project phasing:
- hermetic development phase,
- handover to production, with many interactions between the two areas,
- hermetic production phase.

Two associated systems in development were at a degree of advancement that warranted a *design review*. During the review, questions were posed concerning the development decisions made. The participants questioned the technology choices, the system conception, the assembly breakdown, the materials used, and more. However, the manner in which the questions were posed was such that the development engineers felt they were being attacked. Consequently, even simple questions such as "Why did you choose material A instead of B?" evoked massively defensive responses. It was a profoundly negative experience for everyone concerned. Asked about it months later, one engineer became quite irate and shouted "Why did we use [this material]?! Because!"

The system conception of one piece of equipment was examined so aggressively that the manager in charge of the department that had developed it actually became unable to speak. His superior, the development

[117] The European Space Research Organisation and the European Launcher Development Organisation were the predecessor organizations of the European Space Agency until 1973.

[118] Excerpts from: O'Connor, P. D. T. (1991). Practical Reliability Engineering (3rd ed.), Chichester: Wyley. © John Wiley & Sons Limited. Reproduced with permission.

director, demanded repeatedly—perhaps even intended as encouragement—"Defend yourself! I do not accept that you will not defend yourself!". The equipment had to be completely redeveloped.

The way the review was conducted had the lasting effect that it was nearly impossible to even broach the subject of design reviews with development engineers for a long time thereafter. "Oh, no! Not again!" was a typical reaction to the mere mention of the subject.

This is unfortunately not an isolated case. Seghezzi mentions that this technique was not extended from the defense industry to non-military industries without the resistance of development engineers, who felt hindered within their own scope of competence[xxvii] (1988, 354). Frühauf (1990, 34) provides an indirect confirmation of this issue among his ten rules for conducting reviews of software:

- Help the developer make a better contribution to the project (rule 2).[xxviii]
- Choose your words with care: discuss the topic, not the persons involved (rule 3).[xxix]
- Do not defend yourself: it's a review, not a trial (rule 10).[xxx]

This is in marked contrast to the essential idea that "the emerging purpose of the design review is one of bringing to the designer inputs from the best experts available (on various essential specialties) so that the designer can optimize the design. In the design review meetings, the designer must listen to these inputs and respond constructively. The designer retains the benefit of the doubt and the last word on matters of structural integrity." (Juran & Gryna, 1980, 182).[119]

As Juran (1992, 269) describes, the idea to be conveyed to the design engineers should have been: " 'If you design this way, here is what the consequences will be in our area.' Under the design review concept, the responsibility for action on early warnings rests with the ... designer. The other members of the design review team are there to provide inputs and early warnings, not to make the decisions." There is one distinction to be made in this context, however. If a design review is intended as a revelation of problems, proceeding to the decision-making phase during the review will slow the course of the review even further. A good decision is made after numerous solutions and alternative solutions have been considered that could not be worked out thoroughly during the review. In addition, making decisions during a review restricts the free and constructive discussion of issues. It is more productive to use the review to discuss possible alternatives and leave the decision making to another meeting. Some practitioners quite

[119] Excerpts from: Juran, J. M. & Gryna, F. M., Jr. (1980). Quality Planning and Analysis - From Product Development Through Use (2nd ed.), New York, etc.: McGraw-Hill. Reproduced here by kind permission of McGraw-Hill.

intentionally hold a separate decision-making meeting or a meeting to establish appropriate action plans following the design review (sometimes immediately subsequent to the review), but they ensure a delimitation between the meetings. This not only helps avoid flawed decision-making, it also eliminates the effect of mixing the design review and the decision-making process, which would only restrict the options instead of encouraging the review participants to propose new solution alternatives and discuss their consequences. It also helps concentrate the synergies of evaluating solutions and alternatives.

The idea of conducting design reviews is closely related to the notion of *Critical Testing* established by Albert (cf. page 46). In its implementation, however, many issues arose that counteracted the basic idea. There is often a significant gap between an idea to an implementation; not only did designers fail to listen constructively, criticism was not made constructively either.

According to Juran & Gryna (1980, 182), "A universal obstacle to design review has been the cultural resistance of the design department. It has been common practice for this department to hold a virtual monopoly on design decisions; i.e., these decisions have historically been immune from challenge unless actual product trouble was being encountered. With such a background, it was not surprising that the designers resisted the use of design reviews to challenge their designs. The designers contended that such challenges were based purely on grounds of theory and analysis (at which they regarded themselves as the top experts) rather than on the traditional grounds of 'failed hardware.' " They point out further that "This resistance was further aggravated in those companies, which permitted the reliability engineers to propose competing designs. The designers resisted the idea of having competitors even more than they resisted the idea of design review." (op. cit., p. 182).

There are also problems linked with excessive formality and the consequent inefficacy in making a technical improvement to product quality characteristics as shown by the following case:

Box 34. Case example of avoiding unnecessary formality (case #26).

Two companies working together on a joint program to develop airborne equipment for the defense industry were discussing the inclusion of review meetings to assess reliability and maintainability issues. One member of the more experienced company proposed that the term "design review" be carefully avoided, arguing that this would cause the procuring agency to understand a design review as defined by the military standard (MIL-STD-1521A), in which case it would impose the associated formality. The ensuing "circus" would have to have an "exhibition" character in order for the procuring agency to be satisfied, which would not only be very time-consuming and work-intensive, but would presumably yield doubtful results. It was decided to call the activity a "review meeting" to avoid this confusion. The idea was to conduct a *technical* meeting in which experts could discuss *technical* problems. The need to exchange information and assess the state of the product definition was emphasized, not the need to put on a show.

This is again not an isolated case. The software development standard DOD-STD-2167 referred to in Chapter 2 (page 59) also required many internal and formal reviews. As Coad describes, its "rigid phase-by-phase development approach has adverse cost and schedule impacts. As a major review approaches, the contractor begins to focus its key technical personnel on preparing an award-winning presentation. Documentation is frozen well in advance and presentation is not up-to-date". In its first revised edition (2167A), the formality of the reviews remained a potential problem but where "alleviated by specifically defining the reviews as periodic 'in-process' checks with the customer, rather than the customary major productions" (1988, 49).

Another problem concerning the formality of reviews—even if the customer is not present at the meeting—is reported by Stoff (1990, 23-24). He states that "official 'review meetings' frequently do not proceed with the needed constructive openness. They degenerate to ceremonial events, where conventional forms of behavior are proved and justified.[xxxi] Official conferences create a climate of being exposed publicly[xxxii] in that either opposite interests meet one another, not in the sense of constructive cooperation, or no innovative solution appears because of the tacit avoidance of conflict among the participants".

This shift of focus jeopardized the potential for reviews to improve product quality. Dalton & Perry (1987, 105) describe their findings that, "In many of the high tech firms we have studied, an early indicator that the firm was moving away from radical product innovation occurred in the project review process. The emphasis shifted from reviewing the technical merits of a project to review of a business plan. Once this shift in emphasis occurred, the number of radical product proposals began to drop sharply."

In their survey[120] of selected German companies, Specht & Schmelzer (1991, 83) investigated the efficacy of quality techniques in product development. Design review was the most frequently used technique (89%), but respondents were not very satisfied with the quality of the implementation of the reviews. The issues encountered were:

- determining the evaluation criteria,
- selecting the rhythm and declaring the termination of the reviews,

---

[120] This survey is mentioned previously on page 2. Its authors investigated 40 companies in the electronics, aerospace and automobile industries.

– delimiting the phases after which the reviews would take place, and
– *dealing with the attitude of the participants.*

The authors observe that the low standard deviation in the evaluations[121] suggests that these issues are quite common. In the course of further interviews, many respondents declared among other things that

– reviews are regarded by the developers as a burden,
– requirements concerning documentation and presentation result in additional expenditures,
– the elimination of "unimportant" errors could have been left for later.

Particularly in software development, this formalism was frequently rejected as a "hindrance to creativity"[xxxiii] (op. cit., p. 83).

O'Connor (1991, 161) declares the effect of design reviews is that "techniques often lose credibility, as do reliability staff. The main victim is the design itself, since the protagonists usually prosper within their separate organizations."

To be effective, design reviews must be conducted in a cooperative manner. In many cases, information about latent design problems is already available somewhere within the organization, or problems can be foreseen by attentive employees. Merely gathering this information for discussion already constitutes a considerable task for the development organization. Design reviews provide a good framework to conduct a multi-disciplinary discussion of issues revealed by complementary experiences and viewpoints, but owing to a variety of reasons, this opportunity is often missed.

Box 35. Apollo 13 criticality could have been avoided (Stuhlinger & Orway, 1994, 244).[122]

Michael Collins, the Apollo 11 astronaut who orbited the moon in the command and service module while Neil Armstrong and Buzz Aldrin made the first landing on the lunar surface, recalled in an interview conducted in May of 1987 his contact with von Braun:

"My first meeting of any substance with von Braun was a design review in Huntsville of the Apollo command and service module ... I remember being very much impressed by his ability to go through the circuitry and describe what he thought were shortcomings. One of the things—had it been pursued further, that might have prevented the accident of Apollo 13—was this: In the service module, there were two oxygen supply tanks, and thus there was redundancy for the command module's oxygen supply. Von Braun pointed out during the meeting that the lines from the two tanks came very close together [mutually endangering each other in case of a rupture]. Of

[121] A 5-point Likert scale was used.

[122] Excerpts from: Stuhlinger, E. & Orway, F. I., III (1994). Wernher von Braun - Crusader for Space - A Biographical Memoir, Malabar, FL: Krieger. Reproduced here by kind permission of Ernst Stuhlinger and Frederick I. Ordway III.

course, that may have been what happened to Apollo 13 when one of the oxygen tanks exploded. It broke the line leading from the second tank, and thereby both tanks were lost. It's something I will always remember. Von Braun ... immediately spotted this, [but] no one ever did anything about it." (brackets retained).

The tank explosion could not have been avoided with this information, but the criticality of the situation in which the three astronauts almost died could have been avoided if this issue had been taken seriously. The loss of the two oxygen tanks in the Apollo 13 mission left the command and service modules without electricity, water, life support systems and breathing oxygen. Wernher von Braun was renowned for his cooperative approach to discussions, but for some reason, his warning was not heeded. He was also known to be an effective catalyst in inspiring others to pursue issues within their realm of expertise, thus yielding better implementations, but this was evidently not the case in this instance, with nearly tragic consequences.

In hindsight, it seems obvious that such an issue should have been scrutinized before launch. However, in the sea of details surrounding any given product design, establishing priorities for issues that deserve further consideration is not a trivial task. Nevertheless, concerning the case of the oxygen supply lines of Apollo 13, the fact remains: a highly relevant piece of information was disregarded.

3.7 Competition and its driving forces

The following concept of designing an organization needs clarification.

Box 36. A case of internal competition (case #27).

In a company working for the defense industry in the 1980s, an "innovation" by the company owner was to have more than one team in the same company work on competing concepts to fulfill a product requirement specification. The two development teams worked for the same period of time, independently from one another. Afterward, a competition was held in which the two teams met with the company owner and had to defend their respective solution. The owner then determined which solution was the "winner".

In addition to the duplicate costs, which the company apparently could afford, the participants reported an incredibly unpleasant spirit of competition within the company. Not surprisingly, the respective team members would go to great lengths to hide information from the other team to increase their own chances of winning. In the resulting organizational climate, cooperation was impossible.

Curiously, many development managers consider this adversary approach a good way to achieve better products. They argue that when people are motivated to beat the competition, they are bound to perform better. Indeed, there are many companies whose managers are only motivated to improve a product if a competing company offers significant functional improvements, better quality of execution and/or a lower price. Typically, however, the reaction is late, sometimes too late.

Competition may be healthy, but not necessarily so. The diversity of products available on a competitive market leads to the customer's right to choose according to his or her needs or preferences. Companies providing the market with products have to be aware of these needs in order to succeed. However, the effects of some kinds of competition should be evaluated prior to adopting this approach.

Some proponents of the competition approach point out that the Americans applied it in the missile race against the Soviet Union to achieve a force of intercontinental ballistic missiles. In mutual competition, the Army, the Navy and the Air Force pursued launcher concepts in parallel. This, however, is not so. In reality, the age-old rivalry among the services is what led them to compete with each other, beginning with their core competencies. The Army argued that the rocket was the extension of artillery, the Air Force claimed that the rocket is an extension of an aircraft, and the Navy insisted that as it has both planes as well as artillery aboard its ships, it should be the one to develop missiles (cf. Stine, 1991, 148). This resulted, for example, in a limitation of the maximum range of rockets to be explored by the Army.

For the satellite launcher for the International Geophysical Year in 1950, the three services each made a proposal. The Navy's proposal was the Vanguard, which was ultimately approved. It was associated with much modification work and substantial new developments, and the Army, although its design was mature and proved, was prohibited to launch a satellite, even "accidentally". After the success of Sputnik, a cartoon was published "showing an Army and an Air Force general and a Navy Admiral standing together; a satellite orbited overhead with a hammer and sickle painted on it. The Army general says to the admiral: 'Gosh, am I glad that it is not the Air Force!'; the admiral says to the Air Force general: 'I am so happy it is not the Army!'; and the Air Force general, in turn says to his Army colleague: 'Golly, am I glad it is not the Navy!' " (Stuhlinger & Orway, 1994, 126). This rivalry "overshadowed the perception of joint responsibility for the good of the nation" (op. cit., p. 126). An American delegate to the International Astronautical Congress in Barcelona, Spain, reported a conversation with Professor Leonid Sedov, Soviet delegate to the congress and member of the Soviet Academy of Sciences, who said of the

Vanguard: "One thing we [in the Soviet Union] could never understand, ...is why you chose such a complicated, difficult, and really marginal design for your satellite vehicle. It was an entirely new development! And, furthermore, it had no growth potential. A little device of 20 pounds was all you could hope to launch ... Why, for heaven's sake, didn't you take one of your powerful engines that you have been flight testing for a considerable time now? [That] would be the right thing for your satellite project... We just could not understand why you did not choose this natural, straightforward approach..." (op. cit., p. 133).

In 1955, the Killian Committee suggested that the United States develop an intermediate range ballistic missile with a 1,500 mile range, following the decision to develop the Atlas intercontinental ballistic missile. For strategic purposes, it would be simpler to develop and available earlier than the ICBM. Wernher von Braun proposed that the Redstone Arsenal develop the missile as an Air Force contractor. General B. A. Schriever of the Air Force WDD "turned him down flat, noting, '*It would be naive to think that the Army would develop a weapon and then turn it over to the Air Force to operate*' " (Heppenheimer, 1997, 108).

The Air Force members in charge of the intercontinental ballistic missile also pursued the concept of alternate (or dual) source development, also referred to as parallel development. They actually developed not only two complete sets of subsystems but two complete missile systems: the Atlas and the Titan. This was criticized as "wasteful duplication" but was intended to reduce the risks associated with the uncertainties in the development by providing insurance against delays in the most critical subsystems. In addition, the industrial base for the entire program was too small and had to be expanded to ensure maximum R&D efforts as well as sufficient capacity for later full-scale production (Greenwood, 1974, 193), (Heppenheimer, 1997, 85). They intended to achieve interchangeability of systems and explore alternative concepts (e.g. cryogenic vs. hypergolic propulsion, radio-inertial guidance vs. all-inertial guidance). The resulting products found their utilization and the results of the alternative concepts that were abandoned were used for future concepts. Both missile systems entered service. The Titan system[123] remained in service until 1987. The two systems were not actually competing with each other.

Adversary development might have the alleged advantage of increasing the information available, thus reducing risks and uncertainty. It might also facilitate the concrete evaluation of alternatives, but it has the disadvantage of blocking communication and fostering a climate of suspicion. As Quinn (1985, 79) noted, "perhaps the most difficult problem in managing

[123] A vector for nuclear warheads.

competing projects lies in reintegrating the members of the losing team". This issue has been reported by many developers and development managers: competition inherently means that there is a "winner" and a "loser". As adversary development assumes that only one winner is chosen, a company would do well to uncouple the reward and career progression system from the main product. If the success of a product can be measured and associated with personal commitment, its reward could be based not on winning but on the value to the company of one's knowledge as well as one's potential contributions to future projects. In other words, there should be a compensation for losing.

Quinn mentions an example involving Sony in which, "according to its top manager, the research climate does not penalize the losing team." This might well serve as a good example of compensation, although it might also merely be a characteristic of Japanese employment culture. "We constantly have several alternative projects going. Before the competition is over, before there is a complete loss, we try to smell the potential outcome and begin to prepare for that result as early as possible. Even after we have consensus, we may wait for several months to give the others a chance. Then we begin to give important jobs [on other programs] to members of the losing groups. If your team doesn't win, you may still be evaluated as performing well. Such people have often received my 'crystal award' for outstanding work. We never talk badly about these people".[124] A European manager of product development at Sony declared that this multiple product development path is taken when a consensus regarding product implementation cannot be achieved,[125] and he confirmed that there is a commitment to using the results of the "losing" group in other products (case #28).

A very clear distinction should therefore be made between productive and unproductive forms of "competition". As in sports, the competitor can also be a partner, for example to sustain a rhythm, which can prove to be difficult—sometimes very difficult—to do when practicing alone. Engineering students in an university bicycle race commented that, in periods when they were racing alone, it was very difficult to keep the pace, whereas when another participant was present, it was much easier. The competitor in this case becomes a partner more than an opponent. Working collectively brings the advantage of keeping pace, of exchanging techniques and points of view, of growing.

---

[124] According to the text by Quinn, this declaration is attributed to Sony's Mr. Ibuka.

[125] Readers are presumably aware that Japanese (company) culture is strongly consensus-oriented.

Unhealthy forms of competition and their consequences for the business environment are examined by Kanter (1989). A style Kanter (1991, 33) dubbed *cowboy management* stresses "competition, rivalry, and contention over cooperation (e.g. the popular idea that R&D is improved by 'performance shoot-outs' between competing teams)".[126] Kanter explains: "Cowboy mythology holds that corporate citizens are supposed to enjoy combat, and they are supposed to be tough enough to take it. It glorifies the rough-and-tumble fight-it-out style. Cooperation in turn, is seen as 'soft' as something for 'sissies,' or as something imposed by 'citified' bureaucrats. 'Tough' management styles are regularly glamorized in the press, as in Fortune magazine's listing of the 'ten toughest bosses in America' " (1989, 71). There are five main reasons why competition can become destructive (Box 37):

Box 37. Signs of destructive competition as consequence of "cowboy management," (condensed from Kanter, 1989, pp. 76-82).

- Players pay more attention to beating their rivals than to performing the task well.
- Friendly, respectful competition among people gives way to mistrust, suspicion, and scorn.
- Imitation may preclude competition. Instead of stimulating different alternatives or encouraging variety, more effort is spent on spying on and copying what the competitors are doing rather than on searching for new options.
- The weaker party may give up rather than be stimulated to continue the fight. Kanter also mentions laboratory studies by Kohn (1986, 63), which show that high levels of anxiety in people who compete is correlated with inferior performance.
- The stronger party begins to feel dangerously invincible, thus risking a deterioration in performance or overlooked vital aspects. Kanter argues that "this is why many American companies that dominated the markets before the game became truly global grew lazy and complacent."

Kanter also cites the study by Cohen & Bradford (1985), in which (i) exciting and challenging goals, (ii) some autonomy and ownership, (iii) high visibility, (iv) accountability, and (v) an exciting task were indicated by several thousands of managers as the circumstances in which they performed best. Competition was rarely mentioned in this context.

---

[126] Parentheses maintained.

Moreover, winning the competition can become an end in itself and in some cases, the unfair practice was reported that talented members of a strong competing group were recruited by the "opposition".

Destructive competition can also exacerbate the "not-invented-here" syndrome described in the section on improving communication (page 253). It can also defeat the purpose of generating new ideas if "anything 'tainted' with the 'hue' of the 'others' is automatically rejected" (Kanter 1989, p. 79)

At this point the differentiation between intrinsic and extrinsic motivation should be examined.

As Amabile (1983, 86) proposes, the intrinsic aspect of motivation is essential when high levels of creativity are necessary because, when performance is extrinsically motivated, "attention is directed toward the attainment of the extrinsic goal and away from the exploration of new pathways. ... If intrinsic motivation is sufficiently undermined by extrinsic constraints, task engagement will be avoided". In contrast to this situation, whenever intrinsic motivation is high, "most of the individual's attention will be directed toward the task itself, toward an exploration of its details and of possibly relevant aspects of the task environment. The higher the level of intrinsic motivation, the deeper this concentration on the task."

She also clearly distinguishes between a heuristic and an algorithmic task. The former presupposes learning and perceiving in a guided trial-and-error process, whereas the latter offers a clear and straightforward path to the solution (sometimes one not even considered a creative solution). This differentiation is important to explain an apparent contradiction among study results where the demonstration of seemingly positive effects of evaluation and competition was, "in virtually all of the studies", relevant to tasks that can be classed as algorithmic (op. cit., p. 101). This may also explain the misguided attempts of R&D managers to propose competition as a performance driver—in many cases, they are using the wrong analogy.

Furthermore, the study by Amabile also mentions that behavior (predominantly) under extrinsic motivation is "narrowly directed toward achieving the extrinsic goal that has been imposed, whether that goal be attaining a reward, meeting a deadline, achieving the approval of an observer, or obtaining a positive evaluation from an expert. In order for a creative response to be produced, however, it is often necessary to temporarily 'step away' from the perceived goal, to direct attention toward seemingly incidental aspects of the task and the environment" (op. cit., p. 94). This provides insight into the nature of the observation reported by a development manager concerning his most innovative engineers described in case #11described on page 104.

3.8 Concluding remarks

We examined tentative process modeling and how to cope with the implicit limitation to Cartesian thinking. Although very important in science and engineering, it is not the sole ingredient of success. There are many other scientific approaches that are sometimes more appropriate for product development, such as the approach as convergent – divergent cycles, or models that are closer to the mental processes underlying technical problem-solving.

A good tentative model involving barriers and drivers is also described. It calls for the need to consider the importance of interaction among professionals.

Project management issues presented in Chapter 2 were reexamined in terms of the development environment, especially in its function of protecting workers. Its tendency to become Tayloristic and the related difficulties of time and cost estimation were discussed. In addition, the aspect of uncertainty in development programs was introduced, as well as its apparent contradiction with the manager's need for certainty.

We then approached quality systems, also introduced in Chapter 2, their effects on product development, and the ambiguities they introduce. We analyzed the concept of development control and its ineffectiveness for development activities to approach new problems or achieve innovation. Product specifications and checklists were treated, especially regarding their static – dynamic nature due to market or technology evolution, as well as the difficulty of appropriately transferring these rules.

Control was then presented as a co-relation of controls within the organization.

Autonomy, one of the many widely discussed topics in product development, was also discussed and will be associated with the concept of *creative tension* to be introduced in Chapter 4.

Design reviews were presented, as many factors in their implementation affect the developer's environment. The effectiveness of design reviews varies widely across all the industries that use them or tried to use them. Reasons for this variation are listed, with special attention given to the major objective of improving design.

Competition was then discussed, accompanied by the advice exercise extreme care in adopting this notion from the world of sports to that of problem-solving. It has not necessarily proved to be an effective way to achieve better results in cases where mechanisms of problem-solving and innovation are more relevant than mechanical or physical performance. Although it enjoys continuing popularity as a way to push product development dynamics, its negative "environmental" effects on development are often overlooked

The next two chapters discuss influence factors within individuals as well as development-related aspects within or among groups.

Chapter 4

INTRAPERSONAL FACTORS

4.1 Dirty-hands philosophy and the yardstick

In a hearing before the U.S. House of Representatives in 1964, Wernher von Braun was asked to explain why the Marshall Space Flight Center (MSFC) was spending funds on internal projects. NASA centers were supposed to play an administrative role in coordinating space (and aeronautics)-related research and development, which would actually be performed in industries and universities. Von Braun stated that, "a good engineer gets stale very fast if he does not keep his hands dirty … it is for this reason that we are spending about 10 percent of our money in-house; it enables us to really talk competently about what we are doing. This is the only way known to us to retain professional respect on the part of the contractors." (Levine, 1982, 4).[127]

The so-called "dirty-hands" philosophy was a characteristic of MSFC under von Braun and was accepted "philosophically", because of its effects on the technical personnel involved. It had its roots on the *arsenal approach* followed long before then. There are basically two, strongly interconnected

[127] Cited from Wernher von Braun, Director, Marshall Space Flight Center, briefing of House Science and Aeronautics Committee, subcommittee on Manned Space Flight, 24 Jan 1964, in hearings on 1965 NASA Authorization, 88th Cong., 2nd sess. (1964) p. 803.

elements involved in this statement: motivational aspects and ensuring the knowledge base to discuss technical aspects.

The motivational aspect can be broadly stated and is associated with the vocation for engineering or technical professions.

The aspect of ensuring a knowledge base is an essential element of technical discussions when buying or contracting technical services. It essentially provides a "yardstick," or a reference for comparing and evaluating what is claimed by the experts or persons selling goods and services. Without it, the communication process is hindered considerably, and in-depth discussions of technical details and associated implementations are not possible due to an unbalance between the parties. Taken to the extreme, the person contracting or buying must take the vendor at his or her work, just as any non-knowledgeable person would.

The generalization of this concept is as follows:

Four cases support this philosophy and its effects.

Box 38. Supporting cases for the dirty-hands philosophy.

A young quality engineer in an electronics company[128] was working on testability/maintainability aspects of electronic equipment. He had attended courses on testability and other quality matters but was occupied mainly with problems on paper. Citing the urgency of the case at hand, the managing director requested that this engineer to be trained temporarily to use a test machine for manufacturing printed circuit assemblies and perform the installation, after which he would pass the task to production personnel.

After some time, another engineer returned from a sabbatical-like absence and resumed the function of the young quality engineer's supervisor. He had had direct professional contact with the young engineer prior to the latter's "extracurricular" training and found that, compared to before, the engineer had clearly gained significant expertise from his hands-on experience with the tester. Not only was his command of the pertinent terminology much more precise, he was also able to give exact descriptions of real problems when discussing design aspects of product testability in general. He had gained personal experience with an essential part of the entire process. Later, in a conversation between the supervisor with the managing director, the latter expressed doubts as to whether this temporary detour had served any useful purpose. Apparently, there had been complaints concerning work that could not be accomplished because of the temporary assignment of the young engineer outside of the competence of the department. The returning engineer assured him it had made an essential difference, directly and indirectly, to the young engineer's activities. He stated, "Thanks to that, he now really knows what he is talking about". He

[128] Producing avionics among other products.

praised the director's action, which had actually been intended to give the new engineer an opportunity to gain this practical insight. Moreover, the young engineer was now quick to react to new management measures. His actions were combined with very good performance, especially in dealing with manufacturing and industrial engineering personnel, and this was recognized throughout the company (case #29).

Another case occurred in the aerospace industry. An engineer was successful under a previous manager in an activity that provided a balance between technical activities and paperwork. He was then assigned to a quality manager who believed in creating a division between the quality engineering and quality control departments, where the former would be the "thinking" group, and the latter the "executing" group. The manager explicitly described this approach during meetings. Under this manager the engineer worked exclusively on preparing plans, in other words, on paperwork. A supervisor was then allocated between this manager and the engineer. The latter had not complained about his job, but a certain discontent was apparent. He understood the importance of his task, and when asked about it, he said "It's ok, it's important that it be done". He revealed, however, that most of his manager's requests had involved changing the place of commas and editing text. In an informal conversation, the director responsible for the area asked the new supervisor about the engineer's performance. The director's main concern was that the engineer's work had insufficient technical content and too much emphasis on form. The new supervisor assured the director that this was a consequence of the former manager's style and that it was being worked out.

The supervisor then joined the engineer to evaluate pre-qualification flight tests for a prototype, and realized that essential points were missing in the almost-finished report. A meeting was then held with the flight test group operating the test aircraft to determine the adequacy of systems and to perform a rapid evaluation of some risk hypotheses. The supervisor conducted the meeting in a way that the engineer could follow. However, the latter remained silent throughout the brief evaluation. Back in the office, after a joint discussion of the actions to be taken, the supervisor outlined a summary of points and submitted it to the engineer for comments and corrections. The engineer was assigned to incorporate his comments into the final document on which he had been working for a long time—maybe too long. All this was conducted in a dynamic fashion, partly because of the deadline to present the report to the customer, but partly to make an impact. In the end, the summary was appended to the original document, whereby a substantial difference in content and form was evident.

The new supervisor observed the engineer at his desk, holding the addendum to the original document for a long period of time. The supervisor thought he perceived an absence of eye movement; the apparent point of focus beyond the sheet of paper suggested that the lines of the document were not really being read, although this is only his guess. He knew the engineer was having a difficult time. The engineer's next assignment was to put development test equipment into operation, in other words, a hands-on experience. This assignment was not given with the intention to "drop a bomb" on the engineer. Aware of the necessity of the tasks, the engineer in fact willingly volunteered for it. It was an activity that in the mid-term would require him to discuss extensive technical details with the development engineers, and his own task encompassed numerous smaller technical issues as well. From that day on, things started to change, slowly at first, but this was not surprising. At the end of the process, the engineer approached issues in a much more competent manner. He had improved his technical background based on this experience and was eventually perceived differently by his colleagues, including the director mentioned above. Years later, this engineer thanked the supervisor for the "inputs" that allowed the entire group to work (case #30).

The third case occurred in a design review of a module for a Line Replaceable Unit (LRU) of airborne military equipment. The subcontracted company for the module was experienced in the technology and on market, in contrast to the main contractor that, although well experienced with the associated technical domain, was a newcomer in the market. In the design review, the subcontractor development personnel argued that the specifications were unfeasible within the constraints of size and performance. The engineer in charge of the module at the main contractor, a very good practitioner, was able to experiment a little on his own. He succeeded in building his own laboratory version of the module, although it had been decided that the module was going to be bought and not fabricated in-house. This was actually part of his manager's policy of knowing what they were doing and buying. This hands-on knowledge allowed the main contractor engineer to suggest to the subcontractor engineer what specific component they could use to achieve the required specifications and how to use it. All the arguments cited by the subcontractor concerning the impossibility of building such a module were promptly countered by the main contractor engineer, who described how it could be done, down to the specific component to be used. The main contractor engineer reportedly perceived the subcontractor engineer as arrogant; the former was merely striving for a possible solution, the latter was not. After the review, the subcontractor had to replace the entire development team (case #31). (Note: Another company ultimately provided the module to the proper specifications, including size.)

The fourth case is associated with the technology transfer for the Swiss version of the F/A-18C/D fighter aircraft developed by McDonnell Douglas, described by Surber(1997, 4).[129] "Technology transfer should never be a 'development project' from a 'highly developed' to an 'under-developed' partner.... The company that is taking the technology must possess an existing technology potential, such that it can communicate effectively with the company supplying the technology". Surber stated[130] that he allocates about 8~10% of a given budget for training and in-house R&D to solve this issue. Interestingly, this is in good agreement with von Braun's 10% rule.

These effects are vital to product development, especially when external product development is practiced as a way of optimizing the tension triangle (cf. page 7). The "dirty-hands" philosophy should therefore not be regarded merely as indulging an engineer's desire "to play" or as 10% of useless work that could be better performed by others. It is indeed tempting to see the numbers in the cost spreadsheet and propose their elimination if a manager does not appreciate what these numbers imply.

Lessons unlearned

These lessons are already considered "history". Fries (1992, 177) , in her investigation of NASA engineers of the Apollo era at the time of the Challenger accident describes the effect of losing the dirty-hands philosophy:

Box 39. Losing the dirty-hands philosophy at NASA MSFC.

"Like their fellow engineers at Langley, Ames, and Lewis, Marshall's engineers are restive with a system that relies on contractors for engineering as well as support services. Sam Browning began working for the Army Ballistic Missile Agency in 1956 after earning his degree in chemical engineering. By the mid-1960s Marshall had made the transition: 'We didn't do a great deal in-house in those days, ... which was sad.... I didn't really get to go into the laboratory and get hands-on type stuff, which I would have loved. I had to go visit a contractor's facility, who [sic] was having all the fun. ... I came up with the ideas or picked them up from other people, and we secured funding from Headquarters to go fund the activity, and we'd award a contract to some propulsion company.' Browning [felt] a personal loss from having been denied opportunities to accumulate his own

129 Hans Surber was at the time head of systems management and member of the Board at SF Schweizerische Unternehmung für Flugzeuge und Systeme in Switzerland. The company developed the high-density applications pylon for the Swiss F/A-18.

130 Private conversation during the workshop, in which he presented this case.

experience with advanced propulsion systems—but the loss is not just his. He is currently working on a laser propulsion[131] project that is 'almost more research than technology, because we had to establish that you can, in fact, sustain a stable plasma in hydrogen supported by a high-powered laser.' The work involves an awful lot of high-temperature physics and computational fluid dynamics, and a lot of other good stuff I don't know much about... My frustration with that is that I don't understand enough about it to be able to intelligently guide the people who are working on the program'."

Fries also interviewed Hank Smith at the Kennedy Space Center (KSC) after the Challenger accident and reported that "That event highlighted their concern about the use NASA makes of contractors...".

Box 40. Other effects of losing the dirty-hands philosophy at NASA Kennedy Space Center.

" 'The business of Shuttle processing and launching—I think that's just too critical to turn over to a contractor. Management needs to be involved in that processing work. I don't think any contractor can do the whole thing; it's too big a job.'

Not only is the job too big, but accountability is spread too thin. 'See,' says Smith, the contractor is 'responsible to check himself also'. Now NASA's ultimately responsible, but I don't think [its responsibility goes] deep enough.' "

Smith declared to agree that noncritical work could be assigned to contractors, like operating fire trucks, for example. He continues: " 'But the intricate stuff—the critical stuff—I think NASA needs to be more involved. I just don't think you can say *"OK Mr. Lockheed. Everything wonderful?"* And he says, *"Oh yeah"*, and you walk away. That's just too much' "

"Smith's co-worker, Eleanor Finch, shares his reservations. 'Contractors are in business to make money. ... And they really don't care a lot of times whether the job gets done or not, nor do they even really known what the job is, sometimes. And NASA needs to remind them of what the job is. Day by day. And that is what the contract monitors role was.' Finch says 'was,' because in the late 1970s the management consulting firm of Booz-Allen Hamilton recommended that Kennedy Space Center substantially reduce the number of contract monitors overseeing day-to-day contract activities. Booz-Allen argued that a great deal of money could be saved if NASA were willing to settle for periodic reports from contractors. NASA's euphemism for reduced supervision is 'self-sufficiency', an attribute of Kennedy Space Center's comprehensive mission contracts. Self-sufficiency, explains Finch, ' means that the contractor can make more decision on his own without

[131] Interested readers can find a primer on laser propulsion in Glumb & Krier (1984) and Birkan (1992).

coming to NASA for guidance.' She has managed contracts 'both ways. You can't get much out of a report. You have to go down there and talk to those people and find out what the heck they're doing.' And if you don't like what they're doing, 'once you've turned the contract monitor [role] off, it's very hard to turn it around.' " (op. cit., p. 180, brackets maintained).

The loss of this capacity impairs organizations to handle the technical aspects, not only from the technical, material point of view, but especially with respect to interaction and communication, as the NASA case makes it clear.

In the beginning of the 21st century, NASA is again urging the commercialization of space exploration, thus emphasizing the trend to outsource. This trend is not exclusive to the aerospace industry, and it is necessary to clarify the implications for internal processes beforehand.

4.2 Self-actualizers

In his book "Motivation and Personality", Maslow (1954)[132] describes a type of individual he called *self-actualizing people*. Self-actualization is loosely described as "the full use and exploitation of talents, capacities, potentialities, etc." He describes these individuals as follows: "such people seem to be fulfilling themselves and to be doing the best that they are capable of doing, reminding us of Nietzsche's exhortation, 'Become what thou art!' They are people who have developed or are developing to the full stature of which they are capable" (op. cit., p. 201).

Maslow's work on human motivation makes a distinction between physiological needs, safety needs, the need to belong and to be loved, to be esteemed, and the need of self-actualization.[133] The latter is of particular interest for product development, the activities of which bear a potential richness of possibilities where job content and achievements play a key motivational role.

Especially in the previous century, which was characterized by dramatic technological advances, many technical specialists with a propensity toward self-realization worked primarily in product development. This tendency is due to the "artifact construction" aspect of such work that is implicit in the

[132] Excerpts from MOTIVATION AND PERSONALITY by Maslow, © 1954. Reproduced by permission of Pearson Education, Inc. Upper Saddle River, NJ.

[133] Maslow's approach is of interest for this work mainly because of the characteristics of the persons he investigated to elaborate his theory. The hierarchical organization of these needs is not relevant for this work. (cf. Ulich, 2001, 45-46).

job content.[134] The importance of Maslow's findings with respect to the present discussion has to do with the type of individuals that can be characterized as self-actualizers. They are individuals who are committed to contributing an *added value* through the work they perform. The satisfaction they draw from this is central to the following discussion.

4.2.1 Main characteristics of interest

We shall now explore some of the motivational characteristics of self-actualizers that make an essential difference in the product development environment:[135]

Problem-centeredness: self-actualizers are "in general strongly focused on problems outside themselves. In current terminology they are problem-centered rather than ego-centered. They generally are not problems for themselves and are not generally much concerned about themselves; e.g. as contrasted with the ordinary introspectiveness that one finds in insecure people. These individuals customarily have some mission in life, some task to fulfill, some problem outside themselves which enlists much of their energies. This is not necessarily a task that they would prefer or choose for themselves; it may be a task that they feel is their responsibility, duty or obligation." (op. cit., p. 211)

More efficient perception of reality and more comfortable relations with it: "...they live more in the real world of nature than in the man-made mass of concepts, abstractions, expectations, beliefs, and stereotypes that most people confuse with the world. They are therefore far more apt to perceive what is there rather than their own wishes, hopes, fears, anxieties, their own theories and beliefs, or those of their cultural group." (op. cit., p. 205)

Fear of the unknown: They are generally "unthreatened and unfrightened by the unknown, being therein quite different from average men. They accept it, are comfortable with it, and, often are even *more* attracted by it than by the known." (op. cit., p. 205, italics maintained)

Tolerance of ambiguity and lack of structure: "They not only tolerate the ambiguous and unstructured; they like it." (op. cit., p. 205)

Creativeness: "The creativeness of the self-actualizing man seems rather to be kin to the naïve and universal creativeness of unspoiled children. It seems to be a more fundamental characteristic of common human nature—a potentiality given to all human beings at birth. Most human beings lose this as they become enculturated, but some few individuals seem either to retain

[134] This aspect can also be found in other professions.

[135] Many characteristics would not be reproduced here because of the length of text required.

this fresh and naïve, direct way of looking at life, or if they have lost it, as most people do, they later in life recover it. ... The creativeness appears in some of our subjects not in the usual forms of writing books, composing music, or producing artistic objects, but rather may be much more humble." (op. cit., p. 223)

Philosophical, nonhostile sense of humor: "They do not consider funny what the average man considers to be funny. Thus they do not laugh at hostile humor (making people laugh by hurting someone) or superiority humor (laughing at someone else's inferiority) or authority-rebellion humor (the unfunny, ... smutty joke). Characteristically what they consider humor is more closely allied to philosophy than to anything else. It may also be called the humor of the real because it consists in large part in poking fun at human beings in general when they are foolish, or forget their place in the universe, or try to be big when they are actually small. This can take the form of poking fun at themselves, but this is not done in any masochistic or clown-like way." (op. cit., p. 222)

Resolution of dichotomies: The "age-old opposition between heart and head, reason and instinct, or cognition and conation was seen to disappear in healthy people where they become synergic rather than antagonists, and where conflict between them disappears because they say the same thing and point to the same conclusion." (op. cit., p. 233)

Resistance to enculturation: "Self-actualizing people are not well adjusted (in the naïve sense of approval of and identification with the culture). They get along with the culture in various ways, but all of them it may be said that in a certain profound and meaningful sense they resist enculturation and maintain a certain inner detachment from the culture in which they are immersed." (op. cit., p. 224) "They may be called autonomous, i.e., ruled by the laws of their own character rather than by the rules of society. It is in this sense that they are not only or merely Americans, but also to a greater degree than others, members at large of the human species. To say that they are above or beyond the American culture would be misleading if interpreted strictly..." (op. cit., p. 227)

Own imperfections: Self-actualizers are not "perfect" beings. "They too are equipped with silly, wasteful, or thoughtless habits. They can be boring, stubborn, irritating. They are by no means free from a rather superficial vanity, pride, partiality to their own productions, family, friends, and children. Temper outbursts are not rare. Our subjects are occasionally capable of an extraordinary and unexpected ruthlessness. It must be remembered that they are very strong people. This makes it possible for them to display a surgical coldness when this is called for.... Not only are these people strong but also they are independent of the opinions of other people.... In their concentration, in their fascinated interest, in their intense concentration on some phenomenon or question, they may become absent-minded or humorless and forget their ordinary social politeness.... These people are *not* free of guilt, anxiety, sadness, self-castigation, internal strife, and conflict. The fact that these arise out of nonneurotic sources is of little consequence to most people today [1954] ... who are therefore apt to think them *un*healthy for this reason" (op. cit., pp. 228-230, italics maintained).

All these characteristics are of particular interest because they provide solutions to the "big issues" encountered in product development. One of these issues regarding the mono-disciplinarity of professions or work biographies is that it is relatively easy to become discipline-oriented or method-oriented (in Maslow's terminology applied to science: *means-centered*). Hence one's main concern becomes following the (academic) discipline, or merely applying it as an end in itself, instead of applying it to the issue at hand. Referring to the implications for science, Maslow says that "a methodologically satisfactory experiment, whether trivial or not, is rarely criticized. A bold, ground-breaking problem, because it may be a 'failure', is too often criticized to death before it is ever begun. Indeed, criticism in scientific literature seems usually to mean only criticism of method, technique, logic, etc. ... [The] problem itself does not matter—only so it be well done. In a word, it need no longer to be a contribution to knowledge." (Maslow 1954, 14).

4.2.2 Relevance to development activities

Every professional—competent or not—can argue the importance of his/her contribution to the final product, but this does not necessarily mean that the person is problem-oriented. Indeed, some people tend to do their job regardless of whether it contributes to the overall project.

One of the keys to successful product development lies in the capability to integrate the diverse points of view of various experts and to optimize them to achieve better product characteristics. Of course, an individual's orientation toward a mono-discipline or method-orientational approach allows him or her to strive for better methods and tools. If taken to an unhealthy or even pathological extreme, however, this can work against the successful contribution to the benefit of the whole.

Self-actualizers in product development tend to be more open to the criticism inherent in *falsifying* (cf. Critical rationalism, critical testing on page 46) their design. This quality is due to their problem-centered, instead of ego-centered approach.[136]

---

[136] Readers are surely aware that achieving success in any kind of process depends not only on one's personality traits but also significantly on the quality of interaction among the

Owing to their *enhanced perception of reality*, self-actualizers tend to have a clearer grasp of the nature of the issue at hand, thus compensating for the "blindness" incurred by technical expertise. Instances of this blindness can occur when experts, in their justified efforts to differentiate and be precise regarding the concepts and terms used in their field, achieve confusion instead.[137] The prevalence of this problem is illustrated by the fact that an editorial was dedicated to this subject in the October 1991 issue of "Qualität und Zuverlässigkeit", the official trade newsletter of the German Association for Quality (DGQ) (Krassowski & Treffert 1991).

Self-actualizers' enhanced perception of reality is of particular importance in discussions or decisions concerning the relative importance of the many potential issues to improve the quality of *any* product. This characteristic tends to make self-actualizers more open to *changes of paradigm* necessary in an innovation process.

Fear of the unknown is of particular importance because of the inherent uncertainty in the product development process. Developing innovative designs means coping with this challenge. However, it can be equally important with regard to a technology update of an established product.

Tolerance of ambiguity and a lack of structure is a trait of special interest. An excessively defined structure for solving the problems (cf. Chapter 2) would limit the options to appropriately solve problems. Every product designer is likely to face ambiguity once in a while. Possible causes include the design choices to be made, ill-defined specifications, inconsistencies in the structure defined *a priori* and the interaction with other product components or areas of expertise. Although it is theoretically possible to reduce ambiguity dramatically by means of a thoroughly defined specification, this would be too costly in terms of time and effort. Moreover, it could jeopardize the product implementation, not only because less time would remain for that phase, but especially because of the absence of any degree of freedom to achieve design optimizations (cf. Chapter 1). Furthermore, eliminating ambiguity would require that *all* possible product interactions be known in advance, which is virtually impossible. In its worst form, the avoidance of ambiguity has been known to result in the intentional blocking the communication channels, as new information can potentially have a destabilizing and thus threatening effect.

---

persons involved. Consequently, a universal rule that reliably predicts people's behavior, such as "people of this kind will behave in such a way" cannot be given.

[137] A typical reaction to this is: "He is speaking *technese*. I can't understand him".

The *sense of humor* of self-actualizers is well illustrated by the case of the "egoless" programmer and the case of Wernher von Braun, which will be treated later in this book.[138]

4.2.3 Collective orientation

Maslow's description of self-actualizers did not include the aspect of collective orientation. He did, however, indicate the form of interaction of self-actualizers in dyadic groups, although his discussion is specific to love-based relationships. Some of his works are nevertheless of interest in this context: *Dropping of Defenses* (op. cit., pp. 238-240); *Care, Responsibility, the Pooling of Needs* (op. cit., pp. 248-251); *Acceptance of the Other's Individuality, Respect for the Other* (op. cit., pp. 252-254) and *Greater Perceptiveness* (op. cit., p. 257). No indication was found that Maslow's subjects exhibited similar traits outside of their dyads.

However, let us extrapolate these finding to interactions within working groups. The above-mentioned characteristics of self-actualizers would be highly desirable with regard to the product development environment, although conventional work environments do not foster this.

4.2.4 Attitude and information processing

Resolution of dichotomies, described by Maslow, is important in that self-actualizers tend to resolve the dichotomy of affect, cognition and conation. These terms require further explanation. They make up the so-called "*three-component model*" (Meinefeld, 1988, 121; Reiser 1981, 144) of the *attitude* construct (not to be confused with behavior).

Cognition. Functions involved in synthesizing information, including sensation, perception, attention, memory and reasoning.

Conation. Psychological processes including desire, intuition, volition that lead to purposeful action.

Affect. The emotions behind an action, the pleasantness or unpleasantness of an emotional state, the complexity of ideas involved in an emotional state.

According to Stahlberg & Frey (1988a, 153-158),[139] attitude has four main functional bases that are mainly motivational in nature. They are instrumental in the processing of information that is of special interest to product development:

---

[138] The egoless programmer is described on page 188, and a brief description of von Braun's style can be found on page 293.

[139] Although mainly the original English text is cited here, the German version was also drawn upon (Stahlberg & Frey, 1988b), as both authors at the University of Kiel in Germany apparently contributed to the German text.

Box 41. Motivational and information processing functions of attitude (condensed from Stahlberg & Frey 1988a).

MOTIVATIONAL FUNCTIONS

- *Ego-defensive function* describes how having negative feelings about oneself or about one's own group might lead to an attempt to protect oneself by developing a negative attitude, using for example (Freudian)[140] defense mechanisms.
- *Value-expressive self-realizing function* describes the need to express attitudes that reflect one's own central values or an important component of one's self-concept.
- *Instrumental, utilitarian or adjustive function* describes how the expression of attitudes may help one reach desirable goals or avoid undesirable goals.
- *Knowledge or economy function* describes how attitudes allow one to categorize incoming information such as new experiences along established evaluative dimensions. This helps one to simplify and understand the complex world, as the attempt to deal with every single detail of the (social) environment is of course impossible.

EFFECTS OF ATTITUDE ON INFORMATION PROCESSING

- *Active search for attitude-relevant information.* The most significant theory reflecting this effect is *cognitive dissonance* (Festinger, 1957). It predicts that persons are generally motivated to actively search for (attitudinal) congruent or consonant information and to actively avoid (attitudinal) contrary or dissonant information.
- *Information encoding* (i.e. perceptual and judgmental processes). A study by Fazio & Williams (1986) showed that people tend to distort information about the properties of an object depending on their attitude.[141] Many theories support the influence of attitude on perception. Theories of *cognitive consistency* state that persons seek to organize their cognition in a tension-free or contradiction-free manner. Finding out that one's attitudes are in contradiction places one into the uncomfortable and tension-laden state of cognitive inconsistency or imbalance. This causes one to strive to recover the balanced state by changing one or more of these cognitions. The theory that information is not passively received and stored in memory but selectively encoded and actively organized in

[140] Not explicitly mentioned in the original.

[141] Fazio & Williams (1986) investigated how the subjects perceived the candidates for the 1984 U.S. presidential election according to their attitudes toward the candidates.

cognitive memory structures called schemata also support this information processing function of attitudes.[142]
- *Information recall from memory.* There is apparently no consistency among studies on the effect of recalling attitudinally congruent or dissonant information. However, a meta-analytical study by Roberts (1985) shows a "reliable but modest relationship between opinion and recall". Judd & Kulik (1980) demonstrated that attitudes facilitate the recall of attitude statements in that the persons either strongly agree or disagree, compared to information recall associated with moderate approval or disapproval.

Especially the *active search for attitude-relevant information* has a special meaning for product development. In situations where a person or a group is discouraged from questioning accepted "truths", a maladaptive behavior might be exhibited, for example when a product specification has been frozen upstream and cannot be considered invalid. Changing conditions, the emergence of new findings, or even the attainment of a finer level of details can occur after a project was started, and this can produce conflicting information. It may cause a individual (or group) to ignore the opportunity to perceive significant information and thus avoid failure, or to recognize a major product deficiency, thus jeopardizing the product's chances for success.

Frey (1986, 47) cites the revised edition of the work on cognitive dissonance by Festinger (1964), in which the conditions are specified under which dissonant information is desired:
- When dissonant information is perceived as easily refutable.
- When the dissonant information is useful for future decisions.
- It is assumed that in states of strong dissonance and *when revision of a decision is possible*, dissonant information is preferred.

Research by Frey (1986) investigated not only the conditions under which persons showed no selective preference for consonant information but also those in which they show a strong preference for dissonant information. Results (Stahlberg & Frey 1988a, 155) reveal that persons tend to expose themselves to dissonant information if:
- their cognitive system is robust and consistent (strong), such that they can integrate or disprove the new information with relative ease,
- the cognitive system is relatively weak, such that it appears more favorable to make provisions for changing it in the long term in order to bring agreement with existent (maybe overwhelming) dissonant information, thus creating a new form of consonance.

[142] Cf. Neisser's perception cycle on page 238.

These functions of attitude have a very strong influence on (i) *problem characterization*, (ii) the *search for solution and alternatives*, and (iii) *decision making*. In other words they directly influence three of the four stages of a general problem-solving process also used in product development. The fourth stage is (iv) the *implementation*, and even this stage is also directly influenced because the new information emerging as more things are known about a product can be subject to selective exposure.

Cognitive dissonance and the selective exposure to information reinforces the need for "healthy" persons to correctly perceive the influx of information that accompanies every product development process. Self-actualizers, who excel at balancing the various aspects of *attitude,* can make a vital contribution to the product development environment.

A tendency has been reported of engineers to deny not only the aspects of affect and conation, but also the (seemingly) nonrational, non-logical aspects of cognition. *Intuition*, for example, is a very important form of knowledge that is much more crucial in engineering than in science, as the solution to a problem might lie in taking a good guess. The same principle exists in science, but scientists must endeavor to explain what they see, and intuition is not sufficient for this purpose. *Tacit knowledge* is much more easily applicable to product design than to scientific pursuits, although it cannot be entirely excluded from the latter either.

Interestingly, engineers frequently express themselves in ways that comprise aspects of intuition. An engineer might say, for example, "I just don't *like* this solution", or "I *feel* that we are going the wrong way", or "How do I know? Well, it's an *educated guess*!". Officially, however, few engineers would use such nonrational terms, due in part perhaps to their training, or to what is expected from them: to be exclusively rational. To be accountable, they have to demonstrate logical reasoning in "scientific" terms. A possible consequence is pseudo-rational argumentation that merely serves the purpose of justification.

One point that should be carefully restated here is that, in order to achieve a product, it is necessary to be selective in (technical) information processing (cf. page 3). This leads one to reject certain arguments in order to be able to cope with the main, relevant ones. Failure to do this would cause developers to drown in the flood of information instead of concentrating on the project goals. The process of information selection only becomes a problem when it takes an "unhealthy" turn (e.g. becoming defensive instead of problem-centered).

4.2.5 Problem centering

Maslow's discourse on *problem centering vs. means centering in science* is of particular importance for product development (condensed from Maslow, 1954, pp. 13-21):

- The inevitable [emphasis][143] on elegance, polish, technique, and apparatus has the frequent consequence of a playing down of the meaningfulness, vitality, and significance of an issue and of creativity in general.
- Means centering tends to push into a commanding position in science the technicians, and the "apparatus men", rather than the "question-askers" and the problem-solvers.
- Means centering tends strongly to overvalue quantification indiscriminately as an end in itself.
- Means-centered scientists tend, in spite of themselves, to fit the problem to their techniques rather than the contrary. The beginning question tends to be Which problems can I attack with the techniques and equipment I now possess? rather than what it should more often be Which are the most pressing, the most crucial problems I could spend my time on?
- Means centering tends strongly to create a hierarchy of sciences (or in the scope of this work: engineering disciplines).[144]
- Means centering tends to compartmentalize the sciences too strongly, to build walls between that divide them into separate territories. ... [The expert then becomes the] one who *knows* rather than the one who is *puzzled.*
- Means centering in science creates too great a cleavage between scientists and other truth-seekers, and between their various methods of searching after truth and understanding.
- Means centering tends inevitably to bring into being a scientific orthodoxy that is in its turn creates a heterodoxy, associated with the following dangers:
  - it tends to block the development of new techniques,
  - it tends to limit more and more the jurisdiction of science,
  - it encourages scientists to be "safe and sound" rather than bold and daring.

---

[143] The word "stress" in the original was substituted by "emphasis" to avoid confusion due to the semantical ambiguity in English.

[144] Paraphrased by the current author.

- [Overemphasis][145] on methods and techniques encourages scientists to think that:
  - they are more objective and less subjective than they actually are,
  - they need not concern themselves with values.
- Because of [the] fancied independence of values, standards of worth become steadily more blurred.

It is straightforward to formulate this effect of means-centering on product development, but its use in practice is much more difficult because it implies assessing whether an expert is problem-centered. An example of this is the investigation of possible applications of information technologies to the working world. Experts might take a problem-centered approach by using computers to solve essential problems. However, it is quite easy to switch to the approach of making changes (or *not* making the necessary changes) *because* of the computer, instead of *by means of* the computer. In other words, by asking "What does the computer allow us to do", experts might be taking a meaningful step toward solving a problem, or they might be forcing the use of an available technique. The distinction is often not clear, even if it might be apparent in concrete cases of business reengineering or communication improvement (centered around the computer) (case #32).

Can, for example, a computer expert endeavor to improve computer processing speed by being exclusively problem-oriented (i.e. by succeeding in enhancing the processing speed), exclusively technique-oriented (i.e. by thinking only in terms of the computer science discipline), or both?

Mean centering is not restricted to the professionals who "possess" the discipline. In business reengineering, there are plenty of cases where economists and economy students, not the engineers, have focused their work on the computer. They are the ones notorious for advocating the use of a "docking port, where a vendor would download the actual data to his/her laptop computer" and for overly automated solutions to achieve the person-less handling of goods (case #33).

It is also very difficult—if not impossible—to distinguish between the expert who deals with issues within his/her field of expertise in order to be experienced and ready when this knowledge is to be applied (as this is indeed what is expected from an expert), and the expert who thinks only of his/her own technique or discipline as an end in itself.

---

145 The word "overstress" in the original was substituted by "overemphasis" to avoid confusion due to the semantical ambiguity in English.

4.2.6 Self-actualizers in development

In a survey of attitudes and career expectations, Maitra & Whittaker (1993, 256) requested engineers to indicate the three most important motivational factors: 63% of the respondents gave job satisfaction the highest rating, followed closely by salary (59%), and responsibility (46%). Other factors indicated were autonomy in job (30%), organizational prestige (20%), job security (15%), authority (15%), and creativity (30%).

Investigating the professional needs of engineering personnel,[146] Thamhain (1983) asked subjects to indicate their most important professional need to perform effectively. The characteristics described as "very important" were as follows:

- interesting and challenging work (70%),
- professionally stimulating environment (65%),
- professional growth (62%),
- overall leadership (58%),
- tangible rewards (55%),
- technical expertise (50%),
- assistance in problem solving (48%),
- clearly defined objectives (45%),
- management control (44%),
- job security (42%),
- senior management support (40%),
- good interpersonal relations (35%),
- proper planning (32%),
- clear role definition (30%),
- pen communication (25%), and
- minimum of changes (15%).

The rank-order was roughly the same for all three categories of engineers surveyed. The first two items illustrate the basic motivation intrinsic to engineers.

Not every product developer or person working in product development is a self-actualizer. However, this kind of activity tends to attract people who demonstrate these characteristics. If the evaluating system or the persons using the system have only one orientation, namely that of the *economic man*, than the situation is not only anachronistic but conflicts with the well-being and performance of the development team. A trend has been reported by the International Research Institute that good technical people are hard to

[146] Field study of data collected from 150 non-managerial engineering professionals, 120 engineering managers, and 35 senior engineering managers employed by 75 technology-oriented companies.

find in technical jobs. They attribute this finding to an increasing shift to economical emphasis, which had led to a shift in personal goals and consequently to a variety of attractive offers open to employees, not only in terms of wages. Similar trends have also been reported by leading engineering universities throughout the western world. By assuming the *complex man* (cf. Image of man, Chapter 1), there has to be an evaluation system that considers the entirety of aspects to maintain a healthy environment for the diversity of persons in it, including the persons that are not typical self-actualizers.

There has to be a coherent system of evaluation and progression that reconciles individual freedom and autonomy with commitment and attainment of results, that allows for progress and change. Such a system must also allow for progress in knowledge while guaranteeing the attainment of results for the company. If the situation of increased pressure to achieve shorter times to market persists, there is no opportunity to refresh and enhance one's professional knowledge. Companies will then lose valuable assets that can be difficult to replace as self-actualizers may disappear from the product development market, or at least become *rara avis*.

4.3 Fear and courage

Curiosity, exploration, and experimentation are common elements of scientific pursuits as well as product development. The amount of what is known about the product under development or the disciplines used to address pertinent issues are dependent on these characteristics. Maslow's work on the Psychology of Science (1966) asserts that these elements "can be seen to have the primary goal of allaying anxiety. What looks behaviorally like an interest in the nature of the object being examined or the area being explored, may be primarily an effort by the organism to calm itself down and to lower the level of tension, vigilance and apprehension. The unknown object is now primarily an anxiety-producer, and the behavior of examination and probing is first and foremost a detoxification of the object, making it into something that need not be feared. Some organisms, once reassured, may then go over into an examination of the object per se out of sheer, nonanxious curiosity .... Other organisms may, however, lose all interest in the object once it is detoxified, familiarized, and no longer fearsome. That is to say, familiarization can produce inattention and boredom." (op. cit., p. 20-21).

When cognition is "primarily deficiency-motivated, it is more need-reductive, more homeostatic, more the relief of felt deficit." When it is "more growth-motivated, it is less need-reductive and more a movement toward self-actualization and fuller humanness, more expressive, more selfless, more reality-centered" (op. cit., p. 22). Maslow also made the distinction between the Freudian *defense mechanisms* "for allaying anxiety while still seeking gratification" and what he called *coping mechanisms* "for positive, courageous, and victorious solution of life problems in the absence of anxiety or in spite of it" (op. cit., p. 22). The main difference (for product development) is that Freudian defense mechanisms lead to cognitive inefficiency, in addition to their other effects (op. cit., p. 29). Finally, "growth was seen as an endless series of daily choices and decisions in each of which one can choose to go back toward safety or forward toward growth. Growth must be chosen again and again; fear must be overcome again and again" (op. cit., p. 22)

It is opportune at this point to differentiate between research and development. In product development, the search for knowledge usually reaches the point where the product function can be guaranteed. Very often—although some might be offended by this statement—product developers cannot investigate an item down to the roots of the effects being explored.

Situations in which developers use techniques or concepts that have a different orientation than science research are found for example in Vincenti (1990). Situations in which the complete theoretical explanation is available only after years of utilization by the industry are described in the work edited by Whyte (1975). This differentiation will not be treated in detail here due to space constraints. It is mentioned only to emphasize that even though scientists and engineers may operate at different levels to pursue "knowledge of the object", the contrast between the two approaches still remains applicable.

In product development, the primary goal is of course to end up with a product (not knowledge). This might lead the developer to eliminate any issues that "menace" his/her design to the extent that the status is *dependable*. After that, the designer will approach the next pressing issue, as there are usually many. In contrast, a researcher will typically delve more deeply into a subject and endeavor to develop a theory to explain the phenomena observed.

Finally, the terms "research" and "development" become convergent again in many development programs. Finding an explanation or theoretical foundation of pertinent techniques is an essential part of many development processes and sometimes, this effort characterizes a research activity. On the other hand, the point "menacing" the researcher effort might be omitted in favor of other priorities within the research project. The difference between this situation and the "anxiety-driven" one described by Maslow are better found by emphasizing the word *fear* used by Maslow.

It is not the purpose of this work to investigate cases in which an artificial environment instability is created to provoke changes, especially the rare cases in which fear might be a possible (or maybe even the only) solution to organizational survival. However, it must be said clearly that induced defensive mechanisms (including the Freudian defense mechanisms) lead to cognitive deficiencies as reported by Maslow and weakens decision-making and implementation efforts.

The behavior in the simplest schema can be a result of two forces: problem-centered (coping) and anxiety-allaying (defensive) (op. cit., p. 22).

Maslow also distinguishes between deficiency motivations and growth motivations. The following table summarizes the differences in cognition between the two.

Table 20. Differences in cognition according to primary motivation.

PRIMARILY DEFICIENCY-MOTIVATED	PRIMARILY GROWTH-MOTIVATED
more need reductive	less need-reductive
more homeostatic	more a movement toward self-actualization and fuller humanness
more the relief of felt deficit	more expressive
	more selfless
	more reality-centered

According to Maslow, growth can be regarded as a virtually endless series of daily choices and decisions. At any given point one can choose to return to "safety" or go forward toward growth (op. cit., p. 22). He also states that a safety philosophy or religion or science is more apt to be blind than a growth philosophy, religion or science (op. cit., p. 24).

Many forms of cognitive pathologies might emerge from anxiety-allaying mechanisms. Stemming from a study of brain-injured soldiers performed by Kurt Goldstein, Maslow describes this kind of pathology as follows:

Box 42. Goldstein's brain-injured soldiers.

Their very real injuries and the real losses in capacity that ensued not only made them feel less capable but also made the world look more overwhelming. Much of their behavior could be understood as an attempt to retain self-esteem and to avoid anxiety-producing confrontation with problems from which they could expect only defeat. To this end they first of all narrowed their worlds in order to avoid problems that they were incapable of handling and to restrict themselves to the problems they were capable of handling. Within such constricted worlds, daring less and trying less, being 'modest' about aspirations and goals, they could function well. Secondly they ordered and structured these narrowed worlds carefully. They made a place for everything, and everything was in its place. They

> geometrized their little realms in an effort to make them predictable, controllable, and safe. Thirdly they tended to freeze them into static and unchanging forms and to avoid change and flux. Their worlds were thus made more predictable, more controllable, and less anxiety-producing.
>
> For people who have limited capacities that they cannot trust, who see the world as too much for them, and who can't accept this state of affairs, these are sensible, logical, understandable things to do. They work. The soldiers' anxiety and pain were in fact reduced thereby. To the casual observer the patients looked normal. (op. cit., pp. 23-24)

Managing development is very often a matter of reducing complexity. The classic KISS rule—"Keep It Simple, Stupid"—is frequently cited by project managers and engineers. To arrive at a product implementation, depending on the product's complexity, a project might be doomed to fail if its complexity and thus its level of uncertainty are not effectively reduced. However, an inept project manager may render the development environment hostile to the processes that support its resolution, e.g. by imposing an overly rigid reporting process or defining an inflexible structure. Although the project manager and the group might be able to survive, their efficiency will be significantly impeded, and the work satisfaction of highly competent personnel will invariably suffer.

Maslow also draws a parallel between such measures and the safety mechanisms exhibited by a compulsive-obsessive neurotic who, having lost faith in his spontaneous wishes and instinctive impulses, has to rely on external signals to tell him what to do and when to do it (calendars, clocks, schedules, agendas, quantifications, geometrizations, laws, rules of all sorts). "Since change, flux, and unexpectedness may catch him with his controls down, he must also lay out the future, program it, make it exactly, make it predictable. ...he diminishes the world so that he may be able to control it...he tends to live 'by the numbers', by the rule book, and to rely on external rather than internal cues, on logic rather than on impulse, intuition, and emotion" (op. cit., pp. 25-26).

Other primarily anxiety-instigated, clinically observed expressions were listed by Maslow (op. cit., pp. 26-29). Below is a partial listing of the cognitive pathologies of interest within the scope of this work:

Box 43. Cognitive pathologies according to Maslow (partial list).

- The compulsive need for certainty (rather the enjoyment and appreciation of it);
- the premature generalization that so often is a consequence of the desperate need for certainty;
- desperately and stubbornly hanging on to a generalization, in spite of new information that contradicts it;

- the denial of ignorance (for fear of looking stupid, weak, ludicrous), the inability to say: "I don't know", "I was wrong";
- the denial of doubt, confusion, puzzlement: the need to appear decisive, certain, confident, sure of oneself, the inability to be humble;
- the ability to be only active, dominant, masterful, controlling, "in charge", "masculine", and the inability to be also noncontrolling; noninterfering, receptive. This is a loss of versatility in the knower
- the intolerance of ambiguity: the inability to be comfortable with the vague, the mysterious, the not yet fully known;
- the need to conform, to win approval, to be a member of the group — the inability to disagree, to be unpopular, to stand alone;
- the need to be always and only rational, sensible, logical, analytic, precise, intellectual, etc. Inability to be also nonrational, wild, crazy, intuitive, etc., when this is more suitable;
- intellectualization, i.e. transforming the emotional into the rational, perceiving only the intellectual aspect of complex situations, being satisfied with naming rather than experiencing, etc. This is a common shortcoming of professional intellectuals, who tend to be blinder to the emotional and impulsive side of life than to its cognitive aspects;
- the intellect may be used as a tool for dominating, one-up-manship, or for impressing people at the cost of part of the truth;
- dichotomizing compulsively; two-valued orientation; either-or; black or white.

Maslow points out that "these 'good', 'nice' scientific words (namely: prediction, control, rigor, certainty, exactness, preciseness, neatness, orderliness, lawfulness, quantification, proof, explanation, validation, reliability, rationality, organization, etc.) are all capable of being pathologized when pushed to the extreme. All of them may be pressed into the service of the safety needs, i.e., they may become primarily anxiety-avoiding and anxiety-controlling mechanisms" (op. cit., p. 30). The difference between pathological and primarily growth-motivated behavior is that it is not neuroticized.

Recalling now the section on "Uncertainty and the nature of management" on page 110, let us examine the list of cognitive pathologies (Box 43 on page 182, especially the points – Loss of versatility of the knower, – Intolerance of ambiguity and – Denial of ignorance). We saw that the ability to cope with the issues associated with innovative product development can be jeopardized by the management principle of certainty, if it is not done in a healthy manner.

At this point it must be said that, as with Goldstein's brain-injured soldiers (Box 42 on page 181), making one's world more predictable, more controllable and less anxiety-producing might make life manageable in the short term, but it in fact poses a much greater obstacle to people who attempt to solve actual (technical) problems.

Maslow is not alone

The main reason to adopt Maslow's approach to the product development was triggered by the perception of engineering activities as potentially bearing numerous opportunities for personal achievements.

Maslow's approach should not be used in an isolated fashion for product development. The significant influence of contentual and contextual factors (cf. Herzberg, Mausner & Snyderman, 1959) and their effects on how intrinsic and extrinsic motivation determine quality should also be considered.

A concept by Argyris (1964) , as described by Ulich (2001, 48-49) unites several of these concepts. It proposes that "work motivation, problem-solving competency and mental health are supported by self-esteem based on 'psychological success'." Furthermore "an important condition for the psychological success is the ability to determine one's own goals in accordance with one's own needs and values and to bear the responsibility to exercise control over goals and means. This contradicts the structures of formal organizations, where individuals have only a minimum of control over their own working conditions, or are allowed to contribute only a few specific skills, thus rendering them dependent and lacking initiative.[xxxiv]

Consequently, only when organizations consider and accept that employees wish to apply their capabilities within the scope of the established entrepreneurial goals and to be included in the decision-making process will employees develop their full potential. In other words, only if employees are treated like adults, will they behave like adults. If companies are structured after a different image of man, employees will behave accordingly: dependent, less interested, with a short-range perspective—like in an earlier development stage. Independence in thinking and action-taking are precluded by defensive behavior."

4.4 Orientation to perceiving and solving problems

4.4.1 Problem avoidance

Von Diemer (1994) describes the following conflict of motives that affects an individual's motivation to strive for quality.

Box 44. Conflict of motives after the occurrence of errors according to von Diemer (1994, 1067).

When performance decay, errors or wrong behavior occur, it is unfortunately not always immediately disclosed by the participants, so the possibility of minimizing the consequences or initiating corrective measures is therefore limited. The reason for this is a series of internal, contradictory motives that lead to an internal conflict situation. [...] In this situation, the conflicting motives can be illustrated as follows:

Desire to work properly	←+→	Admit or announce the error
Desire for recognition	←+→	Fear of reprehension
Effort not to cause a negative impression	←+→	Fear of falling from grace
Desire to be popular	←+→	Fear of rejection

Depending on one's personal experience and learning biography, these conflicts are handled differently:

Cover up. Wait and hope the error will never be discovered.

Guilt assignment or justification. When the error is discovered, reasons for the error are sought in order to demonstrate that it could not have been avoided: The material was defective, the previous department caused the error, the documentation came too late, etc.

Rationalization. A suppression pattern is developed to cope with the situation. A determined posture and attitude allows the individual to avoid perceiving the conflicting situation as such.

– *'I just care about the production rate, not the quality. This is the job of the quality department'.*
– *'To err is human'.*

This attitude is frequently interpreted by bystanders as indifference, but it is actually a defense mechanism for self-protection.

Another defense mechanism reported by John McCarthy, then vice president of engineering for the Apollo spacecraft at North American, was described as follows: "A lot of people had developed a fortress mentality — 'us against them'— and the subsystems teams had become parochial. They weren't looking beyond their own fences, and they were saying, 'That's the other guy's problem.' McCarthy proposed a cure. He called it Project Rock, because everybody was hiding behind a rock." (Gray, 1992, 192).

From these descriptions, it appears that the illustration in *Figure 11*, although obviously a joke, indeed contains a kernel of truth. Various translations of it can be seen on the walls of technology companies throughout the (western) world.

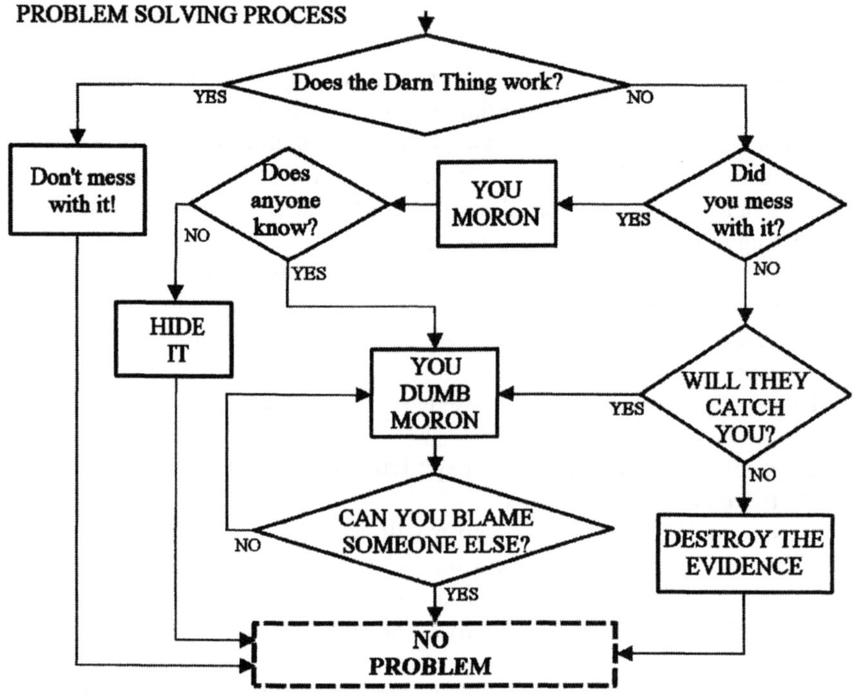

Figure 11. Problem-solving process joke

Of course, jokes are jokes and cannot be generalized to the real world. In the real world, however, the mechanism of guilt or problem avoidance exists indeed and affects the attainment of quality, not only in product development.

4.4.2 Automatic responsibility

Perhaps as a natural reaction to the inability to solve a hopeless situation, some companies simply tolerate the occurrence of the behavior described in the previous section. There is an alternative, however. Recalling the lessons learned from the Apollo program, the Marshall Center used a different coping strategy. According to Tompkins (1977, 11), von Braun declared that the laboratories "have full cognizance and responsibility for all efforts that fall within the purview of their respective disciplines, including active projects, further project studies, and supporting research work. ... [The lab director] is expected automatically to participate in all projects that involve his discipline and to carry his work through to its conclusion." This concept, called *automatic responsibility*, "had long been a part of the organization's method of operation. In practice it meant that, say, a mechanical engineer working in one of the labs, assumed automatic responsibility for any problem he perceived which fell within his area of competence, regardless of whether or not his laboratory had been given a formal task assignment related to it. He was to stay with the problem until its solution was reached. I heard stories from the past about cases such as the young American-born engineer who assumed responsibility for a problem and recommended a solution opposed by his superior and most of the technical personnel at the center. In the end he was right; he was subsequently rewarded with a top management post. If the person who perceived a problem lacked the technical ability to see it through to its solution, he assumed the responsibility for communicating his perceptions up the line. Top management, thus alerted, could direct specialists to give attention to the problem" (brackets and italics maintained).

Tompkins elaborates on this concept calling it "a radical departure from Weber's Germanic 'ideal type' of bureaucracy; ironically the departure was also the product of the Germanic mind. It is such a radical innovation that I have yet to find an executive in a comparable bureaucracy—whether it be business, government or the academic world—who felt he could administer an organization by such a principle. They feared that the result would be anarchy. It is difficult to imagine a Detroit assembly line operating with automatic responsibility. (It could be, however, that in human terms there is something wrong, not with the concept of automatic responsibility, but with the assembly line concept.) The concept assumes that people can be encouraged to aggressively seek responsibility" (op. cit., p. 11). Meanwhile, since the publication of Tompkins work, the Detroit assembly lines might indeed be using this concept originated in the space industry, which in turn might have discarded it in its quest to lower costs.

4.4.3 Independent verification and validation

As mentioned above with respect to software development, it is widely accepted that human beings are limited in their ability to be error-free in complex technical ventures. Conventional programming wisdom holds that software is never debugged. Indeed, it has been published in the computer literature as one of the "Ten biggest lies in computing" to declare that a program is ever error-free. An experienced programmer at IBM once quipped "If you haven't found a bug in your program, you're looking in the wrong place."

In the review process, the lone programmer usually overlooks errors. Hence it is widely accepted among programmers that an independent reviewer can often find the errors in a program better and faster than the program's author. A technique called "software walkthrough" allows error debugging and the critical review of a program implementation and has the effect of shortening maturation time and enhancing the quality of the results.

However, if a programmer feels threatened during the review process to find (hidden) errors, a natural reaction is to defend himself. His efforts would be channeled into surviving the attack, not into finding the errors and its causes. If someone perceives this process as hostile, it is to be expected that this process will be avoided instead of sought. An attack against the program might be taken as a personal attack against the program*mer*, even if the programmer himself denies it. Weinberg (1971, 56-58) describes how an independent software review process can be constructively implemented in what he calls "egoless programming". He traces this approach back to computer pioneer John von Neumann. A personal acquaintance of von Neumann said that "he was constantly asserting what a lousy programmer he was, and that he incessantly pushed his programs on other people to read for errors and clumsiness."

Box 45. Case of an "egoless" programmer's openness to criticism.

Weinberg describes the case of a programmer who matured a code of 13 machine instructions to be used as the kernel of his simulation program and submitted this code to another programmer, who found 17 errors in his 13 statements. Instead of becoming desperate, he became more and more amused as the errors were being disclosed. The reviewer herself had internalized the program and she in turn sought a critique. Three additional errors were found. When the code was finally installed, no errors were found. It is interesting to note is that the original programmer was "giving everyone an enormous laugh at his expense" instead of defending himself in the face of being confronted with his errors, which in fact no one was doing maliciously.

This is not an isolated case. There are many variations on this theme, and many such cases are experienced even by programming students. In his evaluation of the software development process for the American Space Shuttle, Feynman confirms that this technique was used. In an appendix to the presidential report investigating the Challenger incident, he described the software of the general computer system as "highly elaborate" (Lewis 1988, 215). He also praised, as an outsider, the advantages of using this independent verification method in achieving a well-functioning program (Feynman 1988).

However, this does not mean that the verification process is a pleasant situation for the developers, as the next example illustrates.

Box 46. Moment of truth (case #34).

Programmer A is an electrical engineer who worked with software development in 1985. He experienced the problem that his program menus[147] blocked from time to time during development testing. He solved the problem by applying the finite automata theory he had used for hardware development.

Programmer B informally "demonstrated" one of his programs to Programmer A in 1989. As Programmer B spoke about the program implementation, Programmer A (still listening) sat at the computer and started entering "dumb" commands into the program (e.g. entering a '7', when the requested inputs were in the range from 1 to 5). Noticing this, Programmer B became uneasy and asked why he was doing this. Just then, the program blocked. "Voilà," said Programmer A.

A productive discussion followed concerning the incompleteness of the definition of the logic machine. Programmer B had fallen prey to the common mistake of trying to program *the user* as to the "only" correct way to use the program: his way. Users, however, are notoriously creative in using programs in ways other than the programmer intended or assumed. The two programmers discussed the possibility of solving this problem with the technique mentioned above, with which Programmer B, also an electrical engineer, was familiar.

One year later, Programmer B was working with Programmer C, a computer engineer. The latter was discussing the first version of one of her programs with him, and he did exactly the same thing as Programmer A had done with him. Programmer C became not only uneasy, but angry. Moments later, however, she understood the point.

---

[147] Note to users of modern computers: A "menu" at that time was an implementation of a screen with options to be chosen with numbers or letters typed by the user. Pull-down menus were already in existence at that time but the simple menu was the easiest form of implementation.

> Meanwhile, Programmer A was in a software engineering course, in which a program had to be written and formally "delivered" to a "customer" played by the lecturer (Programmer D). Programmer D tested the menus in exactly the same way. Programmer A had tried to conceive a "fool-proof" program, and he of course understood what Programmer D was doing and was in complete agreement with this approach. Later, however, he admitted to having felt slightly insecure and apprehensive. He paid acute attention to what was being entered, and watched the program's reactions to better know how to correct the errors. He knew that, in spite of his care, a program bug can always be discovered, and that it usually appears in this "moment of truth". He knew well that "software is never debugged".

Both in formal as well as informal situations, such software testing evokes uncomfortable feelings, even when the program developers are conscious of the nature of and reason for the test. In this "moment of truth", much is at stake. Feynman's brief evaluation of software development testing for the American space shuttle using similar techniques suggest that the programmer's "reputation is on the line" (Feynman 1988, p. 193).

This "egoless" characteristic can be misleading and has to be further defined. Driskell & Salas (1992) investigated the effects on team performance of the "tendency to attend to task inputs from others in an interdependent manner". They called this *collective behavior*, as opposed to the "tendency toward self-sufficiency and rejection of influence from others", which they called *egocentric behavior.* The way people coordinate, evaluate, and utilize information from others is hypothesized to be dependent on such behavior. In the first phase of their research, they classified their population based on a task with ambiguous results with two possible choices, where the probability of choosing either answer was close to 0.50. The subjects made an initial choice and then received feedback from the other team member. Faced with initial score disagreements, the individuals were confronted with the decision of keeping their initial choice or changing it. The result measured the individual's tendency to stick with one's own choice and provided a measure of the population characteristics. The population was then trisected around the mean value (in their population 0.66). This led to the classification according to (i) an egocentric group, (ii) a collectively oriented group, and (iii) an altercentric group (the latter having a greater tendency to accept the other's input). The third group was dropped and the study proceeded to the second phase, where subjects were given a non-ambiguous task with one correct and one wrong answer. The gain in performance of the two groups was then measured. In a third control group, the individuals received no information. Analysis of variance showed *no significant change of scores for the egocentric group* with the mean gain in performance of 0.36 for 25 trials. There was a *significant gain in performance from initial to final score of the collectively oriented team*, with a mean performance gain of 1.71.

Comparing these two gains, there is a relative improvement in performance gain of almost 5 for the collectively oriented group. For the control group, there was no significant performance gain for the individual performers with the mean gain in performance of 0.39. The low difference between the egocentric and the control groups' performance gains indicates that the egocentric members benefited no more than those who performed the task alone. In addition, by evaluating the latency time between the initial and the final answer, the collectively oriented group took significantly more time to choose a disagreement trial than an agreement trial, reflecting the additional cognitive processing for reevaluation and decision. The egocentric group did not show a significant difference in response latency for agreement and disagreement situations, thus suggesting that they simply ignored the other's input. There was no problem-solving content in the task and there were no requirements for traditional skills such as math or reading in order to dissociate performance with previous experience.

This study has a direct implication on the way a given design is perceived. It is the first step toward improving quality, a central issue of product development. It confirms the basic differences in the verification process that takes place in product development.

Hoffmann & Strickland (1995) reviewed the effects of task orientation and ego orientation as follows.

Box 47. Task involvement and ego involvement (condensed from Hoffmann & Strickland (1995, 496-497).

Task involvement is typified by an approach in which the individual is focused on task mastery and understanding (cf. Nicholls, 1984). It has been typically associated with adaptive attributional, motivational, and performance strategies. Specifically, task involvement, or related constructs, has been associated with the following outcomes:

- a focus on the intrinsic value of learning (Butler, 1987),
- a stronger belief that effort leads to success (Ames & Archer, 1988),
- a preference for more challenging tasks (Elliott & Dweck, 1981; Nicholls, 1984),
- the use of more effective task performance strategies (Dweck, 1989; Wood & Bandura, 1989),
- high task satisfaction/ interest (Butler, 1987; Duda & Nicholls, 1992).

Ego involvement (Nicholls, 1984) is typified by individuals who attempt to demonstrate an ability either to themselves or, more likely, to others. Individuals with this orientation tend to gauge their success relative to a normative reference group. Faced with low performance on a task, people who are ego-involved and uncertain of their ability to perform the task tend to display maladaptive behaviors (Nicholls, 1984). Ego involvement has been shown to be associated with the following outcomes:

- positive feelings after low-effort successes (Jagacinsky & Nicholls, 1984),
- the use of less effective task performance strategies (Nolen, 1988),
- a preference for less challenging tasks (Elliott & Dweck, 1981; Nicholls, 1984), and
- a tendency to engage in task-interfering cognitions (Hoffmann, 1993; Saranson, 1975).

4.5 Stress and performance

4.5.1 A brief field report of stress

Development managers have reported the need to impose stress (they used this word) on development personnel to make them work as desired. Surprisingly, these managers could not specify exactly what they meant by "stress". They evidently simply put pressure on their personnel, even when they knew this would not deliver the desired result. Sometimes they achieved what they wanted, sometimes they were pleasantly surprised, and sometimes they achieved nothing but frustration of themselves and the others involved.

There are indeed cases in which a manager achieved success by placing personnel under "stress", but there are just as many cases in which this approach was not a bit more successful. In fact, the manager often became quite frustrated knowing a person's potential was much higher than the current performance. They overlooked the threshold of exhaustion, the well-known burn-out effect, the symptoms of which vary from a difficulty to concentrate, recognize obvious solutions, express something in writing, sometimes even orally, and hypertension. In one such company, the in-house physician revealed an alarming situation. He declared that, without even leaving his office, he could tell exactly what was momentarily happening with the company, which department was under stress (case #35).

Managers who use the principle of putting pressure on their development personnel do not seem to know how far can they go with it. They seem not to understand the potential consequences of this practice.

Some development engineers and researchers recognize a productive effect of pressure. They put themselves under pressure, for example, to achieve good results, explaining that "when things are too loose, it is very easy to relax and leave the things as they are". Asked whether they would not prefer an absence of stress, they answered that "it would be nice, but it doesn't seem to work".

Box 48. Two cases of too much pressure on technical personnel.

A development engineer put pressure on a colleague to produce a solution to a problem. It happened that this person was already under pressure, and had been so for quite a while. Although the issue at hand was fairly trivial, involving simple computations, the latter was simply overwhelmed. His colleagues were surprised at his inability to provide an adequate solution. Instead of recognizing the adverse effect of the pressure imposed, the first engineer insisted that pressure was necessary in order to get things done. He overlooked that it was entirely counterproductive in this case (#36).

Another case involved an engineer supervising a technician. The engineer noticed that there were many technical details with which the latter was not familiar, information regarding the limitations of the equipment used for development testing. In the engineer's opinion, this knowledge constituted important information for the technician to perform his tasks. Noticing that the latter was evading discussions, he reacted by *increasing* the pressure in an attempt to foster interest in the pertinent information. In fact, however, he did not achieve this because the technician had a different point of view and did not "get his message". With this behavior, the technician was discouraged even more by the engineer's efforts, leaving them both dissatisified. Another person who reported to the same engineer but not directly involved in the situation seriously considered diplomatically maneuvering the supervising engineer out of the picture (case #37).

A potentially catastrophic situation arises for professionals working on problem-solving activities, such as those faced in product development, when their supervisors advocate the application of pressure without knowing exactly what processes this sets in motion. A development-oriented company, for example, had many highly talented, "achievement-oriented" engineers with the finest engineering education in the region. With their higher-than-average standards, they placed their colleagues under stress, who were perhaps not as talented or had a different orientation. The latter insisted that these high achievers could stress themselves if they wished, but should not do this to others. There was no measurement system to assess the extent of the problem when this occurred.[148]

To conclude this brief field report, the degree to which the "mechanisms" underlying performance under stress is frequently unknown to technical professionals and their managers, and this can be quite detrimental. Pressure and conflicts are found to be used more or less indiscriminately in the technical workplace. We shall examine some of the development-related aspects in the following discussion.

4.5.2 Stress and cognition

Bernstein, Roy, Srull & Wickens (1988, 472-503) presented an overview of Stress and Coping that is condensed here for the purpose of the following discussion.

Box 49. Stress and cognition (condensed from Bernstein, Roy, Srull & Wickens (1988, 472-485).[149]

Stress is the process of adjusting to circumstances that disrupt, or threaten to disrupt, a person's equilibrium. It involves a relationship between people and their environment.

Stressors are events and situations to which people must adjust.

Stress reactions are the physical, psychological, and behavioral responses people display in the face of stressors.

"To a considerable degree, the impact of stressors and how people react to them depend on the circumstances in which the stressors occur and on each person's physical and psychological characteristics. This is why stress is defined, not as a specific occurrence, but as a *process*".

There are some reported cases of stresses being induced by desirable events, but one is more likely to find unpleasant events and situations—those involving frustration, pressure, boredom, trauma, conflict or change—associated with stress:

- *Frustrating* situation contain some obstacle that stands between a person and his or her goals.
- *Pressure* situations require a person to do too much in too short a time.
- *Boredom* or understimulation is the opposite of pressure, but it too can be a stressor, especially if it continues for a long time.
- *Trauma*, a shocking physical or emotional experience, can create what is perhaps the ultimate stress situation.

---

[148] No statistical data is available regarding worker dissatisfaction or frustration.

[149] Condensed from: Bernstein, D. A.; Roy, E. J.; Srull, T. K. & Wickens, C. D. (1988). Psychology, Boston, etc.: Houghton Mifflin Company. © 1988 by Houghton Mifflin Company. Reproduced here by kind permission of Houghton Mifflin.

- *Conflict* is almost always stressful. Not only interpersonal conflict can be stressful. Internal motivational conflicts can be equally, if not more, distressing than the first.
- *Change*, whether positive or negative, can be a major stressor.

Cognitive stress responses typically appear as reductions in the ability to concentrate, to think clearly, or to remember things accurately. One of the most common cognitive stress responses is catastrophizing, or dwelling on and overemphasizing the consequences of negative events. Such cognitive stress reactions can be especially severe for those of moderate ability or those who are most uncertain about how well they will do. It can not only interfere with the smooth cognitive functioning necessary for coherent speech but can actually intensify emotional and physiological arousal, which in turn adds to the total stress response and further hampers performance. Cognitive stress reactions to stressors can themselves create additional stress responses, so that having thoughts about failure may help bring about failure.

Other cognitive reactions to stress act to minimize the impact of stressors, at least temporarily. For example, some people automatically (and perhaps unknowingly) respond to stressors with denial, repression, intellectualization, displacement, projection, and other cognitive strategies that Freud called defense mechanisms. Defense mechanisms may cushion the emotional impact of stress, but they do little or nothing to eliminate the source of stress.

Fortunately, many people routinely display rational and constructive cognitive reactions.

4.5.3 Individual tolerance to stress

The concern with individual differences in coping with stress grew after World War II and the Korean War, leading to increased research during the 1950s. According to Lazarus & Folkman (1984, 7), the "psychologists who were involved in this research often cited a universal law propounded by Yerkes & Dodson (1908), the so-called inverted U-shaped curve in which increments of arousal or drive tension improved task performance up to a certain level, beyond which increasing disorganization and performance impairment resulted." Lazarus & Eriksen (1952, cited in Lazarus & Folkman, 1984, 7) found that a "marked increase in variance instead of an average increase or decrease in performance effectiveness under failure-induced stress." This is an important point to remember, especially by those involved in the situation described in the box on page 193.

4.5.4 Coping with stress

Even abandoning the idea that intentionally provoking stress in organizations is a useful strategy, stressful situations nevertheless arise. This reinforces the need to provide individuals with forms of coping with it.

In the framework of stress in organizations, McGrath (1976) considers it the interaction of three conceptually independent "systems": (i) the physical and technological environment; (ii) the social medium, or patterns of interpersonal relations; and (iii) the "person system" or "self-system" of the focal person whose behavior is to be considered. These systems interact or intersect (op. cit., p. 1367) and are of particular interest in studying the R&D organization. *Figure 12* illustrates the model.

According to McGrath's framework, " 'Stress' can arise from various origins, and these origins map in a rather straightforward way to the three embedding systems and the three two-systems intersects." The six identified "classes" of stress or sources of stressful situations (which are not entirely disjoint) are (op. cit., p. 1369):

- Task-based stress (e.g. difficulty, ambiguity, load).
- Role-based stress (e.g. conflict, ambiguity, load).
- Stress intrinsic to the behavior setting (e.g. effects of crowding, of understaffing).
- Stress arising from the physical environment itself (e.g. temperature, hostile forces, noise).
- Stress arising from the social environment, in the sense of interpersonal relations (e.g. interpersonal disagreements, privacy, isolation).
- Stress within the system contributed by the person in question (e.g. anxiety, perceptual styles).

Coping with all these kinds of stress is very important for achieving good results in modern product development environments. Especially as barriers between departments are being dismantled—at least in the organization diagram or room layouts—organizational stresses have to be considered in an adequate fashion.

From the literature review on organizational stress performed by Lester (1983, 16) covering the period 1974-1982, the outcomes of organizational stress,[150] included:

- absenteeism and turnover;
- lowered satisfaction, tension, propensity to leave;
- job-related tension and dissatisfaction, decreased perceived effectiveness, increased negative attitude to role-senders;

[150] The review was not specific to the R&D setting. A particularization is made here.

- anxiety-irritation, depression, job dissatisfaction, psychosomatic symptoms;
- frustration, anxiety, psychosomatic symptoms;
- anxiety, over-assertion of authority, withdrawal;
- emotional distress, medication use, cardiovascular, gastrointestinal, and allergy-respiratory disturbances;
- devalued self-conception.

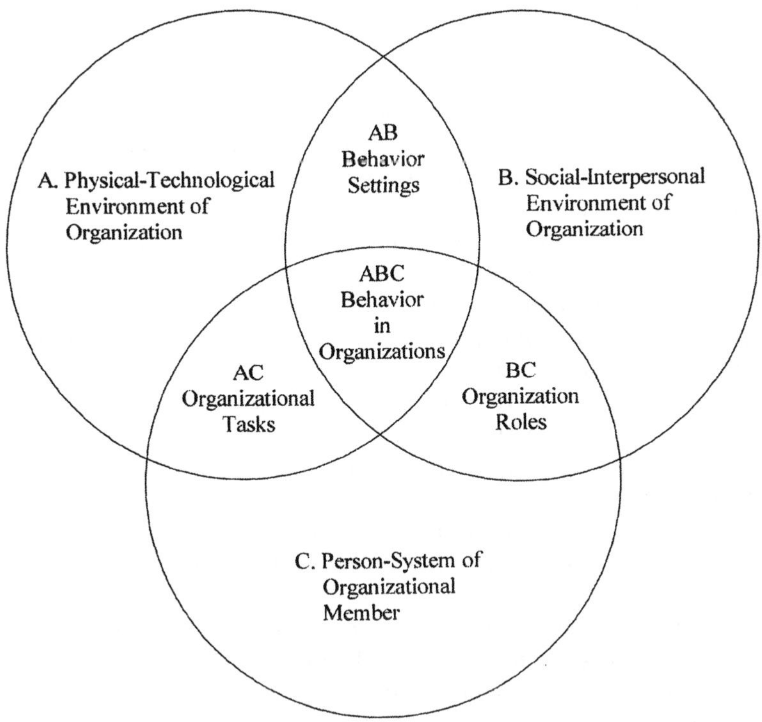

Figure 12. Three overlapping systems for behavior in organizations according to McGrath (1976, 1368).[151]

In addition, the work of Janis & Mann (1977) considers the effects of stress that might adversely affect decision-making, depending on the coping strategy used.

---

[151] From: McGrath, J. E. (1976). Stress and Behavior in Organizations, in: M. D. Dunnette (Ed.), Handbook of Industrial and Organizational Psychology, pp. 1351-1395, Chicago: Rand McNally. Reproduced here by kind permissionof Marvin D. Dunnette.

4.5.5 Concluding notes on stress

The main purpose of this discussion was to describe some situations in the realms of product development that are related to stress, coping with stress, and the problem-solving processes. The extent to which this issue is known to practitioners may deserve a detailed inquiry. In the opinion of the current author, little is known about the practitioner's perspective of the consequences of stress and interpersonal differences in coping with it. It would be valuable to investigate the effects of stress on development and development-related personnel.

In a formal conversation, a specialized consultant was speaking about the concept of "benchmarking" as the best measurable standard. It involves, he explained, taking a result from a worldwide survey, cutting that value in half[152] (e.g. cutting the development time in half), and taking the resulting value as a new goal for enhancing performance. Then the procedure is repeated again, and again. A first curious aspect of this explanation was the idea of performing the cycle indefinitely, although he recognized that this is only theoretically possible (it would tend asymptotically to zero). Second, the explicit mention that this 50% cut is done without any consideration of the limiting factors in the associated processes seemed odd. One merely has to set a reduction goal, and people have to find a way of accomplishing it. Perhaps schedule compression was attempted, but in any case, the stress imposed on individuals to perform the task under such constraints was not even mentioned.

In this context, let us refer to the loss of the Mars polar lander (MPL) at the end of the previous century. Contact with the lander could not be reestablished after descent. The Jet Propulsion Laboratory (JPL) Special Review Board attributed the most probable cause as being a spurious signal on the hall switches during leg deployment. This caused the wrong "interpretation" of ground contact by the computer, and this in turn triggered a premature engine shutoff. In addition, descent telemetry had been eliminated after having been classified as "non-mission-critical." This also eliminated any possibility of knowing the performance of the novel propulsion system that was to be used in future missions.

More important than the technical reasons for the failure, the JPL board mentions that MPL was requested for approximately one-half the cost of Mars Pathfinder, which in turn had been deployed for a considerably lower cost than previous planetary probes. This resulted in minimal project staffing and negatively influenced key technical decisions for which there was "minimal involvement by JPL technical experts." At the same time,

[152] Or doubled, depending on the defined variable.

Lockheed Martin Astronautics "used excessive overtime in order to complete the work on schedule and within the available workforce". The Review Board verified that "much of the development staff worked 60 hours per week, and a few worked 80 hours per week, for extended periods of time. Another consequence of the tight funding constraint was that many key technical areas were staffed by a single individual". All this resulted a "breakdown in inter-group communications, and there was insufficient time to reflect on what may be the unintended consequences of the day-to-day decisions" (Casani et. al., 2000, 6).

Such can be the consequences of undue stress.

4.6 Creative tension

Ewing (1964)[153] refers to concrete cases in which the effective use of tension made the performance and evolution of product characteristics possible. His article considers the consequences of tension for organizations. He cites Harvard psychiatrist Henry H. Babcock's distinction of three forms of tension, "often related, but varying in emphasis". *Table 21* presents his description (op. cit., pp. 72-73).[154]

Table 21. Forms of tension according to Babcock and reported by Ewing.

TENSIONS ARISING PRIMARILY FROM	DESCRIPTION
Inner conflicts	This form is closely related to the concept of anxiety.
External factors	This form may be created by hurdles in the organization structure, for those who seek to surmount them; by management goals, for those who seek to accomplish them; and by similar obstacles that delay the satisfaction of an individual's drives and desires.
A need to be creative	The gifted mind and the special talent need a matrix in which to play out their natural functions. This form of tension can be subdivided as follows: (a) Tensions which prepare for, accompany, and sustain a virtuoso performance of any kind, (b) Tensions of an intellectual or mental nature, which accompany and sustain a more prolonged act of mental creativity, such as analyzing a complicated situation or writing a book. This form of tension operates best when free of tensions from all other sources.

153 Reprinted by permission of Harvard Business Review. Excerpt from: Ewing, D. W. (1964). Tension Can Be an Asset, Harvard Business Review, Vol. 42, No. 5, pp. 71-78. Copyright © 1964 by the Harvard Business School Publishing; all rights reserved.

154 Note from Ewing: Comments in a panel discussion of "Talent and Tension," Fifth Annual Human Relations Forum, The Connecticut Mutual Life Insurance Company, Hartford, October 24, 1963.

It should be mentioned here that this is a fairly old typology that should be taken primarily as a description of the organizational context in which it was used. Ewing refers to the second and third types as "healthy" tensions,[155] and to the first type as "unhealthy" tension. He made explicit in his article that this consideration was taken "from a management point of view". A generalization of this issue will be presented below.

Ewing's findings lead him to associate the constructive side of tension with benefits for the organizations because it takes place namely through:

1. *Learning* "in the sense of learning, understanding and even 'internalizing' knowledge of the problems of other managers. A classical form of such conflict is marketing versus production" (op. cit., p. 73).
2. *Enthusiasm* –"...stated, simply as a matter of common observation, that the organizations and parts of organizations where enthusiasm is greatest are those that feel a strong sense of mission" and that tension would act to ignite enthusiasm to cope with the conflict situation (op. cit., pp. 73-74).
3. *Creative performance* – in that differences in perception, standards, and expectations would act as a stimulus to imagination and performance. "The tensions a manager feels from lack of agreement with other managers and the nagging suspicion that he may be partially right can be effective goads (!) in inducing him to work out new ideas" (op. cit., pp. 74, parentheses added).
4. *Vigilance and self-examination* –"It has to do not with men's relationships with one another but with their adjustment to impersonal conditions and trends." Ewing proposes the tension between the individual and her/his environment as a condition for growth, although recognizing it as being no guarantee that s/he will actually grow (op. cit., p. 75).
5. *Accommodating value conflicts* – "Management decisions repeatedly involve inconsistent needs and principles.... A person who does not see these conflicts – who sees one set of values as completely right and any inconsistent values as completely wrong – will not, of course, feel any tension. Nor will he try to weigh the choices in his approach to administration. On the other hand, the person who does sense the conflicts will indeed feel tension. The result is that he may try to accommodate the conflicting values, weighing the equities of each in his decisions from time to time and letting each influence him. Thus, while he may at times suffer the agonies of the damned, he will develop a broader, more penetrating, more mature outlook. There are many examples of these conflicts: the profit motive versus the community

[155] Both psychological and managerial points of view of this notion of "health" would be questionable today.

service motive, gentleness versus harshness, self-interest versus the organization's interest, and others. One of the most elusive ones is competitiveness versus cooperation." (op. cit., p.75-76).

The generalization of the positive side of tension was investigated in research and development organizations by Pelz & Andrews (1976) , who found that scientists and engineers reached high technical achievements under conditions that seemed antithetical.

Their study, initially published in 1966, involved 1,300 engineers and scientists in 11 research and development laboratories. They used objective measures of performance based on number of patents or patent applications, number of unpublished reports, and number of professional papers as dependent variables. *Table 22* summarizes their findings.

Table 22. Eight creative tensions according to Pelz (1967).[156]

SECURITY	CHALLENGE
Tension 1	
	Effective scientists and engineers in both research and development laboratories did not limit their activities either to pure science or to application but spent some time on several kinds of R & D activities, ranging from basic research to technical services.
Tension 2	
Effective scientists were intellectually independent or self-reliant; they pursued their own ideas and valued freedom...	...But they did not avoid other people; they and their colleagues interacted vigorously.
Tension 3	
(a) In the first decade of work, young scientists and engineers did well if they spent a few years on one main project...	...But young non-Ph.D.s also achieved if they had several skills, and young Ph.D.s did better when they avoided narrow specialization.
(b) Among mature scientists, high performers had greater self-confidence and an interest in probing deeply...	...At the same time, effective older scientists wanted to pioneer in broad new areas.
Tension 4	
(a) In loosest departments with minimum coordination, the most autonomous individuals, with maximum security and minimum challenge, were ineffective ...	...More effective were those persons who experienced stimulation from a variety of external or internal sources.
(b) In departments having moderate coordination, it seems likely that individual autonomy permitted a search for the best solution...	...to important problems faced by the organization.

[156] Reprinted with permission from: Pelz, D. C. (1967). Creative Tensions in the Research and Development Climate, Science, vol. 157. no. 3785, pp. 160-165. © 1967 American Association for the Advancement of Science.

SECURITY	CHALLENGE
Tension 5	
Both Ph.D.s and engineers contributed most when they strongly influenced key decision-makers...	...but also when persons in several other positions had a voice in selecting their goals.
Tension 6	
High performers named colleagues with whom they shared similar sources of stimulation (personal support)...	...but they differed from colleagues in technical style and strategy (dither or intellectual conflict).
Tension 7	
R&D teams were of greatest use to their organization at that "group age" when interest in narrow specialization had increased to a medium level...	...but interest in broad pioneering had not yet disappeared.
Tension 8	
In older groups that retained vitality the members preferred each other as collaborators...	...yet their technical strategies differed and they remained intellectually combative.

Among the findings are reported tensions between:

- science vs. application,
- independence vs. interaction,
- individual vs. organization,
- influence given vs. received,
- security vs. challenge.

This work provides the generalization of the single-case description of the "tension arising primarily from a need to be creative" in *Table 21* to the research and development context. Here, the utopian stress-free environment to create, idealized by many, can be clarified on the basis of two seemingly contradicting statements that a development group should be either "granted extensive freedom" or "kept on tight reins" (cf. Discussion beginning page 13). Actually, this is not an isolated question of autonomy, the characteristics of which were explored in the corresponding section beginning page 137. Instead, it is one of establishing a level of creative tension that serves, among other things, to link individual activities to a whole, a question of tension, which is very often misinterpreted as stress, or pressure. This is the difference between a group "without the ladder" and one "under a whip" (page 13). Care should be taken not to generalize "tension" to "stress" or another of its forms: "conflicts".

4.6.1 Conflict and conflict solving strategies

The perception of product implementation characteristics and strategies varies among professions, specializations and depending on the individual career experiences of the individuals involved. This means that conflicts will inevitably occur. Some argue that conflict is not a bad thing, that creative forces can arise from it and new solutions found. This *can* be the case, but it depends strongly on the conflict-handling strategy used.

4.6.2 Pressure and challenge

Further differentiation between stress, pressure, and challenge and their expected effects on R&D activities is needed. Hall & Lawler (1971) surveyed directors of 117 research[157] organizations, 22 of which were selected for interviews with random samples of engineers and scientists.[158] They identified three kinds of pressure people feet most acutely: (i) time pressure, (ii) financial responsibility, and (iii) quality pressure.

Following the interviews, the respondents filled in a questionnaire to indicate their work attitude concerning satisfaction and involvement. Satisfaction was measured on a scale yielding scores corresponding with the five levels of Maslow's human needs. The needs regarding satisfaction in safety, social esteem (sense of belonging and love), autonomy and self-fulfillment[159] were rated and an *overall lab satisfaction* score was obtained by averaging the scores of the individual researchers. *Job involvement* was also measured by using an attitude scale to reflect the "extent to which the person psychologically identifies with and invests himself in the work" (op. cit., p. 68).

Organizational performance was measured by two variables. *Global laboratory performance* was rated by the research director, who evaluated the performance of the technical employees[160] on a six-point percentile scale. *Objective laboratory performance* was measured by summing the scores of the following indicators: net change in R&D budget during the past year, number of new internally funded projects, percent of project's meeting schedule, number of contracts renewed, percent of project's meeting cost budget, global technical performance, global administrative performance, ratio of contracts won to contracts proposed (the two scores were related to each other).[161] The significant relationships[162] found between the variables are depicted in *Table 23*.[163]

[157] The authors use the term "research", but from their text, it appears that the population studied was predominantly in product development.
[158] Averaging 13 persons per organization.
[159] The category of physiological needs was dropped for this study.
[160] The authors believe this was less threatening than rating the organization itself. They expected that the directors would evaluate others rather than themselves.
[161] No data for this correlation was supplied.

Table 23. Relationships between pressure and performance, involvement, and satisfaction (Hall & Lawler, 1971).

	QUALITY PRESSURE	FINANCIAL RESPONSIBILITY PRESSURE	RESEARCHERS' JOB INVOLVEMENT
Objective lab. performance	0.37 *		
Researchers' job involvement	0.56 ***		
Researchers' autonomy satisfaction		0.40 *	
Global lab. performance	0.40 *	0.54 ***	0.43 **

* $p<0.10$, two tail; ** $p<0.05$, two tail; *** $p<0.01$, two tail

Interesting to note is the correlation between financial responsibility performance and the global lab. performance measure[164] as well as the satisfaction of the professionals' need for autonomy. According to the authors, this suggests that "sharing in the financial responsibilities of the organization increases the individual's sense of freedom from the control of the organization" (op. cit., p. 69).

Time pressure was "virtually unrelated to the effectiveness and attitudinal measures" (op. cit., p. 69). This point was not elaborated upon by the authors. Time pressure was the most common pressure found, reported by 77% of the companies, albeit no assessment of the amount of time pressure was made.

Taking the study by Hall & Lawler and the findings by Pelz & Andrews, Andrews & Farris (1972) noticed that the performance of scientists in the population they analyzed tended to be higher for those working a nine- or ten-hour day on average. Performance declined for an eight- or an eleven-hour day. This yielded a curvilinear relation of time pressure and performance to the extent that working hours are dictated by time pressure[165] (op. cit., p. 368). To avoid potential misunderstandings, it must be made explicit at this point that no evidence was found in the study report indicating the ideal duration of a working day or to physiological

[162] The other findings were dropped from this integration because limitations of space do not allow the pertinent curves to be shown. Summarizing the findings in one numerical indicator would require too long an explanation. In order to avoid a misinterpretation of this study, it was decided to omit them.

[163] Only the correlations significant at the $p<0.10$ level or better were published.

[164] Presumably, the correlation with the *objective laboratory performance* measure was not significant to $p<0.10$ or better.

[165] This is important contextual information present in the subject's working culture.

considerations. The reader is advised not to take these numbers as if carved in stone.

As described in the work of Katz & Kahn (1978, 451-455),[166] the classification by Miller (1960) of the responses to information input overload include the filtering of information: "Filtering or the selective receiving of information can be adaptive if it is set by priorities assigned by the organization and based on an assessment of organizational needs. But without thoughtfully established guidelines, filtering is likely to be maladaptive. People are apt to process the familiar elements in a message, which they readily understand and which do not constitute major problems for them. Under time pressures the parts of the communication difficult to decode are neglected for the more easily assimilated parts, even though the former may be more critical for the organization" (Katz & Kahn, 1978, 453). Complemented with the findings by Pelz & Andrews that, when having too loose a coordination and high autonomy, scientists and engineers were likely to be low performers, possibly due to the low stimulation and/or motivation caused by the absence of pressure (Tension 4, *Table 22* on page 201). Together, these findings suggest an inverted-U-shaped curve for the dependence of time pressure on scientific performance.

Andrews & Farris then investigated the effects of time pressure in a NASA laboratory working on the effects of extreme physical conditions on various materials in 1965 (Time 1) and five years later (Time 2) (the apex of the Apollo program).[167] Work content accommodated a mixture of research, development, and technical services. The information concerning time pressure[168] was ranked as follows:

1. Typical level of time pressure experienced.

---

166 Initially published (first edition) in 1966.

167 The population firstly consisted of 117 scientists and engineers. In the second evaluation, the population mounted to 118 professionals, 78 of whom had also participated in the first investigation. Performance measures were indicated by criteria including innovations (increased knowledge in one's field through lines of research or development which were useful or new), productiveness (increased knowledge along established lines of research or development or as extensions or refinements of previous lines), and usefulness (useful or valuable in helping one's R&D organization carry out its responsibilities) and were independently assessed by an average of 4.4 judges at Time 1 and 7.6 at Time 2. All judges claimed to be familiar with the evaluated person; 2/3 of them were supervisors (including the immediate supervisor) and 1/3 were senior-level nonsupervisors. Based on the interjudge agreement and the average number of judges, the reliability of the performance ratings were estimated to be 0.95 at Time 1 and 0.88 at Time 2 (Spearman-Brown method) and the performance measures were combined into a single score for each quality and each member. The data was adjusted to compensate for the effects of the respondent's length of experience, seniority and formal training.

168 Ranging over a five-point scale from relaxed: "no pressure at all", to extreme pressure: "I'm behind on important schedules".

2. Amount of time pressure the respondent felt would be optimal.
3. Difference between the actual and the optimum.
4. Span or range of time pressures actually experienced.

The respondents' level of experienced time pressure is shown in *Table 24*.

Table 24. Median level of time pressure experienced (Andrews & Farris 1972, 372).

MEDIAN LEVEL OF PRESSURE	TIME 1	TIME 2
relaxed	27 %	39 %
slight pressure	33 %	30 %
moderate pressure	32 %	24 %
great pressure	7 %	7 %
extreme pressure	1 %	0

Almost all professionals indicated that they wanted at least some pressure. Curiously, those who experienced higher pressure were also the ones who wanted more (Andrews & Farris 1972, 372).

A cross-comparison of pressure experienced at Time 1 and the usefulness performance at Time 2 showed a correlation of r=0.49 (Pearson correlation). For the other performance measures, lower correlations were found: r=0.21 for productiveness and r=0.25 for innovation.

A linear relationship between time pressure experienced and the subsequent usefulness measure was also revealed by their data. Scientists who indicated having "relaxed" conditions at least half of their time had an average performance level of 34%. Professionals who reported having "great" time pressure for at least half the time scored an average performance of 74% (op. cit., p. 375). This does not mean that the more pressure, the better one's performance, as the overload data showed. *Figure 13* presents these findings. Here, Andrews & Farris confirmed the inverted-U shaped curve of the relationship between time pressure and performance.

The authors attributed the relationship between greater pressure and better performance to their findings that professionals subjected to higher pressure were also the ones who preferred high pressure (op. cit., p. 379). An interpretation of the results revealed by this article indicate that the challenges faced by scientists and engineers also lead them to invest more to obtain more results from the time spent at their work, thus reflecting the situation of creative tension that allows improvements in cognitive processes and better results for the organization.

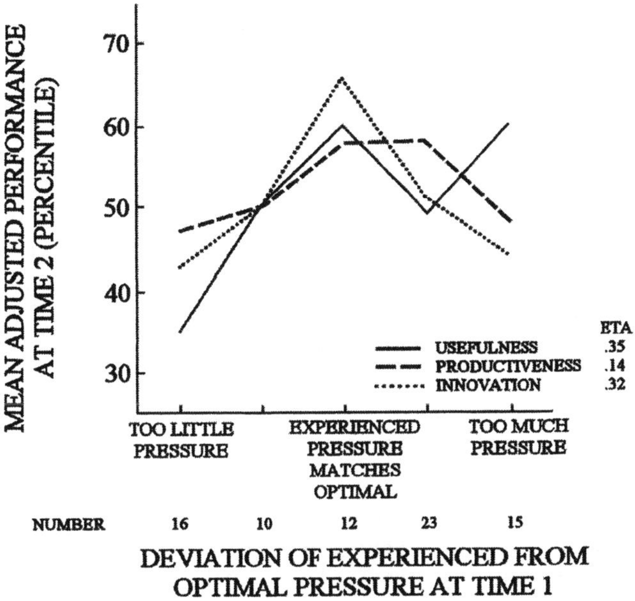

Figure 13. Inverted-U shaped relationship between time pressure and performance (Andrews & Farris 1972, 379).[169]

The authors do not provide an explanation of the curious increase in the usefulness curve. Based on the dates in which the measurements were performed, it is hypothesized here that, in the context of the American moon program, some of the professionals in the population studied were presumably subjected to extreme pressure to accomplish the presidential goal that was achieved in 1969. When the second evaluation was performed in 1971 (Time 2), the scientists and engineers studied might have been benefiting from the hard work performed in the former high-pressure situation. There are no means of verifying this hypothesis from the published data, nor is there any indication from this study that this result might extend indefinitely over time. Indeed, it is to be expected that, over larger periods of time, scientists and engineers may reach a state of exhaustion and their performance would drop abruptly. The duration of the high performance is

---

[169] From: Andrews, F. M. & Farris, G. F. (1972). Time Pressure and Performance of Scientists and Engineers: A Five-Year Panel Study, Organizational Behavior and Human Performance, Vol. 8, No. 2. Reproduced here by kind permission of Academic Press, an imprint of Elsever Science.

believed to be dependent on individual differences and will not be covered in this work, again due to space and time limitations.

At this point, a cross reference should be made to the work of Amabile & Gryskiewicz (1987) described above (cf. p. 143 ff.). It investigated the environmental qualities that foster creativity. Among the 120 R&D professionals interviewed, 12% of them *mentioned* the importance of creative time pressure, following the sense of challenge reported by 22%. Both are coupled to the preceding quality of having sufficient time *to think creatively* indicated by 33% of the interviewed personnel. Moreover 33% mentioned *insufficient time to think creatively, too great a workload within an unrealistic time frame* and *high frequency of "fire fighting"* as environmental qualities that inhibit creativity.

4.7 Concluding Remarks

The discussion started with the dirty-hands philosophy. This notion is related to establishing and updating a measure of reference, and is an essential element of discussing development details with external developers. This approach is becoming increasingly important, as very few companies develop products entirely independently any more. A considerable amount of discussions with experts working externally to a given group or company is necessary. The dirty-hands philosophy also refers to an equally important motivational aspect for engineering professionals.

We then examined the importance of a class of individuals described by Abraham Maslow as self-actualizers and their importance in the product development process. The effects of problem-centeredness, perception of reality, tolerance of ambiguity and lack of structure, ability to cope with fear of the unknown were presented. Attitude and information processing was discussed, in particular the effect of cognitive dissonance at the individual level. Maslow's approach is not the only one to be applied to product development.

We then discussed psychology of science issues to describe the effects on the same processes found in product development as well as the contrast between safety science (or safety development) and growth science (or growth development).

Coping with errors was also approached, and the importance of the concept of *automatic responsibility* was presented. The need for independent verification was also touched upon as a means to improve the quality of development. However, this should not be seen in absolute terms.

The effects of stress and performance were then discussed.

Finally the concept of creative tension was presented. This concept is linked to the concept of autonomy approached in Chapter 3.

Chapter 5

INTERPERSONAL ASPECTS

5.1 Introduction

The following classifications were presented by Ulich in his 1990 lecture.

The functions of the group can be ordered in:
- performance function,
- protection function,
- control or normative function.

Prerequisites for group work are:
- common purpose,[xxxv]
- manageable size,[xxxvi]
- role differentiation,
- common territory,
- mutual understanding among diverse members of equal status,[xxxvii]
- consciousness of membership.

Regarding accessibility, groups can be classified into:
- open groups and
- closed groups.

"Closed" can mean, for instance, that a group disallows changes of its members, stays together, resists external influence, or is homogenous. These characteristics can render a closed research and development team resistant to innovation, for example.

On the one hand it is true that working groups often provide solutions to many problems, in contrast to the individualistic approach. Issues reported in R&D environments include the tendency of individual experts to give insufficient or no attention to problems created elsewhere and the tendency to attach disproportionate importance to one's own discipline. Furthermore, there is an observed propensity of individuals to restrict solutions to a local optimization, thus incurring non-optimal global product characteristics. These tendencies can be compensated by group work.

There are numerous studies in the scientific literature on group performance that focuses on the conditions under which a group (i) outperforms its best member, (ii) provides a solution equal to that of its best member, and (iii) provides a solution inferior to that of its best member. However, these are partial results dependent on task content and context. Consequently, this very promising approach to product development should be applied under consideration of whether a group or an individual would produce the best performance.

Popular business literature is full of success stories that propagate the generally accepted idea that group work is *the* solution to every task. Indeed, many companies and individuals "advertise" the alleged successes achieved in this way. This is a perilous generalization as it implies that *any* group would function better than *any* individual undertaking any task in any situation. Yet the difficulties of implementing group work compared with individual work should not be overlooked.

There are many reported cases (in the late 1990s) where companies have faced difficulties implementing group work because they did not first reflect on whether group work would make sense for the given activity, whether it would contribute to coping with professional challenges. Moreover, it is not uncommon for managers to implement, say, an open space for all employees to work in, believing they are fostering better communication by removing barriers and creating "unity". Although this can of course be successful, it can also be an illusion, with people communicating as before, the barriers remaining invisibly in place, and negative by-products such as increased perception of noise, bothersome distractions, and the uncomfortable feeling that too many eyes are watching one's working habits.

Note that, although the myth of the lone engineer/scientist working in the isolation of the lab is no longer the accepted norm, it should be kept in mind that the need for introspection and concentration to implement a program, to perform an engineering analysis, to write a technical report, to debug hardware, and many other activities in product development calls for a certain isolation that such office arrangements do not allow.

5.2 Advantages of developing in groups

5.2.1 Task typology and group performance

Group vs. individual

The task division of Steiner (1972) as described by Shaw (1976) and Wilke & van Knippenberg (1988) serves as a point of departure for this discussion. Steiner makes a distinction between:

Additive tasks, where the outcome of the group process is the result of a combination of individual products. Examples of this type of task are a group pulling a rope or applauding music performers: the more participants, the greater the effectiveness.

Conjunctive tasks, where every member contributes to the success of the group. An example of this is a group climbing a mountain. If one fails, the group fails—as a group—to succeed.

Disjunctive tasks, where a group chooses a solution from the individual evaluations proposed by its members. The tasks require a choice among several possible alternatives. This may be a simple yes/no choice or the selection of the correct solution from a larger number of possible alternatives.

Wilke & van Knippenberg describe the subdivision of disjunctive tasks into:

Disjunctive tasks with eureka, where the solution, when found, is apparent to all members, and

Disjunctive tasks without eureka, where the solution is not evident to all members, even after it has been found. This distinction is necessary, not because of the solution, but because of the nature of the problem. The difference is that for the eureka type, the solution is clearly recognized by each member as the correct solution,[170] whereas for the non-eureka type, it is frequently the case that the member who proposes the solution cannot "sell it" to his/her group.

[170] An example given by Steiner is the missionary-cannibals riddle, where a group of three cannibals and three missionaries are to cross a river in a boat that can carry only two persons. All three missionaries can row, but none of the cannibals can. For safety reasons, no missionary should be left with a majority of cannibals. The riddle is to find out how many times the river has to be crossed to transport everybody to the other side. The answer, once known, is perfectly clear.

Steiner also specifies the typology in the following terms:

Divisible tasks, where it is possible to subdivide tasks and assign the parts to different persons.

Unitary tasks, where the task cannot be subdivided. Example: two persons are to read a report and draw conclusions about lines of action.

Compensatory tasks, the results of which are the average of the individual evaluations. This kind of task, not present in the typology of Steiner, is of interest in estimation activities, where the individual errors in estimating can be compensated (by the averaging). As Wilke & van Knippenberg (op cit., 348) commented, this averaging approach works only when there is no availability of information about the competence of the members and it can be assumed that the individual errors can be compensated for by the averaging. When members have a greater influence (either hierarchically or technically) or when someone refuses to give his or her estimate, the probability of error grows.

In terms of these typologies, *Table 25* illustrates differences of group performance:

Table 25. Group performance according to task typology (Wilke & van Knippenberg 1988, 351).

TASK	GROUP PRODUCTIVITY	DESCRIPTION
Additive	Better than the best member's	Group performs more than the best member
Compensatory	Better than the majority	Group performs more than the major of its members
Disjunctive w/ eureka	Equals the best member's	Group performance corresponds to that of the best member's
Disjunctive w/o eureka	Worse than the best member's	Group performance can correspond to the best member's performance but frequently is lower
Conjunctive unitary	Equals the worse member's	Group performance corresponds to the performance of the weakest member
Conjunctive divisible	Better than the worse member's	Group performance can reach high levels when the subtasks can be distributed according to the members' capabilities

Characterizing product development activities, one finds all these task types. Examples: (i) An order to implement (separate) drawings for 500 parts could be implemented faster and maybe with better quality by a group than by one individual (*additive*). (ii) The estimation of the performance increase of a given parameter by changing the form of implementation in a product can be assessed by statistical polling of the experts involved (*compensatory*). (iii) If a recognized "star" is present, a more "humble" member might be reluctant or refuse to contribute (*compensatory*). (iv) For cases where solving development problems is finding the optimum by means of calculations and exact reasoning, some of the solutions presented can be mathematically or unequivocally clear to the members (*disjunctive with eureka*). (v) A solution is not clearly definable in terms of a optimization (such as considerations of ergonomy of human-machine interfaces) (*disjunctive without eureka*). (vi) There is a relative "fuzziness" of the choices (*disjunctive without eureka*). (vii) PERT diagrams are a good illustration of the presence of *conjunctive* activities in development programs.

5.2.2 Group performance and task types

Task typology is not unique. For the purpose of product development, another task typology defined by Hackman (1968) in the study of "intellective" activities is of interest:

(Idea) production tasks: where the production of ideas, images or arrangements is performed. This type will be called "production" tasks (in quotes) to preclude confusion with the terminology already adopted here. Readers should be aware that it does not refer to manufacturing.

Discussion tasks: requiring a discussion of values or issues, usually linked to group consensus.

Problem-solving tasks: in that a solution for a specific problem should be worked out within given constraints.

In this study, the 108 groups performing three tasks types with three levels of difficulty revealed the relationship of task type and results as depicted in *Figure 14*. The reader should note the difference between the profiles of performance according to task type.[171]

The same types of tasks appear in the activities of product development. For example, producing design concepts falls under the "*production*" task category, whereas the search for alternatives may fall into the categories of "*production*" and *discussion* tasks. The same may apply to product modifications. Analyzing the adequacy of a proposed solution and its company-wide consequences requires the group interaction of a *discussion* group. Coping with unexpected problems in the convergent – divergent subprocesses (cf. page 93) occurring during product implementation bears

---

[171] *Action orientation* means the degree to which a product states or implies that a specific or general course of action should be, might be, or will be followed.

Optimism measures the degree to which the general point of view or tone of a product can be characterized as positive or optimistic.

Issue involvement means the degree to which a product takes or implies a particular point of view regarding some goal, event, issue, value or procedure.

all three elements. *Problem-solving* tasks are attributed dogmatically (i.e., axiomatically) as belonging to engineering activities.

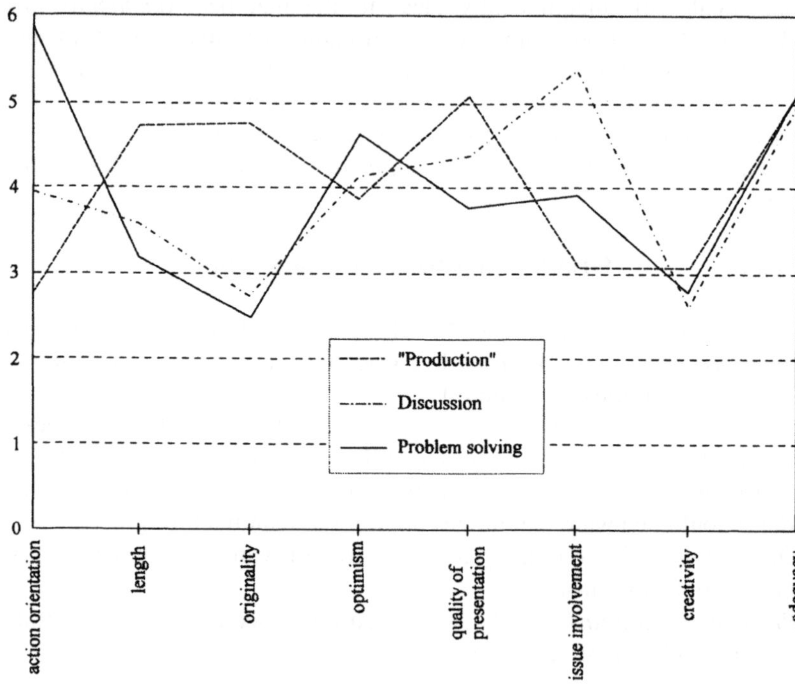

Figure 14. Effects of task type on group performance.

Hackman shows that different task types lead to different distributions of results.[172] As straightforward as it might appear, having success with implementations of group work is much more than just "forcing" it as a management fashion, and task type is just one of the essential aspects of teamwork. A similar misinterpretation of the appropriateness of group work was described by Hackman & Morris (1975, 46): decision makers should not simply *assume* that the additional human resources will lead to a higher-quality product or even reduce the likelihood that the product will be

[172] The operationalization of the study refers to "intellective" tasks performed by undergraduate students on written tasks. This situation is coherent with the essence of the activities in working out ideas, discussing and implementing product characteristics. The particular interest of this study for the purpose of product development is that it could separate the type of tasks, what is not easy to accomplish in a running product development environment.

"grossly defective". Brooks (1982) (the mythical Man-Month) also reports the misconception of time reduction in resource reallocation and the dependence on the task at hand (cf. page 103).

The above-mentioned laboratory studies illustrate the phenomena behind group work. They are an attempt to isolate the influence factors under controllable conditions, but applying this task typology to the product development environment is not straightforward because the task assigned to the group may not necessarily fall into just one task type and may vary with the momentary situation.

One example from the civil construction branch reinforces the difficulty of labeling tasks according to types. A construction company owner reported the following case, which took place in the late 1990s. No sooner had bricklayers finished building a wall with a structural function, an electrician came in and milled a groove diagonally from corner to corner to install an electrical conduit between the two surrounding walls. He was completely unaware that he rendered the wall structurally useless. The manager solved the problem by institutionalizing a daily meeting, at which each group member would tell what he or she intended to do next. With this very simple provision, the group was able to detect this kind of problem early enough for intervention.

Although aircraft and spacecraft are not designed and built the same way as buildings, the anecdote described here refers to insufficient information to practice one's own discipline without interfering with others'—a ubiquitous problem in product development. The task of electrical wiring might be solved by other means in aerospace, but this does not preclude the occurrence of similar failures (even if they might not be technically similar).

A well-known case in point was that involving the incorrect orbit injection of the Mars Climate Orbiter (MCO), which resulted in the loss of the spacecraft. Many people remember the recent case of MCO as being downed by units. The report by the Mars Climate Orbiter Mishap Investigation Board attributed the failure mainly to the use of *lbf-s* units in the ground software file "Small Forces" used in trajectory models instead of *N-s* (named by the report as the root cause). The output of this file was used in the Angular Moment Desaturation (AMD) file (Stephenson et al., 2000, Appendix 1,[173] 16). However, more relevant than the technical error, the Stephenson Report stated that one of the contributing causes of the loss of the spacecraft was that "the operations navigation team was not intimately familiar with the attitude operations of the spacecraft". This "caused the operations navigation team to perform increased navigation analysis to

[173] Appendix 1 is the "Mars Climate Orbiter Mishap Investigation Board, Phase I Report", dated Nov 10, 1999 by the same investigation board.

quantify an orbit determination residual error" that "was masked by the lack of information regarding the actual velocity change imparted by the AMD events. A line of sight error was detectable in the processing of the tracking measurement data, but its significance was not fully understood" (contributing cause #2, op. cit., App. 1, p. 18). Furthermore, the "system engineering process did not adequately transition from development to operations. There were a number of opportunities ... to identify the units problem ... it also resulted in inadequate contingency preparation process to address unpredicted performance during operations" (contributing cause #4, op. cit., App. 1, p. 20). In addition, "inadequate communications between the project elements, including development and operations teams... the project management and technical teams, and the project and technical line management" (contributing cause #5, op. cit., App. 1, p. 21) were cited in the report (cf. Discussion on the importance of communication processes on page 248ff.). The MCO case also emphasizes the dangers of delineating the task types too severely: "The operations navigation team came onboard shortly before launch and did not participate in any of the testing of the ground software ... [and] did not participate in the Preliminary Design Review nor in the Critical Design Review process. Critical information on the control and desaturation of the MCO momentum [caused by the asymmetrical solar array] was not passed on to the operations navigation team." (op. cit., App. 1, p. 18,19). Indeed, this example is quite similar in principle to the building construction example.

Blindly classifying participants along the lines of, say, designing tasks, building tasks, and operating tasks might lead similarly to a failure to "engage operations personnel early in the project, preferably during the mission formulation phase" (text inside quotes is actually Lessons Learned of MCO #2, op. cit., p. 17) and a repetition of the "lack of early and constant involvement of all project elements throughout the project life cycle", e.g., the "inadequate communication between the development and operations teams." ("Specific Inadequacies of Communications #2" of MCO incident, op. cit., p. 21).

Referring again to the case of the bricklayers and the electrician above, even their tasks were not exclusively additive. Recalling the case of the value of the experience gained by operating the Z4 relay computer and its effects on the development of the ERMETH described on page 24, the importance of studying the history of technology should be clear.

Regarding task typology, care should be taken when using multifunctional product development teams, as these groups have the following characteristics:

- high complexity and/or interconnectivity of problems they face,
- usually small in size,

– their task types change many times (sometimes in the same day).

An attempt to divide a group's tasks along these lines of task typology would mean subdividing the tasks into overly basic elements. This would introduce the same kind of problems that Taylorism did in the workplace. Adding the elements of complexity and interconnectivity, problem coping may turn out to be overwhelmingly unbearable. In addition, it would require that the group-forming phase be redone every time the task type changes, and it would cause the number of individuals to grow. In other words, it would produce all the problems that the solution originally attempted to avoid.

Poor application of the task typology to product development job design serves to illustrate that it is not intended here to label-and-divide (cf. Discussion of Cartesianism, page 45; and excessive subdivision of work elements, page 101) with the subsequent task allocation, but to differentiate tasks into a set of activities to be assigned to an individual, while being aware of the diversity of elements present in the teamwork concept. Teamwork should not be misunderstood as *the* solution for *every* development problem. This awareness alone should serve to prevent false expectations regarding the results of group work.

The question remains as to how group work can enhance the product development process and what is needed to achieve this enhancement.

The following comparison between individuals and groups was formulated by Sachse (1995, 70) and is reproduced here as broad statements without the original sources cited by the authors.

- When the activity requires the inclusion of as many different ideas and experiences as possible and knowledge is distributed among many persons, then group decisions are superior to individual decisions.
- The information recall from memory, the exactness and the extent of information is greater in the group than in individuals.
- In situations characterized by uncertainty, groups make better decisions than individuals.
- Individuals make better decisions in cases where each phase of the task requires partial decisions and a deeper approach to the problem.

Again, all these situations occur in the product development process.

5.2.3 Complementarity of viewpoints and experience

Heterogeneity in multifunctional teams might be the most advantageous aspect of group work in development. The different viewpoints and experiences bear the inherent potential of better results compared to single-disciplinary groups that tend to overvalue their (single) view of the issue.

However, one essential prerequisite for success with heterogeneous groups is that the channeling of the different and partially conflicting ways of seeing the issues to be addressed can be realized on a constructive basis, but this cannot be taken for granted.[174]

For example, a professor once tried to bring economics and computer scientist (graduate) students together to evaluate collectively the potential of information technologies to radically reorganize work. A situation evolved in which the economists asked: "What can you do?" and the computer scientists replied: "What do you want to do?". The economists repeated their question, the computer scientists repeated their answer, and group interaction reached an impasse (case #38).

The division of disciplines leads to differing perceptions of an issue, different ways of coping, different approaches (at least regarding terminology), and different priorities. Thus, a design engineer might claim that his or her product satisfies the requirement specifications, whereas a production engineer might say of the same product that its assembly requires too many operations, or that the specified components or materials might have a long lead time, or that no alternative source is available.

The challenge lies in how to reconcile these different views. It is not simply a matter of adding them, because their consequent interaction requires that the results be optimized. If various experts try to optimize their work individually, the global result would not likely be an optimization of the whole. In many cases, contributors must concentrate on the characteristics related to their field of expertise, and still allow room for the contributions of others. At best, this interaction will lead to yet a different solution than any of the contributors originally intended, and the results can be superior. This point was made by Mary Parker Follett as early as the 1920s.

The notion of bringing heterogeneity to product development optimization is therefore not new. In some organizations, product changes (e.g. due to prototype evaluation, new customer requirements, or product deficiencies) have been analyzed systematically by a multidisciplinary group of all pertinent contributors. The process of engineering review changes (ERC) or change requests (CR) has been formally defined, especially in the aerospace industry, and is a careful analysis of the implications of changes in the entire organization, sometimes even including the customer.

The issues inherent in exploring group heterogeneity may have been recognized by the industry long ago, but how to reconcile these different viewpoints remains a challenge. A company with a functional structure is also a "group" in the sense described above, including the criterion of

---

[174] Cf. Section on achieving cooperation, page 232 ff.

manageable size and role differentiation. However, it usually also implies numerous barriers among its subgroups.

5.2.4 Classical efforts to remove barriers

Well known since the publication of *Out of the Crisis* (Deming 1982, 62 ff.), the "barriers between staff areas" constitute an impediment to achieving overall product optimization. Since then, companies throughout the world have been endeavoring to break down these barriers.

New approaches to eliminate barriers between departments have led to experimentation with multifunctional groups, apparently with good results. Here, the focus for comparison is not on individual vs. group efforts but on comparing isolated functional groups working on a product and integrated groups working together toward a product. The idea behind this is that the inefficiencies inherent in functional departmentalization—in spite of its concentration of competencies, its homogeneity and economies of scale—are detrimental to agility and flexibility. Furthermore, the functional organization has a tendency to hinder the mutual understanding among the disciplines and may contribute to an ineffective use of time.

With respect to improving product quality there is an essential difference between the multifunctional group approach (such as simultaneous engineering) and the more conventional process of combining different functional departments to optimize a product. It is not the heterogeneity of a group by itself, but the predominance of object orientation versus method orientation.

As the functional departments contain the experts, the "depository of knowledge" in a given area, they may tend to act not to improve the product as a whole, but to showcase the merits of their group. If under pressure to justify their own existence, they may tend to optimize their methods, instead of optimizing the product as a whole. Multifunctional groups, on the other hand, tend to focus on the product. However, development professionals with an interest in keeping in touch with advancements in their areas, should realize that, when taken to the extreme, the object-orientation approach may encourage members to lose the competitive advantage they once brought to the group. Excessive object-orientation leaves one no time to explore new methods as they appear. This can mean that, even though a multifunctional group can achieve a better product in a shorter time, it is necessary to establish a balance between object-orientation and method-orientation. Of course, participants still work on their areas of expertise, which implicitly helps them stay abreast of their field. Nevertheless, providing the option of dedicating their efforts predominantly to their original domains or to another area of interest might prove very useful in achieving long-term results.

An issue associated with the single-disciplinary approach and departmentalization of organizations is that responsibility is attached to the technical competence of individual persons or departments, not to the ensemble of technical capabilities of the professionals involved. The barriers of the functional organization imply a formal structural conflict when one person or group is interested in the results of another group or department or in the consequences of their activities. In the past, to cross that barrier was very difficult; few dared even try. Instead, the usual way was to "throw it over the (cubicle) wall": then it was no longer one's own problem—the problem "disappeared". Many proponents of multifunctional project teams praise the advantages of eliminating these barriers in order to achieve better product quality.

Boeing instituted the design-build teams (DBT) during the development of the model 777 aircraft. These teams worked on the same principle of breaking barriers down. Again, this is not a new concept, as readers will recall from the section on Lockheed's Skunk Works (page 71). Thus Boeing confronted the throw-it-over-the-wall phenomenon. Sabbagh (1995, 21'25") describes it as follows: "Another innovation on the 777 would be a new method of organizing the 10,000 or so people building the plane. On past projects, airplane design had been bedeviled by communication problems between the many different divisions in the company. For the 777, Boeing would create new work groups called design-build teams or DBTs. They would include people who'd never before worked together."

These barriers are not necessarily within a company's own borders. As cooperation among companies, sometimes spanning various countries, is becoming pervasive in many industries in addition to aerospace, it is becoming increasingly important to reduce or eliminate the barriers between partner companies, customers, and regulating agencies. So it was with the latest development of Boeing: "Underlining the building of the 777 was a new kind of organizational philosophy dubbed 'Working Together'... For the first time, Boeing would be guided by principles that encouraged, even forced cooperation. First and foremost, there will be no secrets. Information will be shared openly, even when things went wrong" (op. cit., 41'). Some forty years separate the development of the 777 from that of the Polaris, but they share the same elements of the philosophy of utter communication instigated with the Polaris (page 67), of automatic responsibility used by Marshall Space Flight Center under von Braun (page 187), and of Lockheed's Skunk Works (page 71).

Neil Standal, 777 assistant general manager, reports: "We had to make some cultural changes...we had to open communications, we had to break down traditional barriers between organizations, we had to come up with forms where the people could speak up, more open communications, more frankness coming from all levels to management or vice-president". (op. cit., 41'30").

Group work can in this sense contribute significantly to overall product optimization. Such optimization can be likened to the "two prisoners" dilemma. Two prisoners are brought to judgment. Owing to a lack of evidence, they were each sentenced to two years' confinement. However, unbeknownst to the other, they are each offered a deal: if one would finger the other as having committed the crime, his own sentence would be halved. There are four possible outcomes, namely: neither prisoner accuses the other; only prisoner A accuses B; only B accuses A; or each accuses the other. It would seem obvious that, acting solely in his own interest, each would accuse the other. But the deal has a catch: the sentence of each one is halved, but the new "evidence" against him adds ten years, meaning a result of five instead of the original two. A crucial point is the absence of communication between the two prisoners. If they would be able to communicate, they could engage in a cooperation to minimize their sentences.

The two-prisoners dilemma is studied in psychology (cf. Grzelak, 1988), sociology (cf. Voss & Abraham, 2000), as well as in engineering optimization (see: originating mathematical formulation by von Neumann & Morgenstern, 1966 and Luce & Raiffa, 1967). All of these approaches indicate the value of communication and cooperation between the two prisoners as the key to solving the problem. Engineering practice also demonstrates that the communication channel in physical form is not sufficient in itself, but that interactions must be cultivated to avoid the tendency of each group to act solely in its own interest.

The two-prisoner dilemma is mentioned here to illustrate one of the issues associated with jointly optimizing a product: If a person or group tries to enhance one's own profit at the expense of others, it is detrimental to the overall effort. The point of removing the physical barriers between groups is but a starting point, not a guarantee for communication. Achieving the necessary collaboration requires much more and will be examined further in the section on Achieving cooperation (page 232 ff.).

5.2.5 Voting and importance of the minority

A development issue widely discussed during the 1980s was related to a programming bug in a computer in a combat aircraft. The development team overlooked the fact that, due to an oversimplification in the coordinate system, there was a mathematical signal change between the northern and southern hemispheres that was not programmed. This "dropping of the signal" would cause the plane to fly inverted upon crossing the equator.[175] The awareness of this event altered the perception for similar kinds of bugs in new systems development. This became apparent, when a development manager, presented with a new kind of lightning detection method to be used for aircraft navigation in adverse weather, questioned "whether the system would also work in the southern hemisphere". He was aware of the navigation bug, and his exposure to this "historical" knowledge evidently triggered the question (case #39).

Following the failure of the Mars Climate Observer (MCO) (cf. page 217), engineers increasingly asked themselves whether the same kind of technical error was a technical "root cause". The pitfall of using an incorrect unit is still familiar from high school physics, an error that can be committed by even the most brilliant student. But the mere knowledge of the MCO case provides a heightened sensitivity to the problem. Without this catastrophic occurrence it would presumably still be wishfully considered highly improbable.[176] Similarly, since the Challenger disaster, o-rings are more closely scrutinized in many industries, not only in aerospace, albeit in very different situations than that encountered in the solid rocket boosters.

Widely publicized and discussed cases like these merely illustrate how one's perception can be affected by experience gained, even passively, from an incident. Of course they have the perceived disadvantage of being common knowledge (*a posteriori*). Moreover, the exaggerated attention they attract may obscure other, less spectacular cases, or many people may be tempted to discount the significance of "what some young engineer experienced in some lab many years ago", as if such experiences were by definition no longer relevant to current problems.

Professionals from different backgrounds and with different experiences solving various engineering problems are expected to approach the same object from different perspectives and thus perceive it differently. For example, navigation or attitude control programmers have a different perspective regarding unit errors than telemetry programmers. *Nota bene,*

---

[175] A confirmation of this case was not achieved for the present work: there were many variations of the aircraft type and error detection (whether an error was detected by actually flying the aircraft, by simulation, or by analysis). The lack of more concrete information requires that this case be regarded for the context of this work as a software development myth. Whether or not it is true is irrelevant here.

[176] There were previous occurrences of this kind of (technical) error in the mid-1980s when the Shuttle rolled unexpectedly toward space to observe a programmed 6500-*mile*-high mountain, instead of the actual 6500-*foot*-high one (Dornheim, 1999, p.41). There was another incident in 1983 concerning the incorrect estimate of fuel in an all-metric Boeing 767 (aircraft #604). The crew, although experienced, took the value of 1.77 lbs/l instead of 0.8 kg/l. The error was discovered only at 14,000 ft. The fuelless aircraft glided to a forced landing.

"different" does not necessarily mean "better", and sometimes the combination and interaction of professionals actually makes the crucial difference.

In another instance, an electronics engineer noticed a problem with the installation of a locking wire of a pyrotechnical line connection used for a space separation system. Experienced mechanical engineers working with the same piece of hardware had overlooked it. The electronics engineer drew on the analogous reliability of coaxial cable connections with which he was familiar. Two different worlds, but with a plausible correspondence: the wire touching the line near the connection might introduce static or dynamic loads, thus jeopardizing the reliability of the connection that it was meant to protect. Heated discussions ensued among the mechanical experts as to whether his arguments could be applicable. A pyrotechnics expert later confirmed the electronics engineer's concern, and the problem was corrected (case #40).

This brings us to the potential significance of what might be known or perceived by a few or even a single person, against a majority opinion.

Although group work is a democratic notion in the sense of (with apologies to Mr. Lincoln) "of the group, by the group, for the group (results)",[177] it is a misconception that the key to group work is majority voting. Experience shows repeatedly that the majority is *not* always right.

One of the difficulties of discussing detailed technical problems is that the solution is not always as obvious as it may seem. Even if the problem is technically characterized, its relevance to the whole might not be clear and might depend on so-called "engineering judgment". This capacity varies from engineer to engineer and, in many cases, is based on opinion instead of hard data. Moreover, differences of judgement can be quite marked, especially in heterogeneous groups. However, people are often aware that there are different functions at work, and this can attenuate such divergences. It is stated here explicitly that the equilibrium between the two forces is not static. Sometimes it is even unstable.

Attempts to settle a dispute over a technical point by majority rule does not solve the problem because it does not necessarily resolve the differences of perception. Instead of heightening the perception offered by an individual, majority rule may actually help silence that person's contribution, and an overlooked detail can eventually turn out to be the difference between a mission's success or failure. This is the situation referred to by many as the

---

[177] Adapted from Abraham Lincoln's, Gettysburg Address, 1863.

"dictatorship" or "tyranny of the majority".[178] This is not to say that decision by voting should be avoided, but development groups should be aware that, in doing so, majority rule becomes but an averaging filter. If a group fails to consider the issues raised by a dissenting minority, then the most advantageous feature of the group is lost, especially if the group is based on the complementarity of disciplines and view points of its members. This is especially the case in multifunctional development groups, as the expert(s) in a specific discipline would typically be a minority, if not the sole voice. For the group to achieve the gain in performance compared to that of the functional organization requires that the minority views be duly considered by the majority.

How might this be possible?

The influence of the minority on the majority perception was studied by Moscovici, Lage & Naffrechoux (1969).

Box 50. Study design for investigating the minority's influence (Moscovici, Lage & Naffrechoux 1969).

The task of the persons was to identify the color presented to them. In their study, groups of six persons were asked to identify the color in a sequence of slides. A minority of two individuals intentionally judged blue light stimuli as being green.

In 36 trials, the 128 individuals in the experimental groups answered "green" in 8.42% of all responses, compared to 0.25% in a control group. The low result on the control group confirms the "blueness" of the stimuli. Even comparing the difference factor of 34 is not as surprising as the result that 43.75% of the subjects made four or more "green" responses.

Following the experiment, another measure was performed on the same set of subjects to determine the influence on the underlying processes, not only the influence on the response.

The experimentors then used 16 disks in the blue-green boundary,[179] and subjects had to identify their color as blue or green. By analyzing the thresholds where 25%, 50% and 75% of "green" responses occurred, they could confirm the expected shift in the thresholds, thus confirming the change the underlying processes of perception.

In product development, subjective factors affect the perception of problems as well as the consequences of a proposed solution. It is in this subjectivity that the experiment with color perception is of interest in the context of this book, as it is dependent on several subjective factors. For

[178] British historian Lord Acton described in his work *The History of Freedom* as "the one pervading evil of the democracy is the tyranny of the majority" (source: Bloomsbury Treasure of Quotations, 1994).

[179] Farnworth 100-hue set perception test.

example, the same color can be perceived differently if a contrast of brightness is presented. Moreover, there are such factors as retention in color perception, the influence of the background color, the dependence on the exposure time to the color, on the colors observed previously, etc.

This study by Moscovici et al. showed how a *consistent* minority can influence the group perception of an external stimulus. Their study also evaluated the effect of an *in*consistent minority that achieved a much lesser influence (1.25%) on the "green" responses.

Of course, this consistency/inconsistency illustrates that the minority is not always right either. Indeed, a stubborn development team member can be quite "consistent". Further differentiation is required here. After the study by Nemeth, Swedlund & Kanki (1974), Moscovici et al. established a clear difference between experimental groups, where the minority merely repeated their position, and groups, where the minority patterned their responses, associating them with changes in color brightness. This study shows the differences in the process associated with rigidity and (situational) flexibility of the minority. The flexible behavior influenced the group more strongly than the dogmatic position of the rigid group. In the real situation of product development, groups have an inherent resistance to stubbornness. Generally speaking, Moscovici (1976, 109-151) finds that "the main factor of success is behavior style".

The importance of these studies for the product development process stems from the fact that the various viewpoints arising from the diversity of training and experience of the participants throughout the process will constitute the difference between a good idea—or good isolated ideas—and a well-conceived product. Their representation in the various groups may not always be balanced because in some companies it is typical to find a predominance of one discipline or another, depending on the product line or the company's orientation. As such, the various areas of knowledge necessary for a successful development are generally not represented by the majority. Particularly in multifunctional development groups, it will be the voice of the minority that adds value to the process, not necessarily the experts who represent the discipline assigned to the group. Moreover, in many situations, important issues are revealed by non-experts for the very reason that they are not as prone to make the assumptions typically made by experts in the field. The minority is very important in such cases because of the inherent diversity they contribute to multifunctional development groups. To circumvent the pitfalls of departmentalization, where specialized groups supported the *dissenting minority,* leading at the same time to *departmental egotism,* care should be taken to avoid the illusion that the majority is always right. In addition, a given issue might not be obvious to all the discussion members, which cannot be achieved without a constructive discussion.

Other effects relevant to product development were explored by Nemeth (1994) in a series of studies investigating the underlying processes of divergent thought and creativity under minority influence. In their first study, Nemeth & Wachtler (1983) confirmed that subjects exposed to a majority view tend to follow that view. However, when exposed to the minority view, they did not tend to follow it but were able to detect novel correct solutions, regardless of the correctness of the minority. These results led them to question whether, when exposed to the minority view, subjects would be stimulated to examine the issue from multiple perspectives, which was confirmed by Nemeth & Kwan (1987). They found that the majority view induced only one of the possible three search strategies (convergence), whereas under the influence of the minority view, all available strategies were used (divergence) and showed much better performance. Recalling the discussion of the convergent – divergent subprocesses model of product development (page 93) reveals the importance of this influence.

It is stated here explicitly that it is not an easy task to be this minority. Such voices are frequently discounted, derided, regarded as detractors, as dissenters, and put under pressure to conform to the majority's opinion. They can be accused of fostering conflict, but if so, such a conflict is not necessarily one between persons. However, once successful in questioning and reevaluating established assumptions, a minority can play a significant role in enhancing the entire group's performance. In such cases, consensus takes non-averaged values into consideration and democracy gains incredible power.

5.3 Group interaction

One of the crucial elements of group work is the form and quality of interaction among the participants.

5.3.1 Critical support

Confronting one's viewpoint of an issue or one's proposed solution with alternative views introduces the possibility of "falsifying" the original viewpoint or solution, potentially leading to a better solution. This is the approach supported by the critical rationalism approach described in the historical perspective (cf. page 46 ff.). Even though an isolated, self-critical individual could achieve this "falsification", approaching this concept in a group makes it more likely that alternatives can be generated, thus providing the mutual cross-fertilization that is often essential to creativity.

A group need not necessarily be heterogeneous in order for this critical support to occur. A situation frequently observed in development labs is that two developers (of the same kind) complement each other by discussing various implementation options and arriving at new ideas or discovering points they had previously overlooked. Heterogeneous dyads possess various advantages, as the kind of issues that can be revealed by fruitful interaction are of a different nature than with homogeneous groups.

This kind of fruitful collaboration cannot be achieved if the participants regard each other as opponents. If that is the case, the participants usually attempt to defeat each other, at least by winning a technical argument, or will avoid the other instead of being open to constructive criticism and self-reevaluation. The position assumed by one group member when discussing or criticizing the position of the other has a decisive effect on the resulting collaboration.

Box 51 contains two questions used in former ASQC[180] Reliability Engineer Certification examinations, published in Quality Progress Magazine:

Box 51. Vital questions for reliability assurance activities (source: Juran & Gryna 1980, 202).

When reliability people say they are breaking the monopoly of the designer, they mean:

(a) they are relieving the designer of responsibility for the design;
(b) they intend to review the design in conjunction with other experts;
(c) they insist on having final approval of the design;
(d) they intend to put the reliability engineer one level higher than the design engineer in the organization.

It is reliability's job to see that all the tasks outlined in the reliability program are:

(a) carried out by reliability engineers;
(b) carried out by the department having the primary responsibility;
(c) defined, making sure that quality control does its job;
(d) done, including those jobs that the primary responsible departments are incapable of doing.

These questions—especially the first—reflect essential issues faced by organizations trying to solve reliability assurance problems. All of the choices listed in these questions have been tried by companies in the aerospace industry[181] contacted during the course of this study, with very

[180] American Society for Quality Control.

[181] The aerospace industry has at least 40 years' experience in reliability assurance activities, especially in the electronics divisions. Companies in other industries have also had quality

different results. They characterize different forms of interaction between reliability experts and designers.

Critical support can be present in design reviews (cf. page 145ff.), provided the participants assume a constructive posture toward criticism. O'Connor (1991, 160-161) emphasizes that "the reviewer should ideally be a reliability engineer who can be respected by the designer as a competent member of a team whose joint objective is the excellence of the design. ... By working as a team, the design and reliability staff can resolve many problems before formal analysis reports are produced, and agreement can be reached on recommendations... Unfortunately, this team approach is frequently not applied, and design and reliability staff work separately in preparing analyses and criticizing one another's work at a distance, either by memo or over the conference table. Design Review techniques then lose credibility, as do reliability staff. The main victim is the design itself, since the protagonists usually prosper within their separate organizations. If the organization is such that it is not possible to set up a team approach to design reliability, it is usually better not to try to apply the Design Review at all, since the effect will almost certainly be too little, too late."

By using Albert's *critical testing* described on page 46, it is possible to "falsify" the design to better know the weaknesses and potentials of the design without proving that the designers are "wrong" or that the design does not work. In this sense, allowing the possibility of errors can be a way to avoid the trap of providing an excuse for sloppiness. It is instead a constructive approach to criticism, from the side of the critics as well as the criticized.

5.3.2 Member expertise and quality of group interaction

Cooke & Kernaghan (1987) investigated the effect of the best member on the group performance. The problem-solving simulation they used to rank priorities is similar to the well-known "NASA–Trip to the Moon" problem. They applied a "Project Planning Situation" to participants who were expected to bring some knowledge of planning and project life cycles to the simulation. Their findings confirm that team performance is dependent on the initial individual abilities measured by the average individual score (correlation r=0.75, N=61 groups. For technically oriented reader, the explanation of the missing 0.25 to achieve r=1 is left as an exercise). *Table 26* summarizes the findings for individual abilities. The average individual

---

and reliability functions, but they were strongly influenced by the former, sometimes achieving considerable technical improvements. Nevertheless, this is one of the errors they replicated.

score and best score together contributed 69% of the variance in team performance, thus confirming the importance of the best member.

Table 26. Ordered regression for effects of initial individual abilities in group problem solving.

	BETA	CUMULATIVE R^2
Average individual score	0.75	0.57
Best score	0.54	0.69

One of the problems associated with utilizing the contribution of the best-member to the group performance is that, very often, the best member is not known in advance by the group. Of course, the criterion for determining the best member is not the person with the most prestigious degree or the most fame, but the person who has the best knowledge of the *specific* problem being addressed by the group. In development groups, the problem at hand can change momentarily when a new detail is encountered or novel information is obtained. Thus the best member in the new situation may change in the new context. In the past, engineering "stars" in management positions often outperformed any member of their team in their area of competence, but the likelihood of encountering such persons today is decreasing, especially in the product development environment because of the rapid evolution of the technologies and methods used.

Another issue related to group work that is of particular relevance to the product development process is that in many (social) experiments, groups interact only during the data collection process, which usually lasts less than one hour, and when working on tasks with which they are unfamiliar. Michaelsen, Watson & Black (1989) circumvented this problem. They investigated 222 learning groups in organizational behavior courses. The groups worked together in the course for at least 32 hours and, although not recorded, most groups reported working an equal amount of time outside of the class. The groups outperformed the best members in 95% of the cases. The authors of this study claim that this setting is similar to the real work environment, although they acknowledge that the groups they studied are more similar to task force groups or project teams, as the groups existed only for the limited duration of the course. Application of their results to product development, however, must be done under consideration of the following points. First, all of the participants were given the same questions to answer, which resembles a situation where a technical problem is approached by professionals with similar backgrounds. Such a situation does not necessarily apply to multifunctional development teams, where professionals with different characteristics approach different aspects of the same overall problem. Second, the same tests were given to the individuals and subsequently to the groups, thus affording the individuals a second chance to evaluate the same problem they had faced individually. This second point resembles typical conditions of group work in product development inasmuch as the issues addressed by the groups have previously been addressed by the individuals. Third, these engineers were participating in an organizational behavior course and were presumably quite conscious of the techniques used to achieve effective group performance.

Collins & Guetzkow (1964) postulated that "effective interaction should allow groups to combine members' knowledge and produce higher quality decisions than would be made by the group best member", what was quantitatively assessed in the studies conducted by Hall & Williams (1970) that revealed that 50% (*N*=30) of groups in which members had received 'laboratory training in group dynamics' outperformed their best members in comparison to only 13% (*N*=30) for untrained groups, and the study by Nemeroff, Passmore & Ford (1976) in that 50% (*N*=18) of the groups instructed in 'consensus decision making' outperform their best peers in comparison to 8% (*N*=12) of conventional interacting groups."

This might be an essential advantage of this study in illustrating the contribution of group work to the product development process. Its real gains can only be achieved with effective and fruitful group interaction.

5.4 Achieving cooperation

Improving product characteristics during the development phase presupposes at least a cooperative environment, but establishing such a thing is not as straightforward as it might seem. Technical people are wont to discuss at length the techniques they use and the limitations thereof, but a barrier automatically appears if the purpose of the discussion is to question the capacity of the other.

Cooperating with experts from other disciplines during product development calls for a free flow of information in order to address the diversity of issues that should be recognized at an early stage when it is still possible to make appropriate changes. In short, cooperation

- opens the possibility of discovering relevant issues that were omitted or overlooked by the developers,
- complements the expertise of a technically competent individual or group with a richness of information that usually is beyond the scope of developers, and
- alleviates the load on developers by distributing the workload.

People trying to influence others' work express a form of criticism. The posture adopted by both sides makes an essential difference in how this criticism is handled. If criticism is perceived as an attempt to demonstrate one's incompetence, people will generally defend their position, and the situation becomes a win-lose confrontation. A frequent consequence is that communication and the flow of information break down. Other defense mechanisms are also possible, such as the refusal to accept potential improvements that a given suggestion might bring. Moreover, one's previous experience with a past critic can affect the way new criticism is perceived, regardless of the way it might be delivered this time. All these factors make achieving cooperation a challenge.

The following case illustrates a possible reaction to information to potentially improve product quality.

Box 52. Difficulty accepting criticism of product characteristics (case #41).

During a workshop on Participatory Design[182] in 1994, the instructor tried to simulate the interaction between customers and programmers in defining program features. The conversations with the "customer" regarding the various concepts were considered by some of the participants to be unfocused and non-pragmatic. A climate of suspicion and distrust developed very quickly, and the simulation eventually had to be stopped. One of the software engineers in the workshop stated later: "It is very difficult for me to talk to customers. They tend to be nasty. They try to prove that our system does not work." (Compare with the case reported in *Box 46* on page 189)

Contrasting with this case, criticism can be invaluable to product development, provided it is constructive and perceived as such.

Box 53. Cooperating interaction (case #42).

In a case of technology transfer of a distributed digital control system, problems arising during the first installation of the established system were discussed between a young development engineer from the company acquiring the technology and a product "gatekeeper" of the company that developed the product. Their long but relaxed telephone discussion ranged from the system down to the component level, including limitations of the built-in test features, weak points of the redundant features, and encountered manufacturing defects. Such issues are still widely present in many sectors of the electronics industry, including aerospace. Instead of engaging in a game of pursuit and evasion, they conducted a thoroughly constructive conversation, starting with the disclosure of the product evolution and encompassing the problems as well as some of the solutions already being working on. It was implicit to both participants that the goal of the

[182] Cf. description of participatory design on page 373.

discussion was to improve the product, the success of which was important to both their employers.

At the end of the conversation, the product gatekeeper said, "*Believe* me, we are *very* interested in your finding out all these points, *really*." The young development engineer was somewhat surprised at the other's emphatic assurances, thinking to himself, "Doesn't that go without saying?"

The point of this example is that the free and open communication was possible of the mutual esteem prevailing between the development and field installation support engineers. The same engineer noted only days later that the same kind of interaction was not—in this case—possible with the manufacturing personnel.

5.4.1 Cooperation as a dynamic process

Contrary to the rather rational expectation of the development engineer in the case described in *Box 53* (page 233), achieving this cooperation is not something that "goes without saying." There is a learning process involved, as well as the awareness that not only technical issues determine success or failure. The following case illustrates how a blocked situation can be resolved.

Box 54. Non-intruding approach (case #43).

The product assurance activities in an electronics company were not being conceived as a control (in the narrow sense) activity. The quality engineer defining the concept was trying to implement it as an activity to *support* the development engineers. Nevertheless, he dubbed the activity "development control", as this was the established terminology in the defense industry.[183] In the defense industry, similar terms were used: project control or design control. Whether these activities were effective in ensuring quality in the development process was a very delicate question that could not be answered adequately at that time. There are indications that these activities were predominantly "on paper", but many companies denied this.

The quality engineer's initial attempts to implement effective activities were unsuccessful, the causes being primarily as follows.

- In the preliminary trials, the non-controlling (in the narrow sense) aspect of his approach was not explicitly communicated to the design engineers. There was interference from another party who emphasized the control aspect (in the narrow sense).

[183] At that time (in the mid-1980s) total quality management was still referred to as total quality control.

- Contrasting leadership styles: controlling, cooperating, participating were not explicitly considered in the conception of the activity.
- The activity was given an unfortunate title. The word *control* suggested from the beginning that it was an intrusive activity.

The activities were suspended for reevaluation. In the next attempt, the quality engineer emphasized that the activity was not intended as an intrusion. He made a conscious effort to:

- Create interest in the general goals instead of the partial goals;
- Concentrate the dialog on technical matters; avoid turning it a personal matter;
- Ensure that experienced developers are not unduly told how to perform their job;
- Demonstrate explicit respect for the experts solving the development problems. Acknowledge[184] their ability to solve design problems;
- Avoid a repetition of activities and competition with the development department;
- Emphasize that the various activities are complementary, not competitive.

These points were not at all new. They had already been performed in the first trial. But the error turned out to be that the quality engineer covered these points *implicitly*, as he somewhat naively thought they went without saying.

Resistance had emerged from the developers' side that had to be overcome. To accomplish this, the quality engineer took the original draft of the activity description and omitted the word "control". At that time, a new product was in the detailed design stage. The product was relatively simple in terms of the number of assemblies required and thus would serve as a good test subject for the new conception of the product assurance activity. The designer in charge of this product was not only a competent engineer, but also a calm and approachable person. When shown the two documents, he said "Oh! *This* is OK!" The quality engineer then found him quite amenable to a discussion of the earlier trials. They examined the misunderstandings, the cross purposes ... the former barriers had disappeared, at least between these two individuals.

They performed a design review, in which the participants were carefully selected. Professionals known to have confrontational personalities were excluded. Managers were also excluded in order to avoid their possible inhibiting effect on the participants (the developers and the other participants as well). This was prompted by various participants' statements to the effect that "it can be very embarrassing in front of the boss to disclose something

[184] An implicit form of acknowledgement is recommended.

we overlook. He has a significant effect on my professional progress and future assignments. I do not want to appear 'dumb' before my superior."[185]

The selected participants were informed of the objective of the quality assurance activity, how were they expected to perform, and what they should avoid. The review took an entire day and went smoothly. The comments were constructive, and some 80 potential quality improvements were listed in the first pass. No decision was made yet regarding their implementation, which was left to the developers or a "superior instance".

The participants found the review's duration to be taxing, and stated that it was difficult to document all the issues revealed. The developer requested a checklist to guide the design.

A series of weekly meetings was conducted with development, acquisition, manufacturing and quality personnel. Prior to the review, they had considered only acquisition and fabrication schedules as well as pending requests for components. After the review, they started discussing technical details of the product implementation and proposing solutions. They covered fewer issues every week, but in more detail. More than 200 changes were protocolled.

A review of the mechanical construction of the same product was announced a few weeks later, but the participants could not be selected as in the previous instance, and the discussions did not converge as intended. Their scope, however, was covered at the weekly meetings.

This case marked a change in the results of a group's quality assurance activities. Efforts had had been launched eight years before to gain access to the development process. The review served to reduce the barriers to communication and cooperation early in the product development phase. It did not allow decisions regarding the implementation of the proposed changes to be made immediately. Many of the issues revealed in the review were not implemented by the development personnel. It was merely in their hands to decide about future implementations. In the weekly meetings, discussions led to such decisions. There was a perceived trend toward sharing responsibility for the decision in implementing the discussed items.

No engineer can expect that the facts will always speak for themselves, independent of how they are presented.

The case described above took place in the early 1990s. Some thirty years earlier during the moon program, the National Aeronautics and Space Administration strove to arrive at a form of cooperation with industrial groups that was based on technical knowledge from both parties and a willingness to discuss the problems. With this knowledge, they could talk to

---

[185] This statement is not a transciption of *one* statement. It was actually composed from different true statements in this specific case.

the contractors about the details of the implementation of the product they were buying. Arriving at this cooperation, however, is no easy task. It took a while for the developers to perceive this kind of openness as a cooperative venture.

The form of cooperation depends critically on the attitude of the participants, but attitude and behavior alone are not sufficient to achieve cooperation between technical parties. The prerequisites for cooperation may be stated as follows:

- a common basis for discussion,
- convergence of goals,
- complementary areas or viewpoints (nonannihilating),
- mutual confidence,
- open communication channels,
- openness to the issues pointed out by the other parties,
- discussion as equals,
- absence of "combat".

The most important prerequisite is the in-depth understanding of the actual product development *processes*. Without it, the chances for success diminish. An essential aspect is that this understanding be not only of a theoretical nature, but is based on experience with similar subjects.

This cooperative environment is not of static nature. Even seasoned technical professionals acknowledge that it can change over time. Expressed in terms of control system theory, the way to actively influence the "state of this system" remains untouched. When a cooperative environment exists, it is generally used (and sometimes abused). When it ceases to exist, there are fewer options to act upon. This also means that the ability to maintain and improve an existing cooperative environment is limited. The following construct clarifies this "system" and how to regulate it.

By *psychological orientation*, Deutsch (1982)[186] refers to a "more or less consistent complex of cognitive, motivational, and moral orientation to a given situation that serves to guide one's behavior and responses in that situation" (op. cit., p. 15). "People orient themselves differently to different types of social regulations, and different orientations reflect and are reflected in different cognitive processes, motivational tendencies, and moral dispositions" (op. cit., p. 21).

He examined the relationships between types of personal interdependence and psychological orientations. The types of

186 Excerpts from: Deutsch, M. (1982). Interdependence and Psychological Orientation, in: V. J. Derlega & J. Grzelak (Eds.), Cooperation and Helping Behavior - Theory and Research, pp. 15-42, New York NY, etc.:Academic Press. Reproduced here by kind permission of Academic Press, an imprint of Elsevier Science.

interdependence in interpersonal relations include the following fundamental dimensions:

1. Cooperation vs. competition.
2. Power distribution (*equal* vs. *unequal*).
3. Task-oriented vs. social-emotional.
4. Formal vs. informal.
5. Intensity or importance (intensity or superficiality of the relationship).

Additional dimensions identified by Deutsch but not considered explicitly were:

- Enduring or temporary nature of the relationship.
- Voluntary or involuntary character.
- Public vs. private nature.
- Licit or illicit quality.
- Number of people involved.

In *Table 27*, Deutsch illustrates 16 types of social relationships assuming the more intense type of relationship. He also emphasizes that the dichotomization is an oversimplification because one relationship may fall into more categories.

This table shows the diversity of types of possible relationships in terms of four of the five cited fundamental dimensions. The labels are taken from day-to-day experience, but can also be applied to the social relationships occurring in the product development, and the reader is encouraged to translate them appropriately. Note especially types 9 and 13 and their effects on efficient product development.

The next step to describe the process of achieving cooperation is based on Neisser's work on cognition, where human perception is dependent "upon preexisting structures, here called *schemata*, which direct perceptual activity and are modified as it occurs." (Neisser, 1976, p. 14, italics maintained). Perception is a constructive process, wherein one continually constructs anticipations of certain kinds of information that enable one to accept it (or reject it) as it becomes available.

The explorations are guided by the anticipatory schemata that act as plans for a perceptual action. The outcomes for the explorations modifies the original schema that then (re)directs further exploration and become ready for new information. This cyclical evolution is depicted in *Figure 15*.

Table 27. Example of the four-dimensional space of social relationships according to Deutsch.[187]

	SOCIAL-EMOTIONAL		TASK-ORIENTED	
	INFORMAL	FORMAL	INFORMAL	FORMAL
COOPERATIVE EQUAL	1 *Intimate* Lovers Love-making	5 *Fraternal* Club members Social party	9 ***Problem-solving*** Colleagues Staff meeting	13 ***Organized cooperation*** Task force members Working together with differentiated responsibilities to solve problems
COOPERATIVE UNEQUAL	2 *Caring* Mother-child Nursing	6 *Protecting* Police officer-child Helping	10 *Educational* Professor-graduate student Working together informally on research project under professor's direction	14 *Hierarchical organization* Supervisor-employee Supervisor assigning employee to do certain task
COMPETITIVE EQUAL	3 *Antagonistic* Personal enemies Fighting	7 *Rivalrous* Divorced couple Custody suit	11 *Competitive* Contestants in informal game Trying to score points against the other	15 *Regulated competition* Business rivals Bidding against one another for a contract
COMPETITIVE UNEQUAL	4 *Sado-masochistic* Bully-victim Tormenting	8 *Dominating* Expert-novice Intimidating	12 *Power struggle* Authority-rebel Guerilla warfare	16 *Regulated power struggle* Guard-prisoner Ordering prisoner to keep in step

[187] Note from M. Deutsch: Each cell characterizes a type of social relationship by labeling the relationship (first entry, in italics), naming people who might be in such a relationship (second entry), and describing an activity that might occur in such a relationship (third entry).

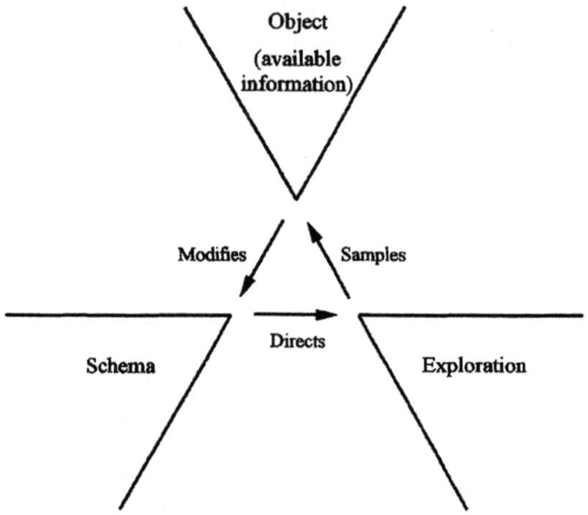

Figure 15. Neisser's perceptual cycle (1976)[188]

This cycle takes place especially when people are mobile within their environment, when they examine it more closely, change their viewing perspectives, occluding or disoccluding entire objects or vistas, with the information gain being systematically related to existing schemata, and to a cognitive map or orienting schema of the adjacent environment (op. cit., pp. 108-110). This *cognitive map* or *orientation schema* is in turn an active, information-seeking structure. Neisser proposes the perceptual cycle embedded in cognitive maps as shown in *Figure 16.* The outer cycle is a more inclusive cycle of exploration and information acquisition that is more comprehensive and takes more time.

[188] From: COGNITION AND REALITY: PRINCIPLES AND IMPLICATIONS OF COGNITIVE PSYCHOLOGY by Ulric Neisser © 1976 by W. H. Freeman and Company. Used with permission.

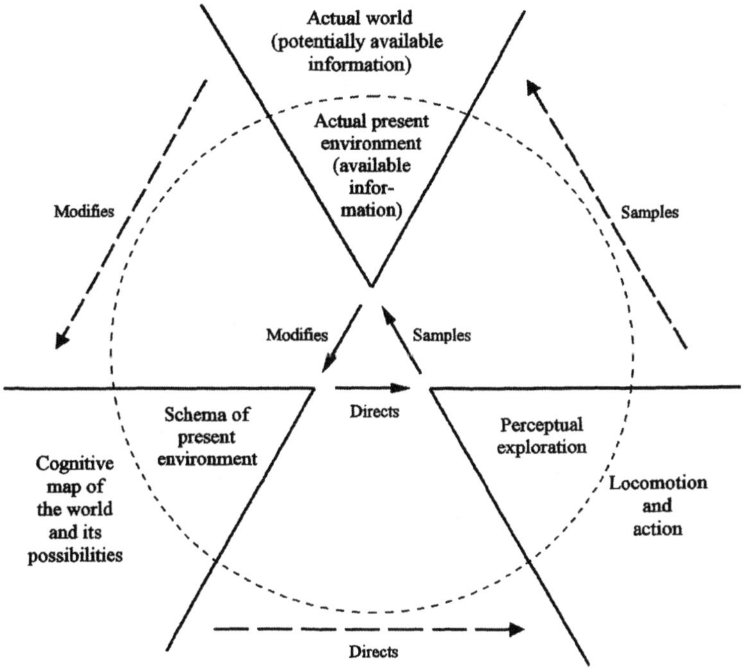

Figure 16. Schemata: Perceptual cycle embedded in cognitive maps[189]

Returning to cooperation, Deutsch presents the dynamic nature of the form of interaction leading to or away from cooperation in a cyclical association between psychological orientation and social relationships (*Figure 17*). Put simply, the basis idea is that one's psychological orientation (to the present relationship) directs its exploration (through actions and reactions) that tests the present social relationship and modifies the psychological orientation. The cycle is thus closed and repeats itself. The external cycle represents the influence of the desired social relationship that affects the normative definitions (of the actions and reactions), thus affecting potential social relationships and so on.

[189] From: COGNITION AND REALITY: PRINCIPLES AND IMPLICATIONS OF COGNITIVE PSYCHOLOGY by Ulric Neisser © 1976 by W. H. Freeman and Company. Used with permission.

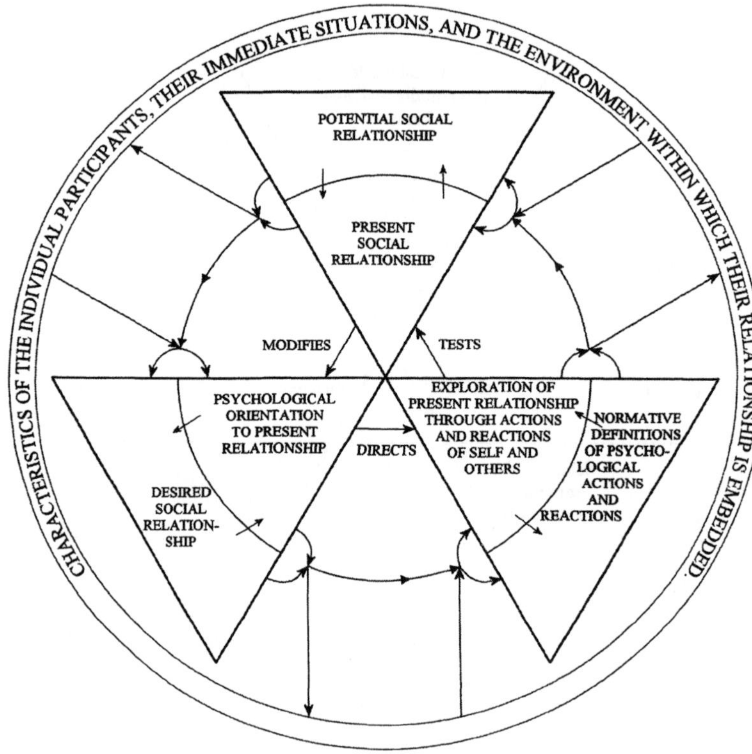

Figure 17. Relation between psychological orientation and the type of social relationship involved (Deutsch 1982).[190]

Now let us consider the following definitions for the discussion of the evolution of psychological orientation.

Box 55. Psychological orientations (condensed from Deutsch 1982, 24-26).

Cognitive orientation

Cognitive orientation, like the terms *schema*, *script* and *frame*, refers to different disciplines (as in cognitive psychology, social psychology, sociology, linguistics, anthropology, and artificial intelligence). It refers to

[190] From: Deutsch, M. (1982). Interdependence and Psychological Orientation, in: V. J. Derlega & J. Grzelak (Eds.), Cooperation and Helping Behavior - Theory and Research, pp. 15-42, New York NY, etc.:Academic Press. Reproduced here by kind permission of Academic Press, an imprint of Elsevier Science.

the *structures of expectations* that help the individual cognitively orient himself according to the situation being confronted.

Schema refers to the constructive and organized features of memory as opposed to the memory as a passive storage.

Script is associated with the notion that people have organized knowledge of a stereotypical form about most recurrent situations they encounter. Abelson (1975, 33) defines it as a "coherent sequence of events expected by the individual, involving him either as a participant or as an observer" and that the "cognitive mediated social behavior depends on the joint occurrence of two processes: (i) the selection of a particular script to represent the given social situation and (ii) the taking of a participant role within that script" (op. cit., pp. 42-43), (cf. also Abelson 1976 and Schank & Abelson 1977).

Frame explains how individuals exchange signals that allow them to agree on the level of abstraction at which a message is intended—for example, whether a message is intended as serious or playful. It can be generalized as follows: individuals, as they face any given situation, ask themselves, "What is going on here?".

Schemas, scripts, or frames may be very concrete and specific—for example, how to work together with a particular person on a given task—or they may be rather abstract and general—for example, what is involved in a competitive as compared to a cooperative relation.

In any society that provides a variety of situations in which different areas in the multidimensional space of social relationships are well represented, it is likely that rather abstract schemas or scripts will develop to characterize the types of relationships depicted in *Table 27*.

The improvisational nature of most social relationships—the fact that given types of social relations occur in widely different contexts and with many different kinds of actors—makes it likely that relatively abstract or generalized cognitive orientations, schemas or scripts will develop for the different types of social relationships.

Motivational orientations

A motivational orientation toward a given social relationship places one within the possibilities of gratification or frustration of certain types of needs in the given relationship. It adds the personal, subjective features arising from the situationally relevant motives or need-dispositions of the individual. It gives rise to the cathexis[191] of certain regions of the cognitive landscape, making them positively or negatively valent, and highlights the pathways to and from valent regions. It gives the cognitive map a dynamic

[191] Cathexis - a charge of mental energy attached to any particular idea or object.

character. It predisposes the individual to certain kinds of fantasies (or nightmares) and to certain kinds of emotions. The individual looks for answers to such questions as "What is valuable in this relationship?" or "What do I want here and how can I get it?".

Moral orientations

Moral orientations provide an individual with a framework in which to evaluate obligations, rights, and entitlements of the people involved in the given relationship. They add an "ought to", "should", or obligatory quality to a psychological orientation. They imply that one experiences a relationship not only from a personal perspective but also from a social perspective, which includes the perspective of the others in the relationship. One can expect that moral orientations, and hence what is considered fair, will differ in different types of social relationships.

Deutsch developed a characterization associated with the four dimensions shown in *Table 27*, a discussion of which is unfortunately beyond the scope of this book. Nevertheless, the following summary is intended to emphasize the differences between the psychological orientations and the associated interdependence in the context of the product development cases described within this book. This summary is very brief, and the interested reader is encouraged to appreciate the work of Deutsch in more detail.

COOPERATION-COMPETITION

Box 56. Differences in psychological orientations in the cooperation-competition social dimension (condensed from Deutsch 1982, 27-28).

Cognitive orientation

The basic schema is that: We are "for" or "against" one another; we are linked such that we both gain or lose together, or such that if one gains, the other loses. This schema leads to an individual's expectation that in a cooperative relationship, the other will be pleased by the individual's success and ready to help him or her achieve it. The individual will expect the opposite to be true in a competitive relationship. If one believes to be in a cooperative relationship and the other is displeased by one's success, insecurity arises: "Is this the kind of relationship I think it is?" or "What can I do to find out the true nature of this relationship?"

Motivational orientation

In a cooperative relationship, one is predisposed to cathect the other positively, to have a trusting and benevolent attitude toward the other; to be psychologically open to the other; to be giving as well as receptive to the other; to have a sense of responsibility toward the other and toward the mutual process of cooperation; to see the other as similar to oneself; etc. One is also predisposed to expect the other to have a similar orientation toward oneself. It is clear that the specific quality of this orientation will be strongly influenced by the type of cooperative relationship involved: social-emotional or task-oriented, equal or unequal, formal or informal, intense or superficial.

In a competitive relationship, one is predisposed to cathect the other negatively, to have a suspicious and hostile, exploitative attitude toward the other; to be psychologically closed to the other; to be aggressive and defensive toward the other; to seek advantage and superiority for oneself and disadvantage and inferiority for the other; to see the other as opposed to oneself and basically different, etc. One is also predisposed to expect the other to have the same orientation. As with the cooperative relationship, its specific quality will be determined by the type of competitive relationship involved, but in addition, it will be colored by one's conception of one's chances of winning or losing.

Moral orientation

Moral orientation connected with cooperation fosters mutual respect and self-respect. It favors equality as a guiding principle to be breached only when inequality brings greater benefits and advantages to those less fortunate than they would have been if treated equally. In contrast, moral orientation linked with competition sanctions inequality and justifies a win-lose struggle to determine who will be superior and who will be inferior in a competitive relationship.

POWER: "EQUAL" VS. "UNEQUAL"

Box 57. Differences in psychological orientations in the power social dimension—"equality" vs. "inequality" (condensed from Deutsch 1982, 29-31).

The basic schema of "relationship power" has to do with the relative power of the participants in a relationship to benefit, harm or persuade one another and, hence, their relative power to influence one another.

In a relationship of unequal power, it is expected that the more powerful member will be advantaged and the less powerful one will be disadvantaged whenever their interests are opposed: Hence, it is considered better to be in the more rather than the less powerful position in a competitive relationship. The competitive context of the unequal power schema highlights the roles of "victor" and "vanquished"; the equal power schema is oriented toward continuing the struggle. In both competitive frameworks, the use of coercion, intimidation, and bluff are rendered more salient. Even in a situation where the more and less powerful members have congruent interests, the less powerful member is expected to be more dependent on the other and, hence, more likely to engage in ingratiating behavior.

The cooperative context of the unequal power schema emphasizes the orientation toward responsibility in the higher power position and of respectful compliance in the lower power position. The equal power schema is oriented toward mutual responsibility and respect. Both cooperative contexts foster the use of positive forms of power: persuasion rather than coercion, benefit rather than harm, legitimate rather than illegitimate power, etc.

Motivational orientation

In an equality relationship, one is predisposed to consider that the other is entitled to the same esteem and respect as oneself. Respect and esteem are more valuable if they are received from those whom one respects; equal status relationships represent the optimum distribution for the mutual support of self-esteem. The need for self-esteem involves the need to have a sense of worthiness of one's goals and a sense of confidence in one's ability to fulfill one's intentions. The need for self-respect involves the need for a sense of one's moral worth, of one's equal right to justice and fair treatment.

In an unequal relationship, one is predisposed either to take a more dominant or a more subordinate role, or to resist the inequality. The need for dominance is manifested in the desire to control, influence, direct, command, induce, dictate, supervise, instruct, or lead. In a competitive situation, the need for dominance will often be fused with the need for aggression and will lead to attempts to coerce and force the other to comply with one's will.

In a cooperative situation, it will often be fused with the need for nurturing and will lead to a protective, guiding, and caring orientation toward the other.

Different needs are associated with the submissive role, depending on whether it occurs in a cooperative or competitive context. In a cooperative context, the need for deference involves a readiness to follow, to comply, to emulate, to conform, to obey, to defer, to admire, to revere, to be suggestible, to heed advice, and otherwise to accept the superior authority of the other.

In a competitive relationship, the need for abasement reflects the tendency to submit passively, to accept blame, to surrender, to seek punishment or pain, to be servile, to be resigned, to acquiesce, to be timorous, to give in, to allow oneself to be bullied.

Moral orientation

In a cooperative, equal relationship one would expect the kind of egalitarian relationship described in the cooperation-competition dimension.

In a cooperative, unequal relationship, the moral orientation obliges the more powerful person to employ his power to benefit the less powerful one, not merely himself. The less powerful one is obliged to show appreciation, to defer to, to honor the more powerful person. These obligations may be rather specific and limited if the relationship is task-oriented or they may be diffuse and general if the relationship is a social-emotional one.

In an equal, competitive relationship, one's moral obligation is toward the value of initial equality among the competitors and the subsequent striving to achieve superiority over others.

In an unequal, competitive relationship the moral orientations of the strong and the weak participants support an exploitative relationship, the weak are apt to adopt the moral orientation of the more powerful and to feel that their inferior outcomes are deserved. Or, they might feel victimized.

FORMAL VS. INFORMAL

Box 58. Differences in psychological orientations in the formal vs. informal social dimension (condensed from Deutsch 1982, 34-35).

Cognitive orientation

The basic element in the schema related to this dimension has to do with whether one expects the people involved in the social situation to let their activities, forms of relationship, demeanor, and the like be determined and regulated by social rules and conventions or whether one expects such people to have the freedom to make and break their own rules to suit their individual and collective inclinations.

Motivational orientation

Informal relationships tend to be more open, more particularistic, more frank, more flexible, more emotional, and more personal than formal ones. They have a more relaxed, improvisational character in which quickly formed, intuitive and impressionistic reactions to the specific other in the particular situation largely determine one's behavior. In an informal relationship, one's motivational orientation is directed toward persons, whereas in a formal relationship it is directed toward rules and authority. Emotion and conflict is more apt to be openly expressed in informal relations and avoided in formal ones.

Moral orientation

Formal relationships go beyond the values of universalism and affective neutrality or impartiality to include a moral orientation to the rules and conventions that guide social relationships. One has an obligation to respect them and to conform to them. One's obligation is to the form of the relationship rather than to its spirit.

In contrast, in an informal relationship one is morally oriented to the spirit rather than the form of the relationship.

It is the relationship to which one is obligated rather than to the rules that are supposed to regulate it.

The work of Deutsch has direct implications for the social interactions leading to or inhibiting cooperation in the product development processes. In addition, Deutsch establishes a relationship of the individual and his/her environment and provides a means to relate the influence of the former experiences of the individual with his/her orientation in a dynamically evolving fashion, not in a fixed form. *Figure 17* depicts this dynamic character.

The interplay among the persons and disciplines occurring in the complexity of modern product design require more than ever a cooperative behavior. Achieving and maintaining this cooperation require much more than technical competencies.

5.4.2 Improving communication

Approaching the product development process as a communication-dependent activity is not new. In a study of R&D organizations by Pelz & Andrews (1976) originally published in 1966, the authors started with the assumption that an R&D laboratory is "a system of interacting scientists (and other components) in which the inhabitants stimulate each other to product high-quality R&D" (op. cit., p. 35), as opposed to merely "a facility which provides services and equipment so that its scientists can conduct (exclusively technical) R&D activities" (op. cit., p. 35, parentheses added). The individual performance of R&D personnel was evaluated by various experts familiar with the work of the individuals and qualified to compare it with the work of others in the laboratory. Half of the evaluators were in supervisory roles. Individuals were evaluated according to their "contribution to general technical or scientific knowledge to the field" and their "over-all usefulness in helping the organization carry out its responsibilities" during a period of the last five years. The frequency of communication with the five most significant colleagues was self-reported by the individuals and rated according to the scale (i) monthly or less, (ii) semi-monthly, (iii) weekly, (iv) semi-weekly, or (v) daily. A positive correlation was found between performance and the frequency of communication summarized in *Table 28.*[192] Other measures of communication such as time spent on the communication, number of close contacts within the group and with colleagues outside the group but within the organization also revealed a positive correlation with performance.

---

[192] Pelz & Andrews also found positive correlations with the more usual measures of published papers, patents or patent applications and unpublished reports, although the findings were not statistically significant.

Table 28. Correlation ratio (eta) of individual performance in R&D activities and communication summarized from Pelz & Andrews (1976).

FREQUENCY OF COMMUNICATION FOR THE CATEGORIES	PERFORMANCE MEASURE SCIENTIFIC CONTRIBUTION	OVERALL USEFULNESS	TOTAL *N*
PhD's in development labs	0.13	0.39 *	73
PhD's in research labs	0.54 *	0.47 *	52
Engineers	0.13	0.28 *	199

* statistically significant to 5% or better (data unclear from the report)

This study shows that the myth of the lone engineer or scientist working on a technical problem to the deepest detail was invalid already more than thirty years ago. This had indeed been broadly stated even ten years earlier by Wernher von Braun as he told *New York Times* reporter George Barrett in an October 1957 interview:[193] "Take a group of people with the same general fund of knowledge, present them with a problem and you will get it done. The idea of a great genius sitting quietly in some corner dreaming up vast secrets is no longer real. So many things in modern science have become so wide in scope, so intricate, that *more and more it takes groups of experts* to do the work" (Spangenburg & Moser, 1995, p. 91, italics added).

But this grouping of experts faces interesting challenges. There is an implicit limitation tied to the specialization of technical experts because it is no longer possible for an engineer to be an expert in all disciplines. This was noted by no less than the philosopher Alfred North Whitehead (1917). According to Pill (1971, 22-23), when "graduate education was then just being developed," Whitehead predicted that "specialists would develop significant new knowledge but that it would become so esoteric that those responsible for integrating it into the real world would cease to appreciate its potential benefit or harm." Pill investigated the contrast between two cases of interdisciplinary large-scale projects. He refers to Whitehead's 1925 Lowell lectures, in which the latter argued against the professionalization of knowledge because it produced "minds in a groove"[194] (op. cit., p. 23).

"The dangers arising from this aspect of professionalism are great, particularly in our democratic societies. The directive force of reason is weakened. The leading intellects lack balance. They see this set of circumstances, or that set; but not both sets together. ... In short, the specialized functions of the community are performed better and more progressively, but the generalized direction lacks vision. The

[193] By Georg Barrett. The New York Times Magazine, October 20, 1957, p. 86.

[194] Further references to Whitehead's "minds in a groove" can be found in (Lowe & Schneewind, 1990, 169) or in pages 275-276 of the 1925 original of Whitehead (1967).

progressiveness in detail only adds to the dangers produced by the feebleness of coordination" (Whitehead, 1967, 197).

Related to this phenomenon perceived by Whitehead, a widely accepted issue in professional training that is not related solely to product development is called here the *generalist-specialist organizational dilemma* and is stated as follows: A specialist (in the extreme) tends to delve deeper and deeper into his/her area of expertise, but in so doing, exhibits the secondary effect of that expertise becoming increasingly narrow in scope.

The result is that various experts working on the development of the same product tend to "dig separate wells". Each expert becomes blind to the issues faced by the other, they may even no longer speak the same technical language. Even worse, they may try to optimize the product only locally, rendering the joint optimization difficult, if not impossible.

A generalist (in the extreme) tends instead to broaden his/her knowledge domain and, in so doing, the level of detail that can be achieved becomes increasingly shallow. Many companies have tried allocating generalists to solve communication problems among experts. In theory, the broader coverage of the generalists should serve as bridges between experts, who then provide the necessary depth in knowledge.

This approach proved unable to solve the problem. Conflicts among different specialists were merely substituted by conflicts between specialists and generalists—the same old power struggles took place. This situation was exacerbated by the fact that generalists and specialists typically follow different professional career paths. Dual-ladder progression systems could not solve the conflicts either, although they provided a means to avoid the so-called "Peter principle".[195] Development managers indicated privately that pure generalists tend to be relationship-oriented whereas pure specialists remain problem-oriented (or discipline-oriented), causing a certain difficulty in integrating the two. Personnel and technical managers complained that many pure generalists could maintain a nice conversation, but did not possess the essential ability to solve problems to the extent their position required (case #44).

This heightened the organizational dilemma: should experts or generalists be hired? The so-called *T-training* strategy partially solved this problem. Basically, the T-training strategy says that the depth of a professional's

---

[195] The "Peter principle" occurs in practice when organizations promote persons based on past achievements and where there is only one path for career progress. In the new assignment or environment, a formerly efficient professional might not perform well. Demoting such an individual is usually not an option due to career expectations. The "principle" describes the deterioration that results from the professional being stuck at the position to which he/she was promoted until performance became inefficient, sometimes even becoming a hindrance to others.

training should resemble the letter "T", i.e. there should be a broad, general coverage of many areas of interest *in addition to* expertise of a particular subset of the whole. With this approach, professionals would possess an overlapping domain of knowledge that, at least theoretically, would allow one to see the others' problems and provide a common base for productive collaboration. This alone cannot solve the entire problem because first, it emphasizes technical competences[196] instead of a balance between technical and social competences, a prerequisite for efficient communication. Second, if goals are still pursued locally, i.e. if the various experts still try to center optimization efforts on their own area of expertise, instead of trying to contribute to the overall product optimization, then communication remains difficult.

One way to correct this problem is the so-called *double-T-training* or *Π-training* (capital π) strategy, in which experts possess detailed competence in more than one area. This is represented by the two vertical bars in the Greek letter Π. An example of this is engineers who pursue graduate training in another engineering area than their original field of expertise, or in business administration. This was proposed as a way to integrate various disciplines *within* an individual, in the hopes of leveraging the potential of leading to the internal mental processes that would result in an opening of the external, interindividual processes necessary to achieve interdisciplinary dialog. The individual who possesses deep knowledge in (at least) two areas can establish a connection that may go much further than the superficiality that the T-training strategy allows (intrapersonal process). It also enables factual communication with experts in other areas, as the perception that one's counterpart possesses sound knowledge in one's own field of expertise (interpersonal process) contrasts with the perceived superficial knowledge inherent to T-training. With the latter approach, having "retrievable knowledge",[xxxviii] that can be recalled as required is necessary but not sufficient.

But the Π-training strategy does not guarantee success. Instead of integrating the disciplines in which they were trained, some individuals have been found to "switch" from one discipline to the other, actually ignoring the important links between their disciplines and hence suppressing the significant internal (intrapersonal) processes. In this sense, an engineer with training in management may become much more of a manager than an engineer, instead of being both. This point is illustrated by the records of the

---

[196] Here, the term "technical competence" means the competence associated with a technique or discipline. The term is quite general, as it abridges every possible discipline and is not limited to the scope of technical professions, e.g. engineering. The equivalent in German is the term *"fachliche Kompetenz"*.

decision-making process regarding the launch of the Challenger space shuttle. In a telephone call the night before the launch, Mr. L. Mulloy, manager of the Solid Rocket Booster Project at NASA, told Mr. R. Lund, Vice President of Engineering at Morton-Thiokol,[197] to "put on his 'manager hat' instead of his 'engineering hat'. "Mr. Lund reversed his opposition to the launch, thus overruling his own engineers" (Feynman 1988, 177).

Another important point concerning the combination of vertical and horizontal orientations of expertise and its effect on communication is the effect of group tenure on R&D professionals. Based on research[198] published in 1962, Wells & Pelz (1976)[199] found a decrease in communication with increasing group tenure. Their data shows a general decrease in average frequency of communication and in the time spent communicating with one's supervisor as well as immediate colleagues with increasing group tenure. There was a general decrease in scientific contribution but the over-all usefulness reached a peak at 4 – 5 years of group tenure. The authors attribute these results to the fact that for "strictly scientific or technical advances, new members of a group do indeed provide fresh viewpoints and stimulate each other to think in new ways, but in order to be useful to the organization…a group takes several years to solidify as an effective team" (op. cit., p. 244). To assess individual characteristics, respondents were asked to declare the extent to which the statements describe the approach preferred: "I prefer to map out broad features of important new areas, leaving detailed studies to others" and "I prefer to probe deeply and thoroughly into selected areas, even though narrow". The combined curves indicating group usefulness, specialization and pioneering depicted in *Figure 18* show the peak associated with a balance between specialization and generalization.

Smith (1970) developed this aspect further, apparently from the same basic set of data. Among the findings from a subset of N=418 scientists and engineers forming 52 groups, the degree of task restructuring (i.e. finding unifying relationships or conceiving means of developing designs) in contrast to accumulating knowledge (such as systematic observation, accumulating ideas, developing or testing specific designs) bears a significant[200] correlation with performance for old groups. No significant correlation was found by the author for young groups.

---

197 The manufacturer of the solid rocket booster.

198 Investigating 83 groups (49 in industry, 34 in government).

199 Excerpts from: Wells, W. P. & Pelz, D. C. (1976). Groups, in: Pelz, D. C. & Andrews, F. M., Scientists in Organizations - Productive Climates for Research and Development, revised edition, Ann Arbor, MI: Institute for Social Research. Reproduced here by kind permission of the Institute for Social Research.

200 No data on significance level was supplied for these findings.

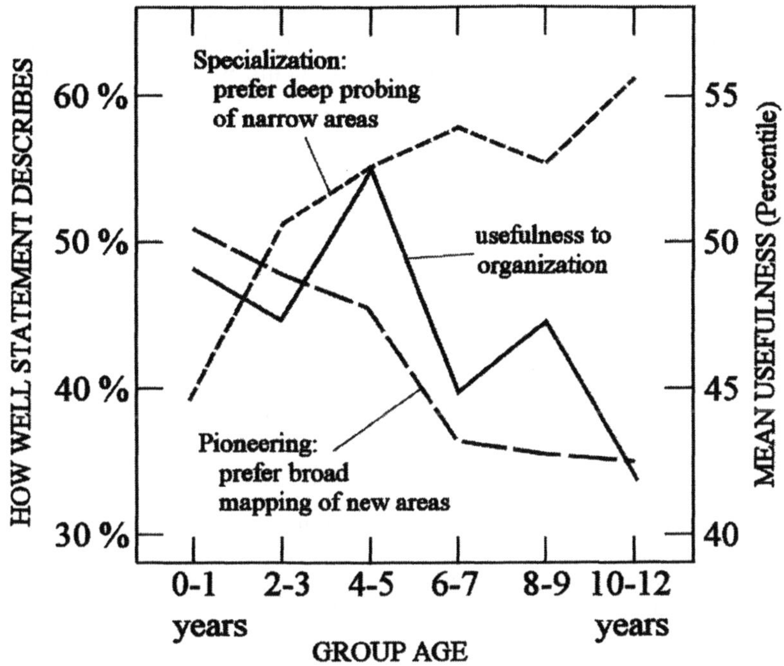

Figure 18. Group usefulness, specialization and generalist tendencies: variation with group age (Pelz, 1967).[201]

Table 29. Effect of generalist-inducing tasks on performance of old groups (Smith 1970, 94).

	DEGREE OF TASK RESTRUCTURING
Evaluated technical contribution	0.48
Evaluated general usefulness	0.44
Average patent output	0.42
Rate of technical papers	0.22

An analysis of group tenure was also performed by Katz & Allen (1982) for a population of N=345 R&D professionals divided into 61 distinct projects. Their findings replicated not only the increase in performance for a group age of up to 1.5 years, but also that it remains constant for a while, before decreasing abruptly after five years of group tenure. This phenomenon was dubbed the *not-invented-here syndrome* because of the

[201] Reprinted with permission from: Pelz, D. C. (1967). Creative Tensions in the Research and Development Climate, Science, vol. 157. no. 3785, pp. 160-165. © 1967 American Association for the Advancement of Science.

observed tendency of R&D groups to develop a monopoly behavior concerning its field of specialization and to ignore the fact that outsiders might produce important contributions that would be relevant to the group. Their study[202] not only confirmed the decay in communication with mean group tenure of at least 2.5 years, but also revealed the decay for various types of communication and project types as shown in *Table 30*. Highlighted in the table are Katz & Allen observations of the highest correlations that show the not-invented-here syndrome affecting negatively the *most relevant* communication types for the related project types.

Table 30. Correlations between performance with group tenure and communication with group tenure Katz & Allen (1982).

	GROUP TENURE BY PROJECT TYPE			
	RESEARCH *N*=6	DEVELOP-MENT *N*=12	TECHNICAL SERVICE *N*=12	ALL PROJECTS *N*=30
CORRELATIONS:				
(a) project performance	– 0.62 *	– 0.39 *	– 0.44	– 0.43 ***
INTERNAL COMMUNICATIONS				
(b) intraproject communication	– 0.26	– 0.14	**– 0.72** ***	– 0.39 ***
(c) organizational communication	0.27	**– 0.53** **	– 0.12	– 0.20
EXTERNAL COMMUNICATIONS				
(d) external professional communication	**– 0.51**	– 0.23	– 0.38	– 0.37 **
	*p>0.10; **p<0.05; ***p<0.01			

The term "communication" in the context of product development means the cognitive processes leading to the encoding, transmission and decoding of a message or a piece of information related, either directly or indirectly, to the product development process. Some authors operationalize their studies by measuring the use of one or more transmission channels. This gives an objective indication of the intensity of the communication process but is not intended as the only element to be considered when facing the issue of effective communication.

202 Project performance was measured by interviewing individually seven department managers and two laboratory directors. They were asked to make an informed assessment based on their knowledge of and experience with the projects. Criteria used included schedule, budget, cost and performance, innovativeness, adaptability and cooperation with other organizational areas. Each project was independently rated by about five managers using a seven-point scale. There was a strong consensus among the ratings (Spearman-Brown reliability = 0.81), and the ratings were averaged to yield overall project performance scores.

Maderthaner's (1989) description of communication processes not only considers the cognitive-informational factors but also the emotional-motivational influence on information processing. He establishes a connection between cooperation readiness and communication in that a depreciation of persons or hard criticism markedly disturb the free flow of information in organizations. Moreover, he mentions the effects on removing inhibitions to aggression and on providing an escalation of quarrels. For these reasons, organizations that endeavor to achieve a cooperative work climate, which is possible only through open reciprocal communication, have to establish on the one hand measures to cultivate fruitful relationships among the participants and to stimulate on the other hand the resolution of existing conflicts (op. cit., p. 496).

The importance of communication to achieve cooperation in the technological arena is elucidated by the work of Elmes & Wilemon (1992). They investigated[203] the determinants of cross-functional cooperation in technology-based organizations. Asked about what they had done to improve their relationships with highly important, difficult partners, the participants confirmed the importance of the additional elements of communication listed in *Table 31*. This reinforces the importance of the social processes associated with technical discussions.

Table 31. Attitudes toward resolution of difficult interfaces Elmes & Wilemon (1992).

	PERCENTAGE
Improved quantity and quality of communication with them	64%
Developed approaches to enhance mutual understanding among groups	24%
Became more responsive to their needs and requests	19%
Other	1%
(multiple entries were allowed) 3 persons provided no response	

The impact of communication on the cooperation processes applied generally to total quality management is also defended by von Diemer (1991).

It is important to consider the non-technical aspects of communication channels in order to enhance them. The availability of a communication channel alone does not guarantee efficient communication. In a report concerning a major automobile manufacturer's experiences with computer-

[203] They interviewed 86 department heads and project managers who regularly interacted with other functional groups. Subjects were not only R&D professionals but included marketing, manufacturing and sales personnel. The main industrial groups represented by their sample were the telecommunications, military/aerospace, electronics and pharmaceutical/chemical branches. On average, the interviewees had organizational experience of 13.6 years, had held their current position for 3.1 years, and interacted with 10.4 departments groups during a typical day.

supported cooperative work in worldwide networking for product development, Hagemann (1996, 116) declares that it places an increased demand on participants with respect to their willingness to communicate and their commitment to teamwork.[xxxix]

The interrelation of communication and cooperation remains therefore a key element of success in efficient product development.

5.5 Effective participation

One problem associated with cooperation is that the advantages of critically appraising the characteristics of a given concept does not provide a way to influence the decision-making process of the product implementation. As Bravo (1993, 11) commented regarding the participatory design of software, there is a difference between making suggestions and making decisions.

Only taking a cooperative approach with the development group, production and quality groups can make essential suggestions on how to improve a product's maintainability and producibility characteristics, but the designers have the final decision. This was an important rule for (internal) design reviews in some non-aerospace companies investigated. Designers decided what was to be implemented, what was to be omitted.

Box 59. Example of the difference between identifying problems, suggesting solutions, and deciding on the implementation (case #45).

In a case of high-reliability electronic equipment, production testing personnel discovered a potential failure mechanism. An unrecoverable hardware latching occurred when a certain switching sequence was performed by the operator. In the subsequent discussions, the chief designer argued that the sequence leading to the failure was quite unlikely to occur. His reasoning explicitly excluded operator mistakes, which many of his colleagues felt was unjustifiable. The chief designer maintained that the highly unlikely mistake did not justify the many major changes that would be required to completely eliminate the flaw.

After much discussion, development decided to mitigate the original design flaw with a component that could absorb the transient, thus eliminating the failure mode, but this was agreed to only in response to considerable pressure from many sides.

A participative approach provides the option to collaborate actively in the development process. This means that members of other departments can develop part of the product or conduct studies or research that would be used for the product development. With this approach, there is a mutual support among groups, which has a very positive effect on the interaction between the different types of professionals working with different types of activities and the associated problems.

Box 60. Implementing a participative approach (case #46).

In a company working for the defense industry, a product was to be developed with extensive built-in test system (BITS) capabilities. This was an essential product quality[204] and was part of integrated logistic support. The equipment was to indicate any failed units to be replaced in order to ensure the fast return to product functionality in case of failures. The product specification called for a very high *built-in test fault location coverage*. The activities of product assurance and logistics were assigned to the quality engineering department. This department did not limit its activities of reliability and maintainability assurance to the performance of MTBF/MTTR calculations. It endeavored instead to influence the definition of product characteristics.

The development manager insisted that the BITS fell to the quality group, and hence took no actions in this direction. He did not even contact the group for pertinent discussions.

The quality group engaged in a long search for solutions for equipment and component testability in many institutions. After having found a solution and working it out, they promoted informational sessions to communicate this information, especially to the development engineers. Two quality assurance engineers met with the development engineers and gave them concrete information about implementation possibilities, what could they use, what technique could be applied, the advantages of the one or the other approach, etc., in order that these product capabilities be implemented.

This approach succeeded in creating an atmosphere of participation between the development groups and a department external to the development groups to perform problem analysis, to conduct pertinent studies and to define the detailed product configuration.

Referring to this participation, the development manager for this product declared, "Before we started working together, I thought your tasks were limited to controlling people. Regarding testability, we had no idea what we should do, how to solve the problems involved. We even assumed it would be unnecessary! Our joint results were in this sense very productive."

[204] The characteristic "extensive BITS" was a part of the marketing mix.

5.6 Pitfalls of the group approach

Despite all the advantages of group work for product development, there are certain caveats. Fortunately, apart from the time it takes to cultivate interaction, undesirable effects of group work are infrequent. Nevertheless, awareness of these effects is helpful to avoid what can turn out to be an embarrassing failure.

5.6.1 Time to interact

Many companies using the multidisciplinary approach to process engineering review changes (ERC) have reported that the entire process is too slow. Piles of ERCs must be processed as every proposed modification is evaluated by many persons, this can have negative consequences for the entire organization.

Box 61. Design by committee.

The case of the British TSR-2 (tactical strategic reconnaissance) plane constitutes a failure to leverage the simultaneity of development and production, another early case of simultaneous engineering. "The initial contract was placed on October 7, 1960, to construct nine development aircraft to be built on production jigs. The disadvantage of this method was that any modifications required as a result of flight tests would have to be incorporated retrospectively to many airframes already on the production line. It was planned that the seventh development aircraft would be the first off the Preston final assembly line. ... The entire project developed into a "design by committee", where even the smallest decision had to be discussed by several people, although it could have been decided upon by one person. Such was the rapid development of aviation in that period, that these delays enabled the Air Ministry to demand changes to meet new roles and requirements, which led to further delays when redesign was needed" (Hygate, 1990, 249).

It must be emphasized that the TSR-2 was canceled mainly because of changes of British government politics under economical problems, not because of technical characteristics of the developed prototypes. However, delays caused by introducing the element of group heterogeneity illustrate that a process can actually be encumbered by the additional interactions implicit in this approach. Much about multidisciplinarity has been learned by technical companies since then. One of the problems at that time was the formalism associated with processing of each modification.

The aerospace industry worldwide continued to use the ERC multifunctional committees approach at least until the 1980s. But in order to avoid slowing down the product evolution process, they had to preclassify every case according to the levels of interconnection required by a given proposed change prior to submitting it to the committees. Sluggish change processes have been mitigated today, the purpose of this example being to illustrate the potential disadvantages of group interaction, one of which is that the group has to be given time to interact.

This case confirms the findings of many social science researchers that—*for the rational solution of complex problems*—group solutions take more time than individual solutions. However, the solution arrived at is usually more correct than that of individuals (Shaw, 1932, cited in Hill, 1982) (cf. discussion on task typology, page 213ff.). Davis & Restle (1966, 104) mention that Shaw's classic study provided evidence that group interaction facilitates individual cognitive processes, a point of particular interest in product development.

Connected to the idea of multidisciplinarity is the reduced interpersonal distances, similar to the situation prevailing in small startup companies. Closeness fosters agility in interconnected problem-solving, thus compensating for the time to interact.

5.6.2 Barriers and boundaries

Another prevailing illusion is that multidisciplinary groups naturally function better, as the barriers between disciplines are removed. In reality, however, real barriers have still to be removed, not only the walls between the departments.

Pill (1971) investigated two interdisciplinary programs in urban development. One of the programs "attempted to collaborate with the relevant components of the urban decision making system; and finally, it was to 'emphasize the use of the operations research and systems approach, including cost effectiveness' " (op. cit., p. 40). But the program failed. Among the many reasons was that divisions along disciplinary lines were followed (op. cit., p. 166) with little interdisciplinary collaboration among the task teams (op. cit., pp. 157, 182). In contrast, the other project had numerous "bridgers" among its consultants and clients (op. cit., pp. 189, 191), was content-oriented, and showed no signs of decreasing commitment to technical objectives (op. cit., pp. 201-202).

Pill found that multidisciplinary in development programs is not a sufficient condition, concluding that the disciplinary lines and how to cross them is a decisive factor.

Hirschhorn & Gilmore (1992, 106)[205] state that "all too often, managers think that getting rid of boundaries also means doing away with conflict. They assume that once the company breaks down the walls that 'get in the way of people working together,' employees like the engineer and the worker will put aside what divides them and unite behind the company's mission. Differences of authority, talent, or perspective will no longer be a source of friction. Nothing could be further from the truth. As traditional boundaries disappear, establishing such differences becomes simultaneously more important and more difficult". Hirschhorn & Gilmore focused on the phenomenon in flexible organizations that the removal of formal boundaries merely makes room for other kinds of boundaries, such as authority, task-related, political, and identity boundaries:

As the function of the boss is no longer defined as the one who issues the orders, collects and channels the pertinent information and whose main task to monitor and control, the *authority boundary* appears. Superiors exerting more effective leadership styles face the challenge of establishing a functioning boundary in that the "subordinates feel trusted by their superiors, and that feeling of trust frees them up to exercise initiative at work. The superiors feel simultaneously supported and challenged by their staffs, which allows them to lead". (op. cit., p. 107)

People working on group tasks are thrown together, and consequently the *task boundary* appears as the former formal relationships between such tasks as development, marketing, production, etc. cease to exist. "Individuals have to depend on others who have skills and resources they cannot control and often don't even understand. To be effective, they cannot simply ignore the work of others in effect, to say 'it's not my job'." (op. cit., p. 108)

The *political boundary* refers to the interaction of groups with different interests. It challenges the groups to "defend their interests without undermining the effectiveness and coherence of the organization as a whole". (op. cit., p. 109)

The *identity boundary* is associated with values as well as with in-group and out-group identification. "The identity relationships are important because they tend to be extremely energizing and motivating". (op. cit., p. 109)

[205] Reprinted by permission of Harvard Business Review. Excerpt from: Hirschhorn, L. & Gilmore, T. (1992). The New Boundaries of the "Boundaryless" Company, Harvard Business Review, May-June 1992, pp. 104-115. Copyright © 1992 by the Harvard Business School Publishing; all rights reserved.

The following case illustrates these issues.

Box 62. A multifunctional team that failed. Condensed from Hirschhorn & Gilmore (1992, 110-111).

A midsize office equipment manufacturer faced serious competition from lower-priced products. It had to counter by developing a new product. Management knew that their functional organization "simply could not respond fast enough". The Chief Operating Officer (COO) proposed a self-managing team approach and appointed a 12-member team composed of representatives from marketing, manufacturing, "engineering" and finance. He also hired a trainer to teach members new skills such as problem solving and group dynamics. As he watched the team grow, he felt extremely encouraged. "The old fighting was gone. He was convinced that he'd fashioned the first successful interfunctional group in what had always been a tradition-bound and turf-conscious company". The team finished the new design on schedule and presented it to the company's executive committee. "Everything proceeded without a hitch. However, when it came time to discuss the team's proposals, the meeting broke down. Company executives asked tough questions to test the task force's design concept... But instead of responding to these legitimate questions, the team members kept defending their original proposal. They didn't think *any* aspect of their design could be modified. And team members seemed ready—too ready—to leap to the defense of fellow team members from different functions. When one senior manager, for example, argued that the product's manufacturing costs were too high to be competitive in the current marketplace, it was a team member of *marketing* who insisted that the costs couldn't be lowered. Similarly, when another senior manager criticized the product as too complex and therefore difficult to manufacture, it was someone from *manufacturing* who defended the complexity as necessary for producing a first-class product". In the end, the company's top management was convinced that the task force had become inflexible. The product design was rejected and the team sent back to the drawing board. Several senior managers insisted that the COO assign different people to the team.

According to Hirschhorn & Gilmore, three boundary mistakes occurred:

- The COO did not create a clear authority boundary.
- The team coped with the absence of authority by enacting an identity boundary that was too strong. All of their energies went into maintaining the group identity that differentiated them from other groups, even at the expense of suppressing the heterogeneity of skills and perspectives that would have enhanced their ability to better cope with their tasks.
- The overly strong identity boundary prevented team members from creating the internal task and political boundaries they needed to do their work. For example, the manufacturing engineer didn't sufficiently

challenge the design engineer's work from the perspective of producibility. The marketing representative rarely questioned the engineer's cost assumptions from the perspective of customer acceptance and pricing. As a result, the team failed to leverage the various skills available to optimize the product design.

In addition, the team failed to represent their original interest groups, and as such could not effectively sell the new design to their own colleagues.

Phil Condit, CEO of Boeing, reported a tendency observed in connection with the design-build team concept used for the development of the 777. "If you are not careful," he said, "organizations, just by the way they behave, want to form little enclaves, so Structures goes off over here and Air-conditioning goes off over there; Manufacturing goes over here and Planning goes over there, and Finance goes somewhere else. And then one day you turn around and you've discovered there's an 'us and them' attitude. 'If only they would do this, then our job would be easier'." (Sabbagh, 1996, 69). It is a problem that appears with scale growth.

5.6.3 Groups and cognitive dissonance

In her investigation of social influence and cognitive dissonance (cf. page 173), Signol (1997, 75) lists circumstances under which exposure to information increases dissonance:

1. accidental exposure, in that the information comes by chance;
2. exposure to an unrelated base,[xl] in that the individual does not expect the information, but it simply appears;[206]
3. forced exposure, in that it is difficult to overlook, even if one is not seeking the information;
4. social interaction.

These circumstances do not always produce cognitive dissonance, but they are expected to be increasingly present in an interacting product development group. This is especially the case for multifunctional or multidisciplinary groups, where the experts might constitute the minority among a diversity of experts. Recalling the discussion of serendipity and its potential stimulation for detecting problems and solutions, circumstances #1 and #2 above are especially appropriate. In a presentation by Jocelyn Bell Burnell concerning her discovery of pulsars, she mentions many interactions with colleagues who played a role in perceiving her exposure to an unrelated base (circumstance #2). Of the numerous legends surrounding the pulsar discovery, it is said that one scientist had previously observed the

[206] Such as a person watching a movie that he/she wants to see and is exposed to information broadcast during commercials.

phenomenon, but attributed it to an equipment artifact and hence disregarded it. Though unconfirmed, this anecdote is included to illustrate similar situations that occur in practice.

Extending the notion of cognitive dissonance to the context of collective actions, Sande & Zanna (1987) recall the necessary conditions for arousing dissonance in that an individual (i) must engage in a counter-attitudinal behavior that (ii) has aversive consequences, for which (iii) he/she assumes personal responsibility.

Sande & Zanna examined three means of reducing the dissonance aroused by collective actions: (i) attitude change, (ii) diffusion of responsibility, and (iii) seeking social support from within or from outside of the group. The first occurs at the individual level and is a simplification—perhaps an oversimplification—that cannot be resolved within the scope of this book. An individual's commitment to one's initial attitude plays a critical role in reducing the dissonance by means of an attitude change. The second is of special significance in a group setting because the diffusion of responsibility is not available to individuals alone. This option can potentially give a group more exposure to negative effects of cognitive dissonance than an individual. Again a simplification, the commitment to the group tends to reduce dissonance by diffusing responsibility. Sande & Zanna also mention the increased public nature and accountability of the collective decision. When neither of the first two is viable, the third may come into play.

At this point it must be emphasized again that this is not expected to happen frequently. Nevertheless, it might occur that one group ignores, say, test results or supplier information or a detail of the design that might or might not be interconnected with other design details. This might mean the inability of the group to catch some "small" detail that may turn out to be a cause of failure.

5.6.4 Group pressure

A well-known instance of group pressure was demonstrated by Asch (1951, 1956). The following presentation of Asch's findings was condensed from the description by van Avermaet (1988, 352-354).

In an experiment, a line of a certain length was shown to test subjects 18 times. The subjects were then asked to choose from among three other lines the one with the same length as the reference line (only one line had the correct length.) The subjects were placed in a group comprising seven persons seated in a semicircle, and were instructed to respond in sequence. But only the 6th person was the test subject; the others were collaborating with the experimenters. To test their influence on the subject, the collaborators intentionally gave a wrong response 12 out of 18 times. Control groups tested the difficulty of the task without the manipulation factor, and found that 2 out of 37 persons (5%) gave the wrong response, corresponding to a failure rate of 0.7%. In the experiment, the failure rate jumped to 37%. Indeed, 75% of the subjects gave at least one wrong response and 28% gave eight or more. During the experiments the experimenter and the collaborators behaved neutrally, expressing neither surprise nor negative reactions to individual responses. As expected, the subjects showed various signs of non-neutral behavior.

Note the complementarity of this study and those after Moscovici et al. (cf. page 223 ff.). The two groups of studies form a duality of influence on group perception in that the majority can have an influence on the minority's perception and the minority can have an influence on the majority's perception.

Asch's findings cannot be generalized directly to problem solving, but they do illustrate a relationship with a *perceived* characteristic of an object overruled by the majority. Application of Asch's results to product development is significant, as the first step toward problem solving is the perception and characterization of the problem.[207] Multidisciplinary product development teams are based on the assumption that one of the members (or a subset of its members) recognize aspects that the others overlook. This complementarity is one of the essential contributions of multidisciplinarity to product development. However, when the group is pressing forward, it is often difficult for an isolated member to insist on addressing an issue that is not perceived by the others as significant. The group may lose one of the essential advantages that heterogeneity is meant to contribute.

It must be said that even though the capability of individuals to work in teams is given considerable emphasis today, teamwork can be misguided or even failure-prone in the context of product implementation if pressure by the majority serves to silence a different perception of the issue considered (cf. Section on the importance of the minority on page 223). What results is a pseudo-teamwork situation in which the true potential of the team is jeopardized. Even functional organizations should be aware of the issues associated with the pressure of the majority.

Other instances of group pressure of interest for product development is the notion of *groupthink*. Janis (1972) defines it as "A mode of thinking that people engage in when they are deeply involved in a cohesive in-group, when the members' striving for unanimity override their motivation to realistically appraise alternative courses of action" (op. cit., p. 8). Famous cases of groupthink are the decision to invade the Bay of Pigs in Cuba, the

---

[207] Cf. first step of general problem solving process on page 175.

decision to engage in the Watergate cover-up, and the last launch decision of the Challenger space shuttle. A revised framework for groupthink in the context of the Challenger is provided by Moorhead, Ference & Neck (1991) and is described in Box 63.

Box 63. Groupthink framework by Moorhead, Ference & Neck (1991).[208]

Antecedent conditions

- cohesive group,
- leader preference (the leader has one preferred outcome),
- insulation from experts.

Groupthink symptoms

Illusion of invulnerability
leads the group to become overly optimistic and willing to take extraordinary risks. It may also cause them to ignore clear warnings of danger.
Rationalization
Group collectively constructs rationalizations that discount warnings and other forms of negative feedback.
Morality
Members believe, without question, in the inherent morality of their position. [They] tend to ignore the ethical or moral consequences of decision.
Stereotyped view of others
[Belief that] the opposition is too stupid or too weak to understand or deal effectively with the problem.
Pressure on dissent
Group applies direct pressure to anyone who questions the validity of the arguments supporting a decision or position favored by the majority.
Self-censorship
Members tend to censor themselves when having opinions or ideas deviating from group consensus.
Illusion of unanimity
concerning judgments made by members speaking in favor of the majority caused by self-censorship and false assumption that those who remain silent are in agreement. The leader and others support each other by playing up points of convergence at the expense of fully exploring points of divergence (that might reveal unsettling problems).

[208] Condensed from: Moorhead, G.; Ference, R. & Neck, C. P., 1991, Group Decision Fiascoes Continue: Space Shuttle Challenger and a Revised Groupthink Framework, Human Relations, Vol. 44 No. 6, pp. 539-550, Tavistock Institute of Human Relations. Reproduced here by kind permission of Sage Publications Ltd.

Mindguarding
certain members attempt to shield the group from adverse information that might destroy the majority view of the facts regarding the appropriateness of the decision.

Decision-making defects
Few alternatives

No reexamination of alternatives
Group fails to reconsider alternatives discarded before by early unfavorable information.
Rejecting expert opinions

Rejecting negative information

No contingency plans
Spending little time discussing the possible consequences; the group fails to develop contingency plans.

Important Moderators
(Leads to the symptoms starting from the group characteristics)
Element of time pressure

Role of leadership
Leadership style is a crucial variable as illustrated by the differences in the cases of the decisions on the Bay of Pigs and the Cuban Missile Crisis where the same group of the Kennedy administration acted in a very different manner.

The possibility of groupthink occurring in product development should not be neglected. First, the antecedent conditions are potentially present. The antecedent condition "insulation from the experts" can occur even when an expert in one area is involved in the group decision, as this by no means guarantees that all areas of expertise are covered. Second, the cases presented throughout this work illustrate many of the symptoms that occur in isolated cases of product development. Third, decision-making defects were also found in product development, particularly in the form of a rejection of expert opinions or negative information. Finally, time pressure (one of the moderating variables that appear to be increased) bears a significant potential to induce a group to avoid evaluating all options thoroughly, thus potentially incurring other decision-making defects. The role of leadership is considered to be critical and may become even more so, particularly for so-called "leaderless groups" in that the function of leadership is diffused in the group elements.

5.7 Penetration and terrorizing

The concepts that affect product development at the environmental, individual and group interaction levels were presented, especially the aspects of autonomy, the dirty-hands philosophy, creative tension, and control in the broad sense. In the following sections, the concepts of penetration and terrorizing shall be examined. They are frequently confused, but the difference between them is significant in product development, which occasionally requires precise measures for specific issues.

5.7.1 Penetration

In the Apollo program, North American Aviation (NAA) developed and built the Saturn V second stage as well as the Apollo command and service modules. For their work on the second stage, they had to interface with NASA's Marshall Center (MSFC). For the command and service modules, they had to deal with NASA's Manned Spacecraft Center (MSC).

The concept of *penetration* denotes the shift from the "arsenal approach" to the "contractor approach" by MSFC in Huntsville (cf. Chapter 1). This occurred in response to the people at Marshall, who were not used to playing the role of merely "accepting" or "rejecting" finished goods. Indeed, they had the substantial advantage of having the *yardstick* or measure of reference for product development also described in Chapter 1.

Their extensive knowledge about the item being contracted equipped them to "penetrate" the contractor's facilities to follow development evolution because they had the necessary updated technical and managerial knowledge associated with the product they were buying.

This penetration did not occur without certain resistance, as the following account by Bilstein (1980, 277-278) describes: "Eberhard Rees[209] admitted that the surveillance of the contractor operations, as well as their management, was 'somewhat sensitive from the point of view of the contractor.'[210] In many instances, contractors felt that they should be allowed to go their own way after the contract was signed. The longing for more freedom of action was evidently a legacy of the experience that most Saturn contractors had previously with Air Force contracts. Huntsville had great

[209] Rees was deputy center director at Marshall under von Braun.

[210] Cf. Section on achieving cooperation as a process on page 234ff.

technical competence. At certain managerial levels of design and manufacturing, grumped one highly placed contractor executive, Marshall maintained a one-on-one surveillance. The Air Force, he said somewhat wistfully, was 'not in your pants all the time.' But Rees maintained that loose reins on the contractor had not always worked out well from the MSFC point of view. 'Consequently,' he said, 'it became clear that close and continuous surveillance of the contractor operation was required on an almost day-to-day basis.' The extent of the surveillance was proportional to the subtleties and problems of the program, its relative position in relation to the existing state of the art, and the extent of the expertise possessed by MSFC. The contractor's reaction to this aspect of NASA monitoring was not favorable at first, but eventually this 'penetration and monitoring' was perceived to be a mutual benefit characterized by the often repeated phrase, the 'government-industry team'. 'Contractor penetration' was an important concept that ultimately involved a contractor's relationship with its own subcontractors. "

"The technique of contractor penetration to maintain high visibility obviously generated some thorny issues in government-contractor relations. Nevertheless, MSFC felt that industry had a strong inclination to take control of the job as well as the funding and pursue the job with a minimum of government intervention. MSFC management believed this inclination allowed too much opportunity for slippage, unidentified problems and poor communications. Vigorous contractor penetration reduced these program difficulties; *in the long run, the contractors seemed inclined to accept the penetration as a mutually useful aspect of completing a successful program. 'The restiveness that stemmed from such close control was gradually dissipated very early in the Apollo program as the benefit accruing from the industry-government team approach was revealed,*' concluded Eberhard Rees" (op. cit., p. 281, italics added).

The Marshall Center was accustomed to discussing the minutest details with its contractors. This was part of their *modus operandi* and they could operate in this way because they had knowledge and experience in developing and building rockets.

NAA, for its part, was accustomed to dealing with the U.S. Air Force, which took a different approach. Gray describes the collaboration from the contractor's perspective (1992, 137): "For one thing, North American was not used to being micromanaged by the customer. When the company built a plane for the Air Force, Storms[211] never had more than a handful of people looking over his shoulder; the Air Force laid down the specs and stood back.

[211] Harrison Storms, general manager of North American's Space Division.

But NASA was actively involved in every detail of the design and wanted to know everything about everything."

Another account from the contractor's side shows the difficulties of working together: " 'The Germans wore their design arrogance,' said Joe Goss.[212] 'They knew all about booster-vehicle design. When North American introduced aircraft-design elements, they didn't think we knew what we were doing.' A typical shootout was the one that took place over the 'kick loads': the bi-directional forces that would be exerted on the joint between the dome and the top rim of the tank cylinder. Bill Parker's designers had come up with a set of numbers that didn't agree with the calculations from Huntsville. Willi Mrazek was head of the design group in Huntsville [...] and he was not easily swayed in an argument. Storms and Rees were quickly drawn into the discussion, and all of a sudden these two huge engineering organizations were nose to nose with the clock ticking. They spent hundreds of hours in tests and analysis, and finally they submitted the results to Professor Court Perkins and the Scientific Advisory Group for independent evaluation. They determined that North American's numbers were correct. This did not improve the relationship between Huntsville and Downey. Eberhard Rees did not take losing lightly, and Storms was not a graceful winner." (op. cit., pp. 160-161).

Despite this reference to "winners" and "losers", these descriptions reveal that technical discussions concerning the Saturn V launcher led to one winner: product quality and its customers. Not always was the contractor the "winner". In another situation reported by Tompkins (1977, 13) a contractor delivered a rocket stage to Huntsville and was promptly quizzed about cracks in it, after which the contractor personnel admitted the existence of *twenty-one* cracks. The customer presented the contractor with evidence of *twenty-six* cracks. The customer was more familiar with the delivered item than the manufacturer at the moment of delivery! This was made possible by the concept of penetration. Tompkins interviewed a top-ranking manager at MSFC, who declared that because of penetration, "a contractor's first experience with the Marshall Center was often 'traumatic'. ... [Contractors were] well known for discouraging the upward flow of 'bad news' ". Tompkins goes on to describe how penetration can also be an internal phenomenon. He reports an interview in which a "thoughtful lab director" declared that "just to get the job done, every man has to penetrate two layers up and two layers down." In the case of penetrating downwards, it was not done to "supervise the subordinates' subordinates, but simply in order to talk and listen." Hence penetration can enhance communication within an organization.

---

[212] Joe Goss worked on the Saturn V second stage.

Once a proponent proved he was technically right, it was accepted. This was done internally as well, but apparently with a completely different style, or at least with a different perception of the style used. Tompkins (1977, 9) describes an internal event where a "lab director whose earlier recommendations about the superiority of one kind of weld over another had been rejected, and who had reported additional evidence which supported his original position" received his reporting note back from von Braun with a comment that read, "Looks like you won after all! Congrats. B". (cf. brief description of Wernher von Braun in the section on the role of the leader, page 293).

A different kind of interaction occurred between NAA and NASA MSC involving the choice of the atmosphere of the command module. It was defined *in the specifications* to be of pure oxygen, which would lead to the dangers of high-pressure oxygen around every fabric and flammable material on board, especially when the capsule was in the earth's atmosphere (in space, the absolute oxygen pressure is lower). The contractor NAA warned of these dangers and proposed a change. MSC personnel replied that it had always been done like that in the preceding Mercury and Gemini capsules, and that they had never had problems with it before. The alternative solution proposed by NAA required monitoring the partial pressure of oxygen (instead of simply controlling the cabin pressure, what is much simpler), so MSC rejected it. NAA alternative proposal was certainly not fail-proof, and that the consequences—astronaut suffocation—were indeed quite grave. The discussion concluded with MSC stating that NAA was the contractor, and the contractor *should do as ordered.* (cf. Gray, 1992, 138-139)

Thus, one day in 1967, a short circuit[213] in the pure oxygen atmosphere in Apollo 204[214] caused the fire that killed the three astronauts, and a nation mourned.

Ironically, NASA Marshall Center had the reputation of being formal with regard to its relationships with contractors, whereas NASA MSC had one of being informal. Contrary to expectations, the "formal" structure proved in these cases to have been more communication-driven and more receptive to meaningful discussions to improve product qualities. This perception of formality could well stem from the German culture permeating MSFC, whereas the culture prevalent at MSC, where people called each other by their first names, for example, encouraged the perception of informality.

---

[213] This was pointed as the most probable cause.

[214] Apollo 204 was later renamed Apollo 1. For a complete account of this tragedy, see Gray 1992, 137-139.

A further discussion of the differences in culture is beyond the scope of this book. It is the intention here merely to emphasize the importance of (i) contextual factors around the discussions on product implementation in the anecdotes presented and (ii) the quality of interaction among professionals dealing with product development. The focus of this section is the concept of penetration and its effects on both contractor and agency in improving product characteristics.

The different perceptions of this form of control (in the broad sense) are also of interest. There is a distinct tendency to perceive penetration as a form of intrusion, which developers frequently try to avoid. There is also an inter*penetration* of knowledge and interaction, which not only enhances one's knowledge of a product's characteristics but also improves the quality of a product's conception. The manner in which penetration is conducted as well as the emphasis on the type of control imposed are instrumental in moderating conflicts. Reaching a common understanding about a product's implementation is vital to whether a product's weaknesses will be discussed openly. If this level of understanding can be achieved, than the discussions will be problem-oriented, instead of being driven by management, politics or individual egos.

5.7.2 Terrorizing

It is important to distinguish between different forms of control. As control can imply taking action toward the people performing a development, it is sometimes the case that managers take drastic measures to exercise control. The management styles appear to diverge: some development managers treat development personnel well, others treat them poorly. Which is the best approach? Let us examine some specific cases.

McGrath, Anthony & Shapiro (1992, 225) cites a case in a "minicomputer manufacturer [where] project managers never left their offices except for 'team beatings' that were based on the assumption that people worked harder if they were constantly harassed and threatened. This company's time to market never improved; in fact, it got longer, and the company was eventually bought out by a competitor."

However, other cases appear to illustrate the contrary: that people respond to pressure by improving their efficiency. Two cases in the computer industry illustrate such instances.

Box 64. Two cases of using threats as a "driver".

A computer company had tremendous success in the 1970s with one of their first machines. It subsequently endeavored to bring an improved successor to the market, but failed. It then changed to an innovative new design with a very promising concept using a graphics interface. In spite of the innovative concept, the computer had a high failure rate in the field, it was priced too high for the market, and almost ruined the company. Near desperation, the company raced to develop a simplified new machine in which the problems of the former one were eliminated. Indeed the new machine allowed the company to regain its position on the market—obviously a success story. But let us examine the aspect of interest in this context: *how* the company arrived at this product. It was reported that the chairman and co-founder of the company continuously yelled at the staff assigned to develop the operating system, threatening to fire them if the work was not completed by a certain date.

Another young computer company (case #47) developed a microcomputer in the 1980s. After a series of failures to build functioning prototypes, the development team was threatened with dismissal if they did not succeed. They ultimately did achieve a functioning product, but their implementation was very inefficient from an ergonomic point of view and had an extremely inefficient hardware structure. In fact it was full of design inefficiencies, but it worked. A major criticism was the excessive component costs, but this had not been given top priority during development because the machine had not been intended to be the cheapest on the market. The machine was not a market success, but it at least constituted a presence on the market.

One theory says that, in order to change, people must be made to feel uncomfortable with a situation. Whether the motivation of the developers was indeed triggered (or driven) by the kind of attitude described in the cases above is highly questionable. As Herzberg (1987, 110)[215] puts it, a "KITA" (kick in the pants) does not constitute motivation. "If I kick my dog (from the front or from the back), he will move. And when I want him to move again, what must I do? I must kick him again". There is no evidence that this kind of attitude was really what made people perform their job. Yet some managers insist that it is necessary because otherwise people will not do what is expected of them.

It might be reasonable to think that there are always some individuals who avoid work and eschew responsibility. Nevertheless, most people try to perform their job as best they can, and object to this kind of attitude from their management or contracting agency. Even if they feel compelled to submit to such treatment for whatever reasons, people's enthusiasm and morale will usually decline over time in such an environment.

---

[215] Republication of Herzberg's "classic" contribution to Harvard Business Review Jan-Feb. 1968, of which more than 1.2 million reprints were sold.

Nevertheless, independent of the morale of the individuals involved, there are cases where a process has reached an impasse, and it takes a certain impetus to get it running again. As described above, Harrison Storms at North American was known to bully people into performing better. An example from his biography illustrates the situation.

Box 65. "Cancelitis" as described by Gray (1992, 62-63).

The missiles were tested at the Atlantic Test Range, an Air Force monitoring operation along the coast of Florida off Cape Canaveral. Here the test object would be carried aloft on a B-52 flying out of Eglin Air Force Base and be fired off into the South Atlantic. Since this operation was all the way on the other side of the country, North American had set up a listening post in Downey.

On the morning of the test, Storms was in his office when Scott Crossfield called from the radio room and said, 'Boss, I think you'd better come down here.' Storms went down immediately, and as he walked in, the project engineer said the flight had already been canceled. Halfway through the countdown they had picked up a red flag on one of the systems and the B-52 was returning to base. Storms said he wanted to talk to the flight crew. He grabbed the microphone and he spoke to the pilot high over the Atlantic. 'How much gas do you have?'

After a moment of puzzled silence, the pilot said he had about five hours of fuel on board.

'Then turn back to the Cape and go back to the top of the checklist and start over,' said Storms.

The people in the room looked at each other—and so, no doubt, did the men in the cockpit 3,000 miles away. The test director said. 'Mr. Storms, we can't restart the test. The Air Force has closed the test range.'

'Open it.'

'But everybody's gone home!'

'Get 'em back,' said Storms. 'And don't land that bomber with more than a teacup of gas in the tank.'

Suddenly, the airwaves crackled with traffic as the operations people scrambled to get the crews back and reopen the test range. And two hours later, on the second pass through the checklist, somebody spotted a switch he'd forgotten to close the first time. They launched the missile, and it flew like an arrow to the target.

'Cancelitis.' Storms said, 'is a disease. Once you start canceling test flights, it gets easier and easier.'

This case required "strong" leadership. If Storms had not intervened, the activities would have stopped, maybe only for that day, but maybe for the rest of the program.

Sometimes, it is not a person but an unpleasant event that shakes up a project. In the moon program, the fire on Apollo 204 had that effect. "Big organizations behave like living organisms, and the shock of the fire went through NASA and North American like a jolt of adrenaline. It seemed as if everybody suddenly remembered what this was all about. Old animosities dissolved. People who had been nose to nose and toe to toe for five years suddenly found themselves in lockstep. 'Down on the working level, there was no finger-pointing,' said Dave Levine.[216] 'Everybody got their act together. We were an impressive bunch.' " (Gray, 1992, 251).

The situations in which such blockages of persons or groups occur are beyond the scope of this book. Suffice it to say that such situations are not rare and apparently the unpleasant way is the only way to get an activity back on track. Carried to an extreme level, however, this situation can deteriorate into what is known as *terrorizing*.

The term terrorizing stems from practitioners themselves. Another example from the moon program describes its origin. Regarding his dealings with NAA,[217] Eberhard Rees wrote in a confidential memorandum quoted in Murray & Cox (1989, 183): "I believe NASA has to resort to very drastic measures, including the possibility of shifting to a new contractor".

In his account of the problems with North American's second stage, Bilstein (1980, 226) cites a comment by Dale Myers, later Associate Administrator for Manned Space Flight, that "there is a need to *terrorize* the contractor once in a while." Terrorizing of this kind was conducted by General S. Phillips.[218] It took the form of a " 'Tiger Team', one of the weapons he had used on the builders of the Minuteman. Separate groups ... would descend on California and rake through Storm's whole operation. Phillips himself would lead the charge" (Gray, 1992, 199). The result of this terrorizing was the so-called *Phillips report,* which "ripped into the company's management, not only on the S-2 matter, but on the spacecraft as well" (Bilstein 1980, 225).

From the contractor's side, the following comment was documented: "Sam Phillips was, by all accounts, an outstanding manager, but his examination of North American was flawed on two counts. For one thing,

---

216 Dave Levine: chief electrician for the Apollo spaceship at North American. In 1966, Levine had 2,000 people working under him. (op. cit., p. 216).

217 The reason for taking so many cases associated with North American Aviation is that they were in charge of the Saturn V second stage *and* the Apollo Command and Service Modules, all of which are very complex and associated with a diversity of implementation problems. Some people at NASA expressed concern that this task was too much for a sole contractor.

218 Gen. S. Phillips was Apollo Program Director at NASA headquarters in Washington. He had previously successfully conducted the Minuteman Missile program.

his auditors examined only the paper trail and paid little attention to the hardware itself. They compiled statistics on missed deadlines, late drawing releases, and cost overruns, but there was apparently no one on the team with a clear understanding of the structural design problems of the S-2, or the chaos in engineering that NASA's own people created every time they asked Bill Parker to carve another hundred pounds out of the thing. And *while it was true the spacecraft was running three months behind schedule, it was cavalier of Phillips to lay that at the contractor's feet and dismiss NASA's complicity.* Years later, one of the key men on the Huntsville team would apologize to Paul Wickham (S-2 chief engineer at NAA) for this witch hunt, but that was long after Wickham was off the program and the damage had been done." (Gray, 1992, 202, italics and parentheses added).

NASA had many reasons for doing this, but it caused much concern, not so much regarding the way in which contracts were awarded, but regarding the feeling it created that things were radically wrong. It was apparently not an extension or direct consequence of the penetration concept described above, but rather an emergency action to get a program back on track. This differentiation is extremely important.

Terrorizing remains a matter of controversy. Some development managers defend its use, others abhor it. Recalling the coping mechanisms under conditions of fear or of courage discussed by Maslow, the reader can imagine the collateral effects that it brings.

5.7.3 Main differences and dithering

The main differences between penetration and terrorizing can be stated as follows. Penetration is used to focus discussions on the tasks at hand. It functions well if distributed over an entire development process as a means to maintain a creative level of tension, thus improving product implementation.

Terrorizing is the last-ditch emergency action to dispel an impasse. It is not necessarily task-oriented, although it might use technical arguments. It can only be effective if applied rarely, but the risk of persons or organizations reacting maladaptively is not to be neglected.

Now, let us recall the previous discussion of creative tension. Penetration, if correctly implemented, can be an instrument at the inter-individual or inter-organizational to maintain a healthy level of creative tension that enables the focus on development goals to be maintained. If applied correctly, it helps detect and correct mistakes, misassumptions, misconceptions, and other coping defects—common to any development—thus enabling an action before it is too late.

At this point of the discussion, the concept of *dithering* should be introduced as well. Dithering is a technique of feedback control systems used to linearize a device's inherent nonlinear characteristic, such as hysteresis, cogwheel backlash or small friction in mechanical instruments. The basic idea is to superpose a high-frequency signal to the desired input to the device. The response of the device then corresponds to a linear reaction to the input. In its simplest form, it consists of a small vibrating mechanism collocated with a mechanical instrument. The best known example of dithering is still found in analog tape recorders, where it is used to linearize the highly nonlinear magnetic characteristic of tapes. In more elaborate recorders, the *bias* level can be changed to adapt to different tapes. Using this concept of dithering for the dynamics of product development is not new. It's need was proposed by Weaver (1959): "we need a certain amount of dither in our mental mechanisms. We need to have our ideas jostled about a bit so that we do not become intellectually sluggish." This was further explored by Pelz (1967, 163) to establish the concept of creative tension (cf. creative tension no. 6, *Table 22* on page 201). In this sense, dithering can act as a means of coping with some of the human "nonlinearities". Penetration bears a resemblance to dithering in that it helps "linearize" responses at the individual and organizational level.

Dithering is quite different from force, and the two concepts should not be confused. Consider for example cases where people take drastic measures when a machine or device is improperly maintained or defective. In their frustration, they begin to pound or shake the malfunctioning machine, apparently in the irrational belief that this will make it work. This is obviously not the best way to operate machinery, and indeed will sometimes jam it even more.

5.8 Concluding remarks

Group work should not be presented as a panacea, as practitioners often take for granted that it is superior to individual work.

In an attempt to describe product development activities, task typology was introduced, but is nonetheless associated with the recommendation of not dividing activities along these lines. The divisions in task typology should be interpreted in terms of the situation currently being faced by a development group, for which a group or individual approach should be more appropriate for problem solving.

The classic factors (i) complementarity of viewpoints and experience and (ii) necessity of removing barriers between staff areas were examined with respect to the product development.

The importance of the minority was discussed, particularly in cases where group interaction supports the minority. Group interaction and the influence of member expertise was also discussed.

The process of achieving cooperation was introduced, for which the approach by Morton Deutsch was followed. Under the topic of cooperation, improving communication—generally accepted as a key factor of improving product quality—was examined.

Achieving participation in product development was presented in contrast to the cooperative approach.

Then, the pitfalls of the group approach were discussed: time necessary to interact, barriers and boundaries, cognitive dissonance at group level, group pressure including pressure of the majority, and groupthink.

Finally, penetration and terrorizing were compared and contrasted.

Chapter 6

SHAPING THE TASKS AND ASSOCIATED ROLES

6.1 Designing the job of designers

In the early 1980s, engineers employed in research and development knew they were in a fortunate career situation. They had the freedom to choose their work; they could propose alternative means of achieving the technical goals of the organization; they could participate in decisions; they had good prospects of influencing strategy; they drew satisfaction from their duties; there were numerous challenges to their technical training, for example those provided by the then-revolutionary technological arena of computers and microprocessors.

At that time, a novel concept was introduced by quality circles to the factory environment. Factory operators are given opportunities to influence their own work. This contradicted the then widely accepted rule of not letting blue-collar workers express their opinions, a rule that was explicitly taught in management courses. This new measure tapped an astonishingly rich source of ideas, and significantly improved motivation. The elements so highly appreciated by engineers were being extended to "simpler" activities at the factory level.[219]

---

[219] Cf. remark on quality circles in Germany in contrast to those in Japan on page 115.

Although quality circles are no longer used in their original form, they and many other concepts have brought significant progress to industrial job design, especially—but not exclusively—in the context of manufacturing activities. The following section describes aspects of both the individual (cf. Chapter 5) and group level. It closely follows the development by Ulich (2001) and will serve to emphasize crucial issues of job design for development activities.

6.1.1 Job characteristics and job design

Hackman & Oldham (1975, 1976) (cited in Ulich, 2001, 104), considered a series of hypotheses on motivation-promoting conditions of the work situation and their effects on the way people experience work as well as on their behavior. The resulting *job characteristics model* states that intrinsic motivation is tied to three "critical psychological states":

- Knowledge of the actual results of the work activities.
- Experienced responsibility for outcomes of the work.
- Experienced meaningfulness of the work activities.

Although it must be emphasized that the models described in this section are not restricted to product development, the reader is invited to reflect on how well product development-related activities stimulate these three critical states.

These three states are determined by five task characteristics called "core job dimensions" that provide the motivation potential

- Skill variety.
- Task identity.
- Task significance.
- Autonomy.
- Feedback (about the job).

The identification of the elements of product development activities is clear, but the most important variable in this model is the *individual growth need strength.* This variable means that, given an opportunity to enhance one's activities in the core job dimensions, a professional with a greater growth need would experience the effects in the psychological states more strongly than one with a lesser growth need. The former is also more likely to react positively to improvements of the core job dimensions. Expected effects on the work are higher intrinsic motivation, better performance, greater work satisfaction, little absenteeism, and lower employee turnover. The importance of this simple model in the context of product development can be enhanced by recalling the discussion on the work by Maslow on self-actualization (page 167) and the psychology of science (page 179). The simplicity of the model facilitates keeping these characteristics in mind.

Moreover, the evaluation method developed by Udris & Alioth (1980) called subjective work analysis[xli] (cited in Ulich, 2001, 108) examines the following.

1. Aspect of alienation with the categories
- decision by others vs. self-regulation,
- meaninglessness[xlii] vs. transparency,
- dequalification vs. authority to take action,[xliii] and
- social isolation vs. social commitment.
2. Workload aspects of
- qualitative underload and
- qualitative as well as quantitative overload.

These aspects are operationalized in a measurement instrument providing the profile indicated in *Table 32*.

Table 32. Task profile in the subjective work analysis.

PROFILE MAIN ITEM	SUB-ITEM
DECISION LATITUDE[xliv]	Autonomy
	Variability
TRANSPARENCY	Transparency of the work task (feedback)
	Social transparency (overview)
RESPONSIBILITY	Responsibility for a common task (status)
	Responsibility fir incidents[xlv]
QUALIFICATION	Qualification requirements
	Application of available qualification
	Chances (psychological future)
SOCIAL STRUCTURE	Peer support
	Cooperation (interdependency)
	Respect of superiors
WORK LOAD	Work quantity (quantitative overload)
	Difficulty (qualitative overload)

For the purpose of this discussion, the main interest in this job analysis tool is the job profile it produces.[220] The reader is requested to peruse *Table 32* to outline the profile for product development activities. Recalling the engineers' perception recounted at the beginning of this chapter, the development activities would probably score highly in many if not all of the profile items, but this is not necessarily so. Compared to simple production activities, which have evolved more or less since the beginning of the past century. The concepts involved in analysis instruments such as subjective

[220] The adequacy of the measurement instrument for development activities was not evaluated in the present work. At first glance, it seems that the method as is would not provide adequate resolution or sensitivity to be useful for job design of product development activities.

work analysis have enhanced job designs for simpler activities than development, which has positive effects for employee and employer.

6.1.2 Task orientation and self-regulation

Ulich (2001, 191-195) states that task orientation plays a significant role in the concept of socio-technical systems (see page 41) and describes a state of interest and commitment that is produced by job characteristics, where Emery (1959, 53) indicates two conditions for its existence:

- "The individual should have control (in the broad sense) over the materials and processes of the task" (op. cit., p. 182, parentheses added).
- "The structural characteristics of the task be such as to induce forces on the individual toward aiding its completion or continuation" (op. cit., p. 182).

It is of particular interest to differentiate between autonomy (exerting influence on the determination of goals and rules serving a given set of goals) and regulating control (influence on the measures to achieve these goals) on the one hand and the concept of self-regulation on the other. The latter means having autonomy as well as control in the sense of influencing a given situation to achieve self-proposed goals (Grote, 1997). This calls for the need to integrate thinking, planning and decision processes into every employee's activity, not only into product development. Grote states: "Making decisions connects motivation to action and, at the same time, appears to act as a consolidator, partially because of the tendency of the individual to 'stand behind his decisions', and partially to 'stand up to the group' "[xlvi] (Lewin, 1982, 282f., cited in Ulich, 2001, 209).

Table 33 presents the goals and possibilities of implementing the task characteristics linked with this concept.

Table 33. Task design characteristics resulting in task orientation or arousing intrinsic motivation (Ulich, 2001, 194).

TASK DESIGN CHARACTERISTICS	SUPPOSED EFFECTS	REALIZATION THROUGH
Completeness	-Employees are aware of meaning and status of their work -Employees get feedback on their own progress from the activity itself	Tasks with planning, performing and checking activities and the possibility to check results of own work against requirements
Variety of Demands	-A variety of abilities, knowledge and skills can be applied -Unilateral strain can be avoided	Tasks with different sensory and physical demands

TASK DESIGN CHARACTERISTICS	SUPPOSED EFFECTS	REALIZATION THROUGH
Opportunities for social interaction	-Difficulties can be overcome together -Mutual support improves ability to cope with workloads	Tasks which suggest or require cooperation
Autonomy	-Promotes feeling of self-worth and readiness to accept responsibility -Enables the experience that s/he is not without significance or influence	Tasks with opportunity for planning and decision-making
Opportunities for learning and development	-Maintains general mental flexibility -Occupational qualifications are maintained and improved	Tasks which present problems, and require application and extension of existing qualifications or acquisition of new ones
Time elasticity and stress-free regulation possibilities	-Avoids unsuitable density of work -Allows time for stress-free reflection and desired interactions	Inclusion of time buffers in scheduling
Meaningfulness	-Gives employees the feeling of participating in the creation of socially useful products -Gives certainty of conformity between individual and social interests	Products of unquestionable social value Products and production processes which can be shown to be ecologically harmless

The orientation of the original table for production activities was intentionally kept. Again, readers are invited to scan the table to identify the elements that pertain to product development activities.

6.1.3 Individual and collective action self-regulation

Action self-regulation refers to the ability of working groups to jointly plan, make decisions regarding the tasks assigned and related resources, and establish agreements concerning the decision principles to be used. (Weber 1997, 56). In this sense, Susman's (1976) typology of short-term regulatory decisions is appropriate for group-work design and is summarized in *Table 34*.

Weber describes them as *essentially "non-material"*[xlvii] *actions* that serve to prepare and organize *material* transformation in the manufacturing process. He states that there is a need to differentiate between *individual* and *collective* action self-regulation because the transfer of a group's action self-regulation decision is not necessarily transferred to every group member.

This observed limitation that one or a few members of the group "occupy" the decision-making roles requires special consideration to bring the advantages of group autonomy to the work results. Unlike individual action self-regulation (with the related effects in thinking, planning and decision-making processes), it results in a *collective responsibility* for the decisions made by the group (Weber 1997, 57). This is not necessarily the case when one or a few persons "hog" the decision-maker's seat. In this context, Weber developed the core task segments listed in *Table 35*.

Table 34. Typology of short-term regulatory decisions (Susman 1976, 127).

SUSMAN'S TYPOLOGY OF REGULATORY DECISIONS	DESCRIPTION
Coordination	Decisions altering the nature, sequence, or timing of conversion activities
Boundary allocation	Decisions conveying raw materials, supplies, group members, etc., to sites where activities are required
Boundary maintenance	Decisions altering the flow of inputs and outputs across the boundary

Table 35. Core task segments related to collective autonomy (Weber 1997, 128, reverse order kept from Ulich, op. cit., in order to emphasize the topmost segments).

	CORE TASK SEGMENTS
7.	Decisions regarding group internal self-management
6.	Planning of personnel qualifications and group internal personnel development
5.	Development of solution proposals to technical-organizational problems
4.	Joint implementation of assignments
3.	Job division and planning of personnel schedule
2.	Internal coordination of fine production schedule and implementation
1.	Planning of internal system and resources

It is interesting to note that the points covered by these segments are meant as a basis for task analysis for work-group or job design, especially for manufacturing activities. They are listed here because they describe essential elements of human activities that should not be overlooked for the sake of organizational efficiency. Actually they are important *because of* organizational efficiency and efficacy.

6.1.4 Job enrichment and wholeness of tasks

Following the two-factor theory, Herzberg proposed the *job-enrichment* concept (1968), the principles of which are listed in *Table 36*. According to Ulich (2001, 197-198), it had been used with success, for example at IBM (cf. Walker, 1950). With the two-factor theory, it has been used successfully throughout the world with the expected results regarding quality and productivity, low absenteeism and employee turnover.

Table 36. Principles of job enrichment (Herzberg, 1968, as developed in Ulich, 2001, 198).

PRINCIPLE	MOTIVATION INVOLVED
Removing certain controls while retaining accountability	Responsibility and personal achievement
Increasing the accountability of individuals for own work	Responsibility and recognition
Giving a person a complete natural unit of work (module, division, area, and so on)	Responsibility, achievement, and recognition
Granting additional authority to an employee in his activity; job freedom	Responsibility, achievement, and recognition
Making periodic reports directly available to the worker himself rather than to the supervisor	Internal recognition
Introduction of new and more difficult tasks not previously handled	Grow and learning
Assigning individuals specific or specialized tasks; enabling	Responsibility, growth, and advancement

Herzberg's concept focused on worker motivation and eventually drove changes in the work place, However, four issues require attention according to Ulich:

- Herzberg called for the need of flexibility or "fluidity" of job structures that would allow the job to be fit to worker development (op. cit., p. 199)
- It does not consider a weak coupling of two factors where, for example, a task devoid of content could be partially compensated for by contextual factors (op. cit., p. 199) (such as a worker feeling prestige in being able to say: "I am working on a space project").
- It is restricted to job enrichment at the individual level (op. cit., p. 199-200).
- The grouping of social relations in "*hygiene factors*" masks the consideration of the potential of self-regulation to improve the performance of a group (op. cit., p. 200).

Specific to the product development environment, good relationships among peers are reflected in their interaction and the flow o f information. This in turn can result in improvements of the developers' results. Hence a task can fall into a gray zone that makes it difficult to allocate it to only one of the two factors of Herzberg's theory or, as a control systems engineer might prefer, it will introduce a strong coupling between the two modes.

Furthermore, Ulich continues (op. cit., pp. 200-201), the effect of "wholeness of task" or "completeness of task" is not clear from Herzberg's model. *Complete tasks* are meant in the sense proposed by (i) Hellpach (1922, 27) for group manufacturing, have (ii) an explicit importance for motivation described by Rice (1958) and Emery (1959), are mentioned in (iii) recent psychology work by Tomaszewski (1981), and are treated in the works by (iv) Hacker (1986 and 1987) as "complete activities"[xlviii] and as (v) "complete actions"[xlix] by Volpert (1987). Ulich condenses all these characteristics of "complete tasks"[l] as listed in *Table 37.*

Table 37. Characteristics of "complete tasks" (Ulich, 2001, 201).

	CHARACTERISTICS
1.	Independent setting of goals, which can be embedded in superordinate goals.
2.	Autonomous work preparation, i.e. planning functions.
3.	Choice of means including interactions appropriate to goal attainment.
4.	Performance functions accompanied by feedback of performance, enabling corrections to be made.
5.	Checking with feedback of results and the possibility of checking results of one's own actions against the goals set

This *wholeness of tasks* is currently being proposed by work psychology to improve the characteristics of work. The evolution of this section was intended to give an overview for the following discussion.

6.1.5 Product development job analysis and design

Recalling that the elements developed so far in this chapter are particularly applicable to job design at the *factory floor level*, the reader is encouraged to browse through the entire chapter to appreciate the power of the various concepts for product development activities. The engineers at the beginning of this chapter can be left in the context of twenty years ago, or can be translated into today's context, bearing in mind the current situation in the development environment: companies are trying to improve the efficiency of product development processes. The goals include the gain in time, the enhancement of quality implications during later stages of the product lifecycle, the optimal use of available resources, the improvement of product characteristics (including costs) leading to user preference, or a combination of these reasons.

In this attempt to enhance performance, it is to some extent understandable that production techniques are being tentatively applied to product development by establishing standard procedures, introducing cost accounting, error reporting and correction, manufacturing line discipline, and emphasizing the importance of the (shorter) time span in the development lab. What is not understandable is that in adopting these concepts, the importance of the job characteristics, which are essential to a sustainable situation in product development, is sometimes forgotten.

Let us recall the tension triangle quality – deliverability – costs or quality – time – cost mentioned above (page 7). It consists of an optimization operation, not a naive minimization of the time and cost with simultaneous maximization of quality. Tension exists among the three variables that makes the process complex. Sometimes, one factor must be given room to increase slightly in order to reduce the others.

At present, NASA is applying the faster – better – cheaper approach. It can indeed show a certain success with this approach, but has also suffered some considerable setbacks. Space professionals are often heard to quip: "When asked to implement faster – better – cheaper, choose two!"

Not seldom, managers tend to exaggerate the validity of the rule "cost reduction + time acceleration + perfect results." This has led to the problem in present space programs mentioned by Wertz (1996, 42ff.) where, although schedule compression is a major tool for cost reduction, it is not sufficient in itself, but requires the ability to trade off requirements, performing rapid and accessible decision making. Wertz also mentions the use of small-group work (stated "empirically" from his experience with space companies) and simultaneity of activities. Similarly, Fleeter (1996, 170f.) criticizes the dogma of schedule aggressiveness in space projects and voices strong reservations against acceleration as a means of cost reduction. He recalls the need to include plenty of contingency, delegate responsibility, allow autonomy of team members, and improve communication. All these issues are covered both theoretically and in practice by the concepts illustrated in this work. They are all aspects of optimizing the product development process. Optimization, however, cannot be achieved (also from the perspective of industrial engineering) by the inadequate application of the Cartesian principles of dividing to the extreme, optimizing the part, and rejoining them. There are managers in aerospace who, trying to meet today's challenges, exclude persons that could make an essential difference in improving product characteristics. Under pressure to reduce labor costs, for instance, they cut activities and thus risk forgoing cost reductions that could be realized by avoiding problems downstream. Or they might make the mistake of discouraging project members from analyzing an issue perceived by them as a potential threat to the program's success. Some may go so far as to punish their staff for automatically pursuing a problem. They may even try to impose a production line practice on development activities.

Such managers seriously risk suddenly finding themselves in the paradoxical situation that manufacturing activities benefit from the advantages mentioned in this chapter, whereas the development groups progressively assumed the disadvantages that manufacturing once had. They might experience the dilemma of seeing their development groups suddenly reaching the yield point—a very poor analogy to materials strength—manifested by a sudden decline of performance that could prove quite costly. At the very least, the result will be a work environment where experienced staff will be loathe to stay, and failures that could have been avoided are lurking in every detail.

Wertz (op. cit., p. 43) presented a drawing by Jack Friedenthal created for the TDRS program with the caption: "Schedule compression without changing the rules or requirements is a lot like getting a cat to move a piano up the stairs. The whip doesn't help much." The drawing shows a dead cat and a man holding a whip, complaining: "...damn cat died!..."

Forgetting these lessons is not an option.

6.1.6 Need for integrative functions

The issues described so far illustrate the need for integrative functions. The role of stimulating groups and individuals to consider the myriad of details can make the essential difference in product design quality. Encouraging individuals to consider the whole without losing focus on their own work is another crucial factor. Integrating not only individuals into groups, but work packages into larger work packages (i.e. not merely combining tasks but leveraging product interactions), fostering group interaction and communication, promoting internal and external cooperation, stimulating the critical appraisal of one's work, looking for effects of "interdisciplinary nonlinearities", designing the design job (i.e. not merely putting things together but also people), finding overlooked aspects, cultivating the group balance are all additional issues to consider.

Many of these functions are in the hands of system integrators or program managers, and sometimes of quality personnel. We shall explore these roles in the next sections.

6.2 Roles of leadership

6.2.1 Isolated geniuses or teams of experts?

In descriptions of development leaders, especially those who established milestones in their areas of expertise, we find many references—direct and indirect—to a notion known a the "great man" theory. In essence, it refers to brilliant, bold men whose knowledge outshone others', who were more clearly focused than those around them, and had the personal qualities that enabled them to lead a veritable army of engineers and technicians to "get the job done". As historian Thomas Carlyle declared: "They were the leaders of men...the modelers, patterns, and in a wide sense creators, of whatsoever the general mass of men contrived to do or to attain" (Lambright, 1995, 2-3).[221] However, the great man theory has a strong tendency to attribute an individual's success to a set of generalized characteristics inherent in the "leadership traits" theory described above (page 100). A further tendency is to gloss over if not completely ignore the ways such leadership icons interacted with others within an enterprise. A comprehensive discussion of leadership theories is beyond the scope of this book, but we shall examine selected cases applicable to product development where the *interaction* of the "great man" was much more the determinant of success.

Willy Messerschmitt. Aircraft designer and aerodynamicist.

On having conviction:

"It is obvious from his many statements and writings that Messerschmitt had very little doubt that he was usually, if not always, right in what he said and thought. When his views were challenged he was quick to spring to his own defense and to stave off any criticism made against his decisions. No doubt, if he had not possessed those qualities his name would still be unknown to the world at large today." (Vann, 1993, 143).

Technically criticizing the work of others:

Messerschmitt was concerned about possible structural failure of the Fieseler Fi 103[222] due to the 15 g catapult launch (cf. letter to Robert Lusser, op. cit., p. 172). Indeed the aircraft exhibited this problem later.

Messerschmitt objected to the implementation of the wing-to-fuselage attachment and the avoidance of automatic slots (op. cit., pp. 245-247) of the early European Airbus, which he preferred, but were deemed not adequate for the aircraft.

Reluctance to pass his creations to others:

The Me 264 was going to Dornier: "Messerschmitt was never pleased to see any other designer interfering with his work and he was obviously not

[221] Lambright refers to the original source in the footnotes: Thomas Carlyle, On Heroes, Hero-Worship, and the Heroic in History (1840), cited in Jameson, W. Doig & Erwin C. Hargrove, Eds., Leadership and Innovation: Entrepreneurs in Government (Baltimore: John Hopkins University Press, 1990), 1.

[222] Renamed by Goering the "V-1".

happy about the modifications to the design that Dornier was intending to introduce" (op. cit., pp. 129-130).

Me 323G modifications were going to Zeppelin, then part of Dornier: "Messerschmitt was once again confronted with the unpleasant prospect of another designer tampering with one of his projects" (op. cit., p. 139).

Ability to admit one's own failures:

A report from an inspector for the German Air Ministry sent to check Messerschmitt's production and development of fighters mentions that: "In this respect, there was no show of arrogant infallibility on their own part, but faults were freely admitted which the management and workers had committed during the improvisation of the dispersal" (op. cit., p. 164).

Refusal of solutions other than one's own:

During the development of the HA300 jet aircraft in post-war Spain, Messerschmitt exhibited an unwillingness to accept a diverging point of view. "The alternative suggestions put forward by the Spanish designers were not welcomed, and bad feeling was an inevitable result" (op. cit., p. 240).

Desire to control a situation

"[Messerschmitt] was always anxious to control and vet every detail of every blueprint. If necessary, he would correct calculations himself and often made improvements to drawings submitted by his engineers or draughtsmen. Sometimes, Messerschmitt would quietly place in their hands a new sketch or calculation, and frequently a designer, when proudly showing Messerschmitt his solution to a particular problem, would realize that in fact Messerschmitt had found a much better solution himself. When it came to technical problems, he would work on regardless of time"[223] (van Ishoven, 1975, 164).

Vann pays tribute to Messerschmitt as a technical genius of a kind that will probably never appear again because of the predominance of the computer in an engineer's education and the limited amount of real experimentation that new generations of engineers are exposed to.

[223] After losing management control of his company because of the Me 210 catastrophe (cf. page 81), Messerschmitt returned to his "Konstruktionsbüro" (design office) where he "could remain his old self".

Clarence "Kelly" Johnson. Lockheed, creator of the Skunk Works concept.

Bad temper

"[Johnson] had designed more than forty airplanes over his long life, many of them almost as famous in aviation as he was, and he damned well only built airplanes he believed in. He was the toughest boss west of the Mississippi, or east of it too, suffered fools for less than seven seconds, and accumulated as many detractors as admirers at the Pentagon and among Air Force commanders. But even those who would never forgive Johnson *for his bullying stubbornness and hair-trigger temper* were forced to salute his matchless integrity. On several occasions, Kelly actually gave back money to the government, either because [Lookheed] had brought in a project under budget or because he saw that what we were struggling to design or build was just not going to work." (Rich & Janos 1994, p. 9, italics added).

"When Kelly rolled up his sleeves, he became unstoppable, and the naysayers and doubters were simply ignored or bowled over. He declared his intention, then pushed through while his subordinates followed in his wake. He was so powerful that simply by going along on his plans and schemes, the rest of us helped to produce miracles too.

Honest to God, there will never be another like him. He was a great boss if he liked you and a terrible boss if he didn't. Once he was down on an employee, the situation was usually terminal. We would kid, 'The only way out of Kelly's dog-house is out the door.' Unfortunately, that was true. I was annoyed by things he did at least half a dozen times a week — but I loved that guy." (op. cit., p. 299).

Technical skills:

"Back in 1932, then 23 years old, [Johnson] contradicted positive findings of established experts about the Lockheed Electra and warned that the design was inherently directionally unstable especially with one engine out" (op. cit., p. 113).

"Hall Hibbard, who was Kelly's first boss at Lockheed, watched Kelly work for three days during the war to transform Lockheed's Electra into a bomber for the British called the Hudson. The transformation was so successful that the RAF ordered three thousand airplanes, and Hibbard was so awestruck by Johnson's design skills, he claimed 'that damned Swede *can actually see air.*' " (op. cit., p. 114. Johnson was the son of Swedish immigrants. Italics added.)

"I had never known anyone so expert at every aspect of airplane design and building. He was a great structures man, a great designer, a great aerodynamicist, a great weights man. He was so sharp and instinctive that he often took my breath away. I'd say to him, 'Kelly, the shock wave coming off this spike will hit the tail.' He would nod. 'Yeah, the temperature there will be six hundred degrees.' I'd go back to my desk and spend two hours with a calculator and come up with a figure of 614 degrees. Truly amazing. Or, I'd remark, 'Kelly, the structure load here will be…' And he would interrupt and say, 'About six point two p.s.i.' And I'd go back and do some complicated drudge work and half an hour later reach a figure of 6.3" (op. cit., p. 129).

John Warren. B. F. Goodrich.

Reaction to criticism:

One of B.F. Goodrich's most capable engineers, Warren is noted for the scandal concerning the A-7D aircraft brakes referred to in Chapter 2 (page 54). "The happy-go-lucky manner he usually maintained belied a temper that exploded whenever anyone ventured to offer criticism of his work, no matter how small. On these occasions, Warren would turn red in the face, often throwing or slamming something and then stalking from the scene. As his co-workers learned the consequences of criticizing him, they did so less and less readily, and when he submitted his preliminary design for the A-7D brake, it was accepted without question."

The brake compact conception with four disks turned out to be unfeasible, with disastrous results. In the end, it had to be redesigned with five disks, a concept that was initially refuted by Warren (Vandivier, 1972).

"Dutch" Kindelberger. North American Aviation.

Intolerance of error:

"By March 1950 the first version of Bollay's[224] engine [for the Navaho cruise missile] was ready for testing at full thrust at Santa Susana. … Several officials were present, including von Braun. If all went well, a bright sword of yellow-white flame would stab downward from amid the girders, shaking the rocks with its roar, a roar that would grow louder as it reverberated off the canyon walls.

The test conductor called, 'Oxygen valve open!'—and the rocket blew up in a sudden explosion. A designer had specified that a key part be made of mild steel, not knowing that this commonly used metal would become brittle

[224] William Bollay led the effort of North American Aviation's interest in cruise missiles as a springboard for the ICBM. "His work established the main line of American rocketry" (Heppenheimer, 1997, p. 50).

when chilled with liquid oxygen, and would shatter. *Kindelberger, who now was chairman of North American, sought out the luckless engineer and chided him: 'You should have made the blasted thing out of solid gold!'* The specification changed to stainless steel, which stood up well when very cold. Later that spring, the engine was working well in tests." (Heppenheimer, 1997, p. 50, italics added).

Walter Dornberger. Director of Peenemünde.

Tolerance of error:

"The mistakes we made then [in connection with the static firing of rocket motors] may, of course, cause smiles now, but we trod with fine freedom of ignorance in a new field of technology and learned from failure and bitter experience....

Three weeks after the test stand explosion, the first rocket motor design was ignited and burned successfully. 'Burned' is the literal description of what happened. After a few seconds, a dazzling white light appeared in the bluish-red rocket exhaust and grew brighter. The combustion chamber had been made of aluminum and was on fire. The Kummersdorf team had encountered its first cooling problem." (Stine, 1991, 31).

Wernher von Braun. Director of Marshall Space Flight Center.

Involvement of participants:

"Von Braun's strongest point was always getting things done. He could bolt down great quantities of work in a single gulp. He could command a meeting of people in a way that generated dedication and enthusiasm, that led rather than brow-beat. Those who worked with him respected the wisdom of his experience and his ability to state his case coherently and cohesively.

'Von Braun,' his former personal assistant, Dieter Huzel, once wrote, 'didn't conduct a meeting, he led it with perception and incisiveness. He knew most problems at first hand, and those few that he didn't, he knew by instinct bred of long, *intimate* and successful experience in the field.' " (Spangenburg & Moser, 1995, 97-98, italics added).

According to Konrad Dannenberg, a veteran project manager under von Braun, he "would talk as easily with the 'top brass' as with the 'tin benders'... He would solicit advice and suggestions from workers in the shops, taking into account the realities of fabrication and manufacture as the design evolved... In this way, von Braun avoided the pitfalls of having top-level managers making critical decisions among themselves, and arriving at assumptions about fabrication that might not approach reality." (Stuhlinger & Orway, 1994, 223).

When von Braun "visited a laboratory or a test stand, his questions and debates would frequently continue far beyond the comfort limits of his associates. He would go from room to room, up stairs and through narrow passages, always inquiring, listening, suggesting, playing with valves and switches, and immersing himself fully in the problems of his hosts. 'He can really wear you out,' Was a remark not unusual among his companions" (op. cit., p.216).

Stimulus of new ideas:

"Von Braun was not a creative genius—to that extent, the public image misrepresented him. It is not possible to think of 'von Braun concepts' that changed the development of rocketry in the same way that 'Faget[225] concepts' changed the development of spacecraft. Instead, he was a natural leader and technical manager. Karl Heimburg, chief of the Testing Lab, explained how von Braun went around looking for new ideas, which he would then take to his associates. Heimburg would listen unimpressed and explain to von Braun why something wouldn't work. And then Heimburg would find himself saying, 'But we could do it in this other way,' and an innovation that he had not considered would have opened up before him." (Murray & Cox, 1989, 52).

Tolerance of error:

"One of our early Redstone missiles (in the 1950s) developed trouble in mid-flight. The telemeter records indicated that the flight had been flawless up to that instant, and permitted us to localize the probable source of trouble. However, the suspected area had been very carefully checked in numerous laboratory tests so that all explanations sounded highly artificial. Several theories were advanced. Finally one theory was accepted as most likely and remedial action based on it was initiated. At this point an engineer who was a member of the firing group called and said he wanted to see me. He came up to my office and told me that during pre-launching preparation he had tightened a certain connection just to make sure that there would be good contact. While so doing, he had touched a contact with a screwdriver and drawn a spark. Since the system checked out well after this incident, he hadn't paid any attention to the matter. But now that everybody was talking about a possible failure in that particular apparatus, he just wanted to tell me the story for whatever it was worth. A quick study indicated that here was

[225] Max Faget.

the answer. Needless to say, the 'remedial action' was called off and no changes were made. I sent the engineer a bottle of champagne because I wanted everybody to know that honesty pays off, even if someone may run the risk of incriminating himself. Absolute honesty is something you simply cannot dispense with a team effort as difficult as that of missile development." (von Braun, 1956, 41, parentheses added).[226]

Not hunting for the culprit:

Dieter Huzel reports a situation after a failed launch in a control bunker at Peenemünde: "I watched the firefighters move into action. A great hubbub, meanwhile, had begun in the control center. A guidance engineer and a powerplant engineer were in a heated discussion... There seemed to be some idea that an erroneous emergency radio cutoff signal had caused the malfunction. The guidance engineer kept shouting that some switch had been tripped by the console operator. The latter was an elderly man whom I knew to be extremely reliable. Obviously still under the shock of the catastrophe, he was unable to speak a coherent sentence in his own defense... All this time von Braun had not moved from his periscope. Now, he put down his headset and stepped over. In the background, Muenz[227] was still busy with instructions, messages for headquarters, and answering questions over the intercom, all in his calm, steady voice. Von Braun likewise appeared unaffected, a little solemn, perhaps, but entirely collected. His voice was just a trifle louder than usual when he cut into the argument. 'This missile will never fly again,' he said firmly, and the effect was complete ...the room became quiet. 'That's an obvious fact. We didn't accomplish what we hoped we would. But since it has happened this way, it is extremely important that we obtain every bit of information as to the cause of the failure. And by all means, let's not conduct a fault-finding investigation. No one is going to be hanged. If a human error occurred with this experienced crew, it can only mean that our control equipment has to be made more foolproof.' He smiled. 'Don't forget, we have to place these monsters in the hands of Army units within a few months.' There was relieved laughter. The tension subsided, and the man at the console was able to tell his story." (Huzel, 1962, 90-91).[228]

---

[226] There is another, slightly different record of this event, but it also involves a bottle of champagne. This version was chosen as it was reported by someone present.

[227] Willi Muenz was lead test engineer.

[228] Excerpt from: Huzel, D. (1962). Peenemünde to Canaveral, Englewood Cliffs, NJ: Prentice-Hall. © 1987 by The Huzel Family Trust Agreement. Reproduced here by kind permission of Irmgard Huzel.

Yet inspiring commitment to rigor and quality of work:

In 1954, during the launch of a Redstone, Samuel Hoffman[229] experienced the following failure of an engine. "It was indeed a sobering experience when the missile rose only a few hundred feet, faltered and exploded in a white ball of flame... Colonel H. N. Toftoy[230] looked [at Hoffman] and said, 'You *will* be here for the next launch, won't you Sam?, It was of course a rhetorical question. Von Braun, on the other hand, did not say a word. 'But his silence spoke volumes and impressed me more than anything he could have said. I returned to Rocketdyne instilled with a new sense of dedication and purpose' " (Bergaust 1976, 502).

Sense of humor:

Von Braun was well known among his peers and contemporaries for his sense of humor. His biographies recall Maslow's description of self-actualizers described in Chapter 4 (and the above recommendation to refrain from labeling individuals). Bergaust (1976, 487) describes a situation after the failure of one of the Explorers, where the top project people apprehensively "expected to [catch] hell from the boss. All of them had worked hard for several days on their analyses pertaining to the failure, but were hesitant to suggest remedies and corrective measures for their next satellite shoot. ... They [appeared at a meeting] like a group of schoolchildren in trouble with their teacher... there was absolute silence in the room.... Von Braun studied the Explorer flight pattern diagrams for a long while ... [made a witty comment, and]... they all roared with laughter. Before they went back to the drawing board, everybody had been relieved of their tension." Bergaust also cites a description by Bonnie Holmes, von Braun's secretary: "[His humor] kept the team together. Not only that—it also made it the best of teams. Von Braun's sense of humor and his human ways were among the greatest assets of the whole program."

Sense of teamwork:

Stuhlinger & Orway published a transcript from a speech in which von Braun said "First and foremost, let's give more appreciation to the value of the team in modern research and development efforts! Successful R&D teams are the most valuable assets this country has in its continuous struggle for scientific and technological leadership; therefore, let's protect the integrity of those teams ... , and I wish to include all our missile competitors in this plea..." They go on to explain that "One of the reasons, [von Braun] said, why the significance of teams is often overlooked is because people

---

[229] Samuel Hoffman was head of Rocketdyne.

[230] Col. Holger N. Toftoy was the Commanding General of Redstone Arsenal.

tend to identify achievements only with individuals; actually, those individuals who get the publicity rarely deserve more than a fraction of the credit they receive. 'I myself am a case in point,' von Braun added. 'There may be differences in the degree of ignorance among rocket engineers, but there simply is no such animal as a "rocket genius". Neither at the Army Ballistic Missile Agency, nor in the Navy, nor—mark my words!—in the Air Force.' " (Stuhlinger & Orway, 1994, 133)[231] (see Inter-service rivalry, page 153).

Alexander Lippisch. German aircraft designer, recognized as the father of the delta wing

Sense of teamwork:

Lippisch pioneered not only the delta wing for high-speed flight but also other innovative concepts of flying machines such as the aerodyne and the airfoil (ground effect plane). He is also known as the "father" of the Messerschmitt Me 163 Komet, the first rocket plane to enter service. He protested the latter credit in his memoirs, declaring that it was not true and insisting that his associates had contributed equally to the development of the machine. (Lippisch, 1982).

Concluding remarks about the biographical excerpts

The applicability of these cases of strong leaders is still valid in the present context of product development. In fact, several examples presented in this work serve to demonstrate the old truism that history does indeed repeat itself. Although the leadership style of some of these men would certainly be labeled "out" today, this does not mean that such cases no longer exist. Nevertheless modern leadership approaches presumably enjoy less tolerance to certain behaviors described. It is therefore important to evaluate these cases in terms of the elements—whether present or absent—that can impede the product development process.

This work refrains from labeling a given style or method as "in" or "out", "old" or "new", as they all have elements of validity regardless of their historical context. Before drawing conclusions from descriptions of development leaders, care should be taken not to examine an isolated characteristic or property out of context. This is especially important if a

[231] Citing a paper by von Braun entitled "The Lessons of the Sputnik", presented in Washington on October 29, 1957, two weeks after Sputnik's launch and three weeks before receiving the assignment of launching the American satellite.

generalization of their work methods or styles is intended for application to other product development groups, either in a different time or domain of knowledge other than aerospace. This section was intended to provide insight into the way these leaders *interacted* with their development personnel. Any attempt to infer their personal characteristics (e.g. personality) would be potentially misguided and is considered unfair practice.

6.2.2 Leader orientation and problem solving

This section concentrates on the way interaction processes take place. Nevertheless, it does not constitute the full approach referred to as the *interaction theory* of leadership psychology, because then it would have to integrate (Gibb, 1969, 268):

1. The personality of the leader,
2. The followers, with their attitudes, needs, and problems,
3. The group itself: its structure of interpersonal relations and its syntality[232] characteristics,
4. The situations determined by physical setting, nature of task, etc.

As the information available regarding personality and syntality is limited, I will refrain from using *interaction theory* in its entirety for the cases cited in the previous section.

The focus of this section is on interactions of leaders with their development personnel, and whether they facilitate the flow and processing of information such that productive discussions take place and appropriate action taken. The product development process is generally an information-intensive endeavor, and essential information is typically scattered throughout an organization. Leadership styles influence the way information carriers (here: the specialists) succeed in optimizing their contribution.

Let us return to the influence of the minority examined in the previous chapter on interpersonal factors. Again, the perception of this problem is not new at all. Maier & Solem (1952)[233] investigated the contribution of the discussion leader to the effective use of minority opinions. Thirty-three of the 67 groups were provided with an observer and 34 with a discussion leader.[234] "The discussion leaders were instructed to encourage the

---

[232] Syntality: the equivalent of the personality of an individual when referring to a group.

[233] Excerpts from: Maier, N. R. F. & Solem, A. R., 1952, The Contribution of a Discussion Leader to the Quality of the Group Thinking: The Effective Use of Minority Opinions, Human Relations, Vol. V, No. 3, pp. 277-288, London: Tavistock.. Reproduced here by kind permission of Sage Publications Ltd.

[234] Their subjects were 353 male and female students divided into groups consisting of 5 or 6 persons.

participation of all members and to ask questions so as to cause the group to think together rather than as individuals. They were requested strictly to avoid expressing their own views on the problem. Instead, they were asked to be permissive and to accept the various points of view expressed. Finally, they were asked to try to get group agreement on an answer" (op. cit., p. 280). All participants, including the leaders and the observers, were asked to give their individual answers before and after the discussion. Then they were assigned the so-called horse-trading problem, which goes like this: "A man bought a horse for $60 and sold it for $70. Then he bought it back for $80 and sold it again for $90. How much money did he make in the deal?". The correct answer is $20, but the answers ranged from –$10 to +$30, $10 being the most frequently given incorrect answer.

Table 38 summarizes the raw effect on the two groups. Both groups enhanced their performance by means of the discussion, but the groups with a discussion leader showed a greater improvement.

Table 38. Influence of discussion leaders on quality of group problem-solving (Maier & Solem 1952).

	CORRECT ANSWERS BEFORE DISCUSSION	CORRECT ANSWERS AFTER DISCUSSION	RELATIVE INCREASE IN CORRECT ANSWERS
Leader groups	45.8%	83.6%	82.6%
Observer groups	44.3%	71.6%	61.6%
chi squared	0.0744 (non-significant)	7.3459 (significant at less than 1% level)	

The influence of minorities was demonstrated in this experiment (*Table 39*) as follows. In the absolute minority groups, where only one person had given the correct answer before the discussion, 75.9% of the persons who had given an incorrect answer in the groups with a discussion leader changed their responses after the discussion. In contrast, only 36.4% of the persons who had given an incorrect answer in the groups with an observer corrected their answers after discussion. From the cumulative percentual change, the groups with a leader showed better performance than the groups with an observer, especially when the minority had originally given the correct answer.

Maier & Solem (1952) also investigated the effect of the discussion on the responses of the leaders and observers, *Table 40*. From this data, the authors conclude that intellectual leadership was not among the forms of influence the leader exerted on the group members. Note that the leaders had given fewer correct answers before the discussion than the observers had.

Table 39. Effect of discussion leaders on the voice of the minority (Maier & Solem 1952).

		LEADER GROUPS			OBSERVER GROUPS		
TYPE OF GROUP	RATIO CORRECT-TO-INCORRECT ANSWERS	NO. OF EACH TYPE OF GROUP	% CHANGE	% CHANGE CUMULATED AS MAJORITY VS. MINORITY	NO. OF EACH TYPE OF GROUP	% CHANGE	% CHANGE CUMULATED AS MAJORITY VS. MINORITY
Majority groups	(a) 5:0	1	0	64.5	2	100	56.3
	(b) 4:1 4:2	5	33.3		4	100	
	(c) 3:2 3:3	11	72		10	41.7	
Minority groups	(d) 2:3 2:4	8	80.8	72.3	7	78.7	45.5
	(e) 1:4 1:5	7	75.9		8	36.4	
	(f) 0:5	2	40		2	0	

Table 40. Influence of the leaders' and observers' response (Maier & Solem 1952).

	LEADERS			OBSERVERS		
	% CORRECT BEFORE DISCUSSION	% CORRECT AFTER DISCUSSION	% CHANGE	% CORRECT BEFORE DISCUSSION	% CORRECT AFTER DISCUSSION	% CHANGE
Majority	58.8	88.2	71.4	75	87.5	50
Minority	29.4	88.2	93.3	41.2	52.9	20

The influence of the minority on the majority took effect in this experiment when the minority had chosen the correct answer. Only in these cases did the majority change its views (op. cit., p. 286).

The pressure of the majority (cf. Asch's study on page 263), the influence of a consistent minority on group perception (cf. section on page 223) and the leader influence described above constitute a very important triad affecting product development performance.

According to the Maier & Solem, "The reason a clearly led discussion is superior to one not so guided was shown to be due primarily to the fact that the leader permitted minority opinions to exert a constructive influence. Ordinarily persons who change opinions are influenced by the social pressure that is exerted by the majority view. It follows that when a majority opinion is wrong social pressure has an undesirable effect on the outcome of the discussion. With the presence of a discussion leader this social pressure is reduced, allowing minority opinions to exert an influence. In protecting a minority opinion from the social pressure of the majority, the leader allows the minority to have enough influence to make a possible contribution to the quality of a group's thinking." (op. cit., p. 285).

Maier & Solem conclude that the following three forces were present in discussion that led test subjects to change their opinion (op. cit., p. 287):

- The discussion leader's contribution of ensuring that the minority be heard.
- The social pressure of the majority.
- External reality or the facts supplied by discussants to support their opinions.

Particularly the latter two factors are present in the product development process. The first is not always present.

Compared with the effect of group pressure, especially the case of *groupthink* in the Kennedy administration reported by Janis (cf. page 264), the study by Maier & Solem gives the leader the additional function of preventing the group from engaging in false assumptions, or overseeing a good solution to a problem.

This is not the only way to ensure that vital issues are not overlooked by the majority. A strong leader may perform this role too. A case that occurred during the development of the Hound Dog missile at North American is described by Gray (1992, 61-62) as follows:

Box 66. Strong leadership can ensure that a solution is not overlooked by the majority.

"The next test flight ended in failure, [and] Storms went down to the engineering department and sat in on the postmortem. The chief engineer and a dozen assistants were in a little conference room going over the data, and as Storms looked at the printouts, something caught his eye. For some reason the autopilot was making corrections to the flight path about twenty times a second —a nearly constant rate— literally flying every inch of the way.

'Why've you got your autopilot working so hard?' asked Storms. The project engineer patiently explained that the missile was tail-heavy, so the autopilot had to correct for it constantly.

Storms said. 'Why don't you put a few pounds of lead in the nose?'

The suggestion was met with stunned silence. *Add* weight? These people were trained to take weight out, not put it in. But Storms was an old model airplane builder, and he knew that a few sinkers from the tackle box often did wonders for flying performance. They put a fifty-pound lead weight in the front end of the missile—sacrificing a few miles in range—and the ship was balanced so the autopilot wouldn't have to work itself to death" (italics maintained). After some operational incidents, the missile was successfully flown.

Storms' proposal was presumably entertained in part because of his influential position. This may well work, but in numerous other cases described throughout this work, it certainly will not always be so successful. In some of the cases described, it may even result in disaster. Even the most technically capable development leader is not free of mistakes.

Interaction theory holds that not only leader characteristics themselves are important, but also how they are perceived by others. But there is a reciprocal relationship, and it was investigated by Maier & Hoffman (1965). They demonstrated that "when the foreman most often reported having problem employees, solutions were least likely to be innovative and acceptance was relatively low. Conversely, when the foreman perceived his subordinates as idea men, innovative solutions increased and satisfaction with the solution was greatest for all concerned, despite the fact that all persons involved had to change their initial positions" (op. cit., p. 373). The main idea behind this hypothesis is that the "leader's perception of his subordinates will influence the type of discussion he will conduct, and the nature of this leadership will determine the degree to which they can constructively (vs. negatively or passively) contribute to the discussion" (op. cit., p. 374).[235]

This data strongly supports the hypothesis that disagreement leads either to resentment or to innovation, depending on the discussion leader's perception (op. cit., p. 380).

The innovation that resulted from this experiment is the product of conflict resolution, which Mary Parker Follett described some four decades earlier (cf. page 38). In the latter experiment, Maier & Hoffman concluded that, "disagreement can serve as a stimulant to further exploration, but unless it is respected it tends to become the source of interpersonal conflict" (op. cit., p. 384).

---

[235] The situation they studied is called "changing work procedures." It consists of a group of four persons, one assigned the role of the foreman and the other three the roles of workers with different production rates. The role-playing experiment requires the subjects to engage in the following conflict: whether to maintain the current work method that avoids monotony issues or to accept a new solution proposed by the foreman that saves time but incurs monotony. The group also has the option of "proposing integrative solutions that are not obvious or given as background information, but must be generated in the discussion" (op. cit., p. 379). The foreman classifies the members as "problem employee" or "idea man". The members are asked to estimate whether the solution adopted would increase, maintain, or decrease productivity, and this served as a measure of their acceptance of the solution. The test subjects were 600 executives in middle-management positions from several industries.

6.2.3 Leadership and the learning organization

Not only day-to-day problem-solving is affected by leadership. Developing the organizations in which the development group is embedded is another area in which leadership is crucial. (McGill, Slocum & Lei, 1992) describe the contrast between two types of organizations. There are companies that emphasize generative learning, which calls for continuous experimentation and feedback concerning the way problems are defined and solved. Then there are companies that emphasize adaptive learning, where problems are solved as they have always been, in a "traditional" fashion without examining the appropriateness of current learning behaviors. Generative and adaptive learning are also called double-loop and single-loop learning (Argyris, 1982). *Table 41* summarizes the differences.

These contrasting characteristics play a crucial role when organizations are also interested in improving product development processes while ensuring that the development groups will learn to cope with technological advances and update their current technological standpoint.

Table 41. Characteristics of types of learning organizations (McGill, Slocum & Lei, 1992, 14).[236]

	ADAPTIVE	GENERATIVE
MANAGERS' BEHAVIORS		
Perspective	Controlling	Openness
Problem-solving orientation	Narrow	Systemic thinking
Response style	Conforming	Creative
Personal control	Blame and acceptance	Efficacious
Commitment	Ethnocentric	Empathetic
HUMAN RESOURCE PRACTICES		
Performance appraisal system	Rewards stability	Flexibility
Reward basis	Short-term financial	Long-term financial and human resource development
Focus of rewards	Distribution of scarcity	Determination of synergy
Status symbols	Rank and title	Making a difference
Mobility patterns	Within division or function	Across divisions or functions
Mentoring	Not rewarded	Integral par of performance appraisal process
Culture	Market	Clan
STRUCTURAL CHARACTERISTICS		
Structure	Bureaucratic	Network
Control systems	Formal rules	Values, self-control
Power bases	Hierarchical position	Knowledge
Integrating mechanisms	Hierarchy	Teams
Networks	Disconnected	Strong

[236] Order of groups in table altered to emphasize leadership aspects.

	ADAPTIVE	GENERATIVE
Communications flow	Hierarchical	Lateral
STRATEGIC CHARACTERISTICS		
Core competence	Better sameness	Meaningful difference
Source of strength	Stability	Change
Output	Market share	Market creation
Organizational perspective	Compartmentalization	Systemic
Developmental dynamic	Change	Transformation

6.3 Roles of quality professionals

Figure 19 serves as a starting point for this discussion of the role of the quality department in the product development process. It was obtained from a training course, where the instructors required a group of managers to draw a picture describing the way that new product decisions are made at their company.

This picture is intended only as an illustration. It describes a problem encountered by many companies: involving quality personnel at too late a stage of development. Usually, quality control does not possess an approach to participate in the process from the very beginning. Hence development personnel in many companies refer disparagingly to quality control as "those guys who come after everything has already been defined or is already done and criticize it, demanding changes", as if the quality control had somehow mysteriously appeared out of thin air. As we've seen, this has many causes from both the development and the quality sides, meaning that the role of the quality organization had to be reviewed. To control (in the narrow sense) a process, the quality team has to wait until the process is ready to be controlled, otherwise they would be barred in trying to attempt it, and so they wait…

Companies that successfully implement effective product assurance activities at the development stage have found a way to get the quality organization involved in the product definition process. This means that product assurance personnel have had to change their roles.

Figure 19. Design decision process of a company (McGrath, Anthony & Shapiro 1992, 47).[237]

Seghezzi (1992a) classified quality management into four technical functions described in the following table:

Table 42. Four quality management technical functions (Seghezzi 1992a).

FUNCTION	DESCRIPTION
Quality planning	Determination of needs and implementation into products or processes
Quality steering or control	Carry out all processes and be very proficient in them
Quality assurance (in the narrow sense)	Determine quality risks and carry out measures to minimize or eliminate them
Promotion of quality	Conduct efforts to increase product or process quality

These technical functions cannot be assigned to specific departments but are distributed among many groups inside a company. Hence they are

[237] From Product Development: Success Through Product and Cycle-time Excellence by M. E. McGrath, M. T. Anthony, A. R. Shapiro. Reprinted by permission of Butterworth-Heinemann.

functions acting at the interface between disciplines. In order to deal with interface issues, Seghezzi proposed a fifth function: *quality leadership*.

The role of quality might be decentralized if, for example, multifunctional groups are to be implemented. However, it is still necessary to have a central quality function as described below. Acting on a centralized *and* decentralized basis, the overall quality organization can be efficient (due to hierarchical independence) and fast (due to decentralization).

6.3.1 Power of the law, the error and the culprit

Recalling now the differences between quality control and quality assurance described on page 56, the former focuses on controlling and finding errors (and consequently rewarding or punishing those involved). The latter concentrates on determining which problem is generating a given error (and consequently finding the cause(s) of the problem). Unfortunately, quality controllers have been known to hunt for the culprit when an error is found. Some even threaten with apparent relish that "heads will roll" in an effort to instill fear of the consequences and achieve discipline amongst workers.

Box 67. Read me my rights before you arrest me (case #48).

Referring to a former quality assurance manager, colleagues said that "everything you say to him can be used against you". He had since transferred to a different company, and was attending a meeting at which participants were discussing the characteristics a product should have and the possibility of implementing them. In the course of this discussion, he proudly mentioned the notorious reputation he had had in his previous company. One engineer, who had known him for many years but was not aware of this reputation, later reported having thought, "So that's why I always want my lawyer when talking to you! Don't you realize that your behavior discourages me from saying anything, when I should be discussing with you freely?"

In contrast, consider the example of a teacher who conducts tests to evaluate deficiencies in the learning processes of individual students. A good teacher uses the results to address any deficiencies and to correct them. A poor teacher uses the results only to classify students by grade and students are fearful of failing an exam. The quality assurance principle would dictate the former approach of taking action to achieve improvement, thus acting on the basic causes of the problem. Of course "errors" must be assessed, and cases of negligence and sloppiness cannot be ruled out. But to assume this as the default situation can have many disastrous consequences. This is further illustrated by the contrast between the "vicious circle of theory X" and the "reinforced cyclical effect of theory Y" (Ulich, Baitsch & Alioth, 1987, 18-19). McGregor (1957, 23) described as Theory X the belief that "the average man is by nature indolent—he works as little as possible, ...lacks ambition, dislikes responsibility, prefers to be led, ... is inherently self-centered, indifferent to organizational needs,... is by nature resistant to change, ... is gullible, not very bright, the ready dupe of the charlatan and the demagogue." As a consequence, people must be directed, ordered, controlled in their actions, punished. McGregor's classic work emphasizes that this notion of human nature does not cause such management actions, but is a *consequence* of them. Ulich, Baitsch & Alioth depict this as a *vicious circle* in that the assumption of Theory X leads to stringent procedures and control. This gives rise in turn to passive work behavior, leading to responsibility avoidance and lack of initiative, thus confirming Theory X, and the circle continues. In contrast, Theory Y proposes completely different assumptions of human nature and motivation, namely that "people are not by nature passive or resistant to organizational needs...[, that] the motivation, the potential for development, the capacity for assuming responsibility, the readiness to direct behavior toward organizational goals are all present in people" (op. cit., p. 88). In contrast to Theory X, which relies exclusively on external control of human behavior, Theory Y "relies heavily on self-control and self-direction" (op. cit., p. 89). The *reinforced cyclical effect* holds that the adoption of Theory Y improves the margin of action and self-control that enables more commitment to work. This enhances initiative and readiness for taking responsibility, which in turn reinforces Theory Y. Again, the circle is closed, although in a different sense.

Assuming the need for "police" to arrest the "culprit" of quality problems, there is first a need for a legal system that tells citizens what is *to be* done and what is *not to be* done. A Roman legal principle still used today states that "there is no crime if it is not foreseen by a law". Approaching quality with a police approach without a good legal system is a sure recipe for injustice. An inherent deficiency of this approach is that if a given issue is not on the books, then no "crime" has been committed, even if something clearly wrong has occurred.

Regardless of whether this approach is the most efficient, for production purposes it is straightforward to define what is to be done, in a form quite reminiscent of a law. It is easy to say, "This is what you must do! This is what you must not do!" (cf. standard operating procedures, workmanship standards), and to guarantee production homogeneity, there has to be a normative system. Extrapolating this approach to product development leads only to a partial solution that should not be applied blindly. Construction rules such as: "it is prohibited to use potentiometers", "it is prohibited to use solder paste", "it is prohibited to wash with water", "it is prohibited to use plastic components", "it is prohibited to use the GOTO instruction", "it is prohibited to use CMOS components," for instance, have their justification, but all these rules make sense only in context and subject to technical reasoning. They all have concrete underpinnings, and are not a question of right or wrong, but of appropriateness. Moreover, approaching quality like a legal system leads to the "opening of precedents" (jurisprudence term, esp. in English law) and a resistance to deviate from the rules, even when this would produce a better, more appropriate solution to a given problem.

Box 68. Opening precedents in a quality "legal" system (case #49).

A company manufacturing high-reliability equipment in the early 1990s followed the rule for such equipment still valid at that time, which prohibited the use of plastic-encapsulated components. The components were consequently specified as encapsulated in ceramic or metal. The rule was about two decades old. Meanwhile, advances made in plastics and the encapsulating techniques used did not render the rule invalid, but made its application non-mandatory. Production planning then inquired of the quality and reliability organ of the company whether, due to the unavailability of one of the components of the ceramic casing, its plastic counterpart could be used. A brief analysis showed that the components in this case were not critical for the functioning of the equipment in the environment in which it was to be operated. Therefore the quality manager approved the use of the component *in that particular case*. Nevertheless, the experienced quality manager was concerned that, because he had approved it once, production people would use it again without asking.

The interpretation of the activities to achieve quality in product development can lead to "interpreting as law". Three contrasting cases are presented in Box 69.

Box 69. Three contrasting cases of interpreting quality activities as a legal system.

Case 1. In an electronics company (case #50), quality assurance engineers had called a meeting with development and production staff. The purpose was to discuss recommendations for the conception of a product that was in detail design. They were aware that the competence for the final decision was in the hands of the development personnel, and they did not intend to encroach on that turf. They were intent on avoiding compulsory dialog styles such as, "You should do this", "You should not do that". The meeting proceeded haltingly. At one point, a development engineer, clearly apprehensive, abruptly asked, "Are we obliged to implement these measures?"

The quality manager tried to reassure him by explaining, "These measures are not obligatory. They are recommendations we are trying to discuss. We believe that..." but the development engineer interrupted him. "What is the purpose of this meeting, then?" The quality manager explained the short-term and mid-term goals: "The goal of this meeting is to talk about the implementation of our product. In the long run, a list of guidelines for implementation could be developed. It would contain several recommendations for the development of our products." At this the design engineer, clearly upset, retorted, "Recommendations! That doesn't work! You should *give orders* to us regarding what should be implemented in the design. Recommendations mean nothing! You should *command us* what we have to build!" The meeting ended unresolved. Neither the development engineers nor the production engineers grasped the purpose of the meeting.

Case 2. In an electronics company (case #51), a development engineer asked the quality assurance department for standards for power supplies. Showing him the standards file, a quality engineer tried to help him find the appropriate standard. The latter noticed that the development engineer was consulting standards for power supplies that were intended for very specific equipment quite different than the application he knew the designer was working on. The quality engineer asked what exactly the other was searching for. The engineer answered, "I just want a standard. There must be a standard somewhere about the power supply I should use." He was not looking for general standards, such as test methods for power supplies, nor for specific standards applied specifically to the product in which his power supply would be installed. He was simply looking for a standard, any standard, in the mistaken belief that compliance to some random standard would bring some kind of a benefit. He was of course not wrong to seek information on the *available* standards, but *any* standard?

Case 3. In the definition of a product (case #52), a development director and a quality assurance engineer were working together to revise product specifications. The product was to be developed for the defense industry. Therefore, the first version of the specification included an extensive list of standards "inherited" from the *platform* on which the equipment was to be installed. This revision was planned, as it was expected that the knowledge gained during the development would lead to modifications of the product specifications. After the first version, the company's knowledge of the applicable standards increased considerably. Consequently, it was in a position to discuss such issues as the electrical and mechanical specifications, the level of environmental stresses, and test procedures, as well as to introduce changes and propose the omission of non-appropriate standards. Discussions focused on the *applicability* of the standards, enriched by the knowledge gained about the *platform* that was also in development. In this case, the collaboration with the development staff proceeded smoothly because it centered on the reasons behind the standards, not on whether compliance was obligatory.

In the problem-solving processes used by development, complications are frequently incurred by defining *a priori* what constitutes a good or a bad solution. A legal system to guarantee product development quality would solve such conflicts only in cases where there is an obligation to implement a given rule (e.g. by law, an automobile must have safety belts). All other rules for implementing products are potentially subject to technical scrutiny. This is particularly due to the dynamic nature of technological development and the contention that "law, police and punishment" can impede innovation and the evolution of the product and persons involved.

Generally speaking, correction and improvement are crucial processes in any mature legal system. Therefore the implementation and update of a law system can be slow, depending on its complexity and the human resources available to the task. Society has considerably slower dynamics, especially compared to the fast dynamics of product development. This significantly limits the application of "legal" system concepts to product development, which tends to be expensive to update at the appropriate pace.

6.3.2 Role of control

The evolution of the concept "quality" from "the best" to "the most appropriate for a given purpose" made it an interdisciplinary discipline (Birolini 1994, Seghezzi 1992b). It considers not only its own techniques, namely statistical quality control, environmental tests, cause-and-effect diagrams, FMEA, quality audits, among others, but also the techniques of other participants in the product lifecycle, from marketing (customer needs) to customer support (customer satisfaction) and product disposal.

This integral consideration was one of the essential elements of the ISO 9004 standard (ISO 1987b). Nevertheless, it posed a formidable challenge to quality organizations worldwide. The diversity of issues and techniques quality assurance personnel would face if they attempted to control everything would be overwhelming. Therefore, in order to interact effectively with other kinds of experts, it is highly useful to be familiar with their terminology, their perspectives, and their culture.

Assuming a quality assurance professional has achieved the necessary command of all these other disciplines in order to collaborate effectively with experienced professionals, he/she will notice that this knowledge-based position is very difficult to maintain. In practice, partial achievements in this direction can be made by reallocating development personnel to product assurance activities because then they would know "how to talk" with designers. People, however, are not always willing to make this change. Moreover, although encouraging this redirection gives a person enhanced familiarity, it detracts over time from that person's original state of expertise. For example, a former development engineer no longer working in his/her field will find it difficult if not impossible to stay abreast of the field. Hence the original benefit is not implicitly sustainable, for the criteria he/she would be using might no longer be valid.[238] Regarding an issue exclusively from the vantage point of the technical knowledge needed—if the approach to be taken were to control the work of development engineers—this controller would have to know *more* than all the engineers working on the project. As one good "control" engineer cannot in the long run know more than many good development engineers, there have to be many good control engineers to control the many development engineers. Moreover, it would make more sense to have these "good" ("control") engineers work actively *with* the development engineers—a simple matter of resource efficiency.

Thus, whereas multidisciplinary is a problematic aspect of the approach control, it provides problem-solving mechanisms and hence better way to improve quality than by controlling it.

The approach control (in the narrow sense) is self-limiting, although it does solve a few of the issues. The purpose of this section is to reconfirm the trend observed by Seghezzi in 1992 that "quality specialists would change their roles from controllers and inspectors" (1992b, trend #11.) This is a must for handling the quality of a development.

Closing this discussion with a comment about control, it is sometimes deemed necessary. Recalling the sections entitled Achieving cooperation on 232 and Effective participation on page 256, we have seen cases where the function of a quality group extended much further than the role of controlling.

6.3.3 Quality leadership

One of the managerial myths mentioned by Juran (1992, 428) is that "the work force is mainly responsible for the company's quality problems". He

---

[238] In some cases, the result of this experience was not at all negative because the former development engineers eventually returned to development with a different perception of the activity and a better understanding of the problems associated with it. Others who stayed evolved to suit the new activities but had to leave the development in the background.

also claims that "80 to 90 percent of the damage done by poor quality is traceable to managerial actions".[239]

Owing to the quantity of details, practicing quality leadership is not as trivial as it might seem. Project managers should rely on additional staff for support in performing this function. Even, or perhaps *especially* in decentralized organization structures, there is the need of a central quality function that practices this fifth function (cf. page 306). The conditions under which decentralized groups or divisions would accept someone else in this role are not entirely clear, owing to the stereotypical image of the quality department as controllers in the narrow sense and of the emphasis on paperwork. The strong interrelationship and partial overlapping of duties with the project manager function causes *potential* conflicts. If not appropriately resolved, this might render any activity of the quality department useless. Another source of conflict is that managers are evaluated for their ability to push a project forward. This can conflict with attention to detail. Quality personnel—still belonging to the project team—can help resolve this apparent ambiguity. It must be emphasized that a clear division between "pushing forward" and "attention to detail" should not be made between the project manager and the quality function and that the responsibility for both duties should be carried by both functions.

The quality of the decision process is not unconditionally better in decentralized structures, as a group might engage in maladaptive behavior as discussed above (page 258ff.). A centralized quality function is necessary to ensure the relative independence of the group's approval and peer evaluation processes. A very delicate effect of a central quality function can be to improve the quality of a group decision, to act as a moderator to enhance the degree of vigilance or hypervigilance of a group in order to avoid group pressure effects.

The quality function must accommodate concepts of leadership in order to achieve company-wide effectiveness. This goes far beyond the basic idea of leadership as "command and control". This is not new. Juran (1992, 455) lists participation as a managerial tool that is widely applicable to quality planning. Although by "participation" Juran means "the process of securing

[239] At this point, it must be said that this kind of indicator is believed to be dependent on the industry and on the specific work culture of the enterprise(s) in question. It is used here to indicate the general presence of the problem over the "population" of companies in the experience of Mr. Juran. This kind of information has different attributes when compared with statistically polled data, but it indicates the general presence of the issue, which differs from the information in such statements as "in this specific company, 80 to 90 percent...". The latter kind of data was used in this research only when the informational content *as well as* the context were unambiguous and that it was described sufficiently precisely to allow analysis.

inputs from those who will be impacted by an intended action," i.e. cooperation in the sense used in this book, these concepts can be extended in encompass cooperative and participative leadership styles in the concept of quality leadership. The leadership style will determine the effectiveness of actions toward developing quality *with* the products.

As with individual and group leadership concepts, the differences in results achievable for the quality of product development are highly dependent on the leadership style practiced by quality personnel.

6.3.4 Support

An experienced development engineer with a background in quality matters can improve the quality of his or her work by virtue of the enhanced perceptions of the potential problems he or she can incur downstream (despite performing his or her own tasks quite brilliantly). However, even when a development group relies on a professional with ample previous experience with quality issues, it would be difficult for the latter to stay abreast of the quality field. Hence, there is the need for a function to support the professionals working with development, not only on quality disciplines. This role might range, for example, from operating and maintaining central testing facilities with the necessary knowledge to perform thorough evaluations, to keeping a study group informed of developments in the area of quality in order to accelerate the (usually slow) learning process for group members, to extending the activities to provide specialized technical support for dealings with suppliers or customers. One has to be able to react quickly when the analysis of specific product problems is required. In addition, new methodical approaches to quality issues could be learned by the development engineer.

Sometimes, even a control (in the broad sense) instance is an effective way to support a group assigned to assess a product's strengths and weaknesses.

6.3.5 Independent evaluator

6.3.5.1 Independent test laboratories

An independent test lab can assess the quality of a development product in a neutral way, at least theoretically. An outside group can provide information that was not available to the groups developing a product, or perform a critical examination of results without, for example, being affected by company politics. More than just an independent appraisal, a significantly more interesting approach is offered by an independent test laboratory equipped with the necessary specialized equipment and experienced personnel to perform technical evaluations that would not be possible for the majority of companies owing to financial constraints. Thus external consultants can complement the knowledge and experience of product development groups. This can be done for isolated characterizations such as electromagnetic compatibility of electronic equipment, environmental testing, or failure analysis. More elaborate possibilities are the complete reliability assessment of an entire equipment or system, or the qualification of an equipment or system for its intended use.

Another example of implementations of this approach are consumer test societies that publish test reports independently from vendors.

6.3.5.2 Certifying, evaluating and regulating organizations

There are many examples of organizations whose function is to independently evaluate quality-related aspects of products to be sold on a given market. Examples of independent organizations in the aeronautics industry are the American Federal Aviation Administration (FAA) and the National Transportation Safety Board (NTSB). In Europe, there is for example the Joint Aviation Authority (JAA), which federates the European National Aviation Authorities. There are numerous examples in industries other than aerospace as well.

Government regulating agencies increasingly assume a role in product development in that they have to react rapidly in order not to hinder the speed of the development process. Sometimes, the success of an industrial sector can be dependent on this agility. Taking for example the role of the FAA in the case of the problem with DC-10 cargo door latches, the agency was criticized for having conflicting goals: both to promote the aviation industry and to provide for its safety (Fielder & Birsch, (Eds.), 1992, 4). According to Eddy, Potter & Page (1992, 122) "most of the blame for the failure to do something about the extreme vulnerability of jumbo-jet floors[240] must ... lie with the FAA".

However, without this two-pronged organizational goal, regulating agencies would tend to halt the development process like the independent evaluating group described above (cf. *Box 24* on page 121). These agencies play an important role in the success of a product and in improving its quality for the benefit of the public they are serving. In the case of the DC-10 cargo doors, the FAA might have failed to realize the ambiguous goal

[240] The abrupt difference in pressure made floors collapse (cf. case of DC-10, p. 112).

described above (page 112), but this does not mean that is not achievable or beneficial to set complex goals, even in an ambiguous, conflicting situation.

Several years ago, there was a controversy over the evacuation requirements for large civil aviation aircraft. They were criticized for being outdated and hence partially invalid. A joint working group from industry and regulating agencies was established to examine the issue.[241]

Another case that involved the certification of the successful Boeing 777 was the approval for extended twin engine operation (ETOPS). This involved extensive testing and monitoring of aircraft systems and intensive involvement of suppliers such as the turbine manufacturers. It also demanded additional training and the demonstration of the proficiency of the flight and maintenance crews. Moreover, it required a revision of the criteria for approved routes (such as maximum time to divert to another airport). ETOPS was achieved from "Day 1", at least for one type of power system. In earlier twin-engined aircraft, ETOPS approval was obtained only after airlines had demonstrated minimal quantitative reliability levels. It was never issued from the first day of operations. According to Proctor (1997), the entire certification of the 777 was the "most thorough—and public—in aviation history." The FAA admitted that "it invested a record 130,000 hours in the certification effort". The entire process of certification, the ETOPS approval in particular, was an important determinant of the success of the aircraft. Had the FAA merely assumed a controlling (in the narrow sense) position, the 777 would probably not have been featured this quality advantage.

This case of cooperation and participation illustrates the need to consider several aspects of an issue, not only the side representing each facet of the problem, i.e. industry considering *only* commercial "success" and the FAA considering *only* safety.

Without joint efforts such as those, famous historical cases might be repeated:

The "unsinkable" Titanic oceanliner had sixteen wooden lifeboats and an additional four collapsible ones, which was sufficient for only one third of the passengers (Büscher, 1984, 0:15') The requirements regarding lifeboats established in 1899 by the British Board of Trade was also recognized in America. It stated that, for ships displacing more than 15,000 tons, a minimum of sixteen lifeboats is to be provided. The Titanic displaced 46,000 tons (Schlotterbeck 1984, 0:29'). Thus, the manufacturers had actually "exceeded" the formal requirements.

The Ford Pinto automobile had the fuel tank mounted behind the rear axle. This rendered it vulnerable to fire in case of a collision from behind at

241 The results could not be examined for this integration.

30 km/h (20 mph). A protective part costing $6.65 each would have made the tank safer. The Pinto was produced from 1971 on without the part because regulations did not specify this safety characteristic. A memo recommended that Ford wait until 1976, when the government was expected to introduce fuel tank standards. By delaying making changes, Ford saved $20.9 million, based on the average cost per automobile of about $10 (De George 1991, 177). The federal standard for gas tanks was not introduced until 1977; 1.9 million automobiles were on the roads at that time. Ford issued a recall only in of August 1978 (op. cit., p. 176) as some 50 lawsuits were filed against the company in connection with rear-end accidents (op. cit., p. 175). In the end, it cost Ford considerably more than what they had intended to save.

Another example of insufficient cooperation from a regulating agency (case #53) occurred in the telecommunications industry. Electronic telephones were available with keyboards for the use with the conventional pulse-type telephone switching. They were designed to replace rotary-dial telephones and to function with the same telephone exchange as its mechanically dialed counterparts. One of the regulating agencies contacted the telephone manufacturers to impose a specification for the tolerance in the pulses sent by the electronic telephones to be narrower than that for the mechanical counterparts. The new telephones would be connected with the same telephone exchange, which worked well with the wider tolerances of the older telephones. When asked why a tighter specification was being issued, the regulating agency declared that their request was simply because a tighter control of the pulse specifications would be *possible* with the new electronic-pulse generation of telephones. The new specification made no apparent sense because it produced no quality improvement of the telephone system nor for the telephone manufacturers.

Some obstacles to good interaction with regulating agencies are associated with

- a relationship based on formality,
- excessive aggressiveness from one or both sides,
- each side wanting to impose its will on the other,
- emphasis on justification instead of on a critical appraisal,
- time constraints to characterize a problem.

The relationship between the organizations concerned is perhaps the most delicate of all, as the occurrence of political, economic, technological and public opinion influence factors can be expected to be more frequent in such cases.

6.3.6 Normative role

The normative function is still necessary in order to provide an internal reference for development personnel. External standards or old rules of practice or checklists that are usually not readily accessible to development personnel must be "decoded." These professionals need rapid and usable information and usually cannot stop their work to question these rules. Establishing standards, in addition to the control function, is probably one of the oldest forms of quality assurance in the development phase. This should not be neglected, as it helps avoid repeating errors, and hence is part of the natural learning process.

Care should be taken when quality personnel operates in a multimode (or changing mode) role because this can generate ambiguity in that the developers do not know when a quality measure is a normative function and when it is a supportive or controlling function. Another issue associated with this ambiguity is illustrated by the following case (Box 70).

Box 70. A case where an obsolete mentality prevailed (case #54).

In an aerospace company, the product in development was in the prototyping phase. One of the goals was excellent maintainability characteristics on which the logistics was based. In a meeting of the development and quality assurance engineers involved, the internal aspects of the diverse *shop replaceable assemblies*[242] and their interconnections were being discussed.

The discussions went smoothly. Quality engineers asked questions such as, "Is it possible to implement it this way or that way?". The development engineers asked, "Is it necessary to have this or that characteristic?" All the participants were satisfied about the quality of the meeting. A list of implementation suggestions was established. Most of them were small changes in the initial conception to improve the product maintainability characteristics.

One of the issues raised by the quality engineers questioned whether it was possible to provide connectors in both sides of a connecting cable. The development engineers requested time to evaluate this point. At the next meeting, they expressed concern that it would not be possible due to the space constraints of the compact design. The quality engineer who had raised the issue said, "Ok, no problem. It would improve the maintainability, but we do not *have* to have it. In this case [due to space limitations], we can

[242] Denotes the division of the product in assemblies or modules to be maintained or substituted with the aid of additional equipment. It is in contrast to line replaceable unit, which refers to modules that can be substituted in the field, usually without the use of additional test equipment.

have this connector in only one side." He sensed that the developers were silently staring at him. He added, "Hey, I never said we *must* do it. I was only asking about the technical feasibility." The development personnel replied that they had understood it as such and reproached him by saying, "But you said we *should*." "No, I didn't," the quality engineer responded. "Yes, you did." "No, I didn't"... The development personnel concluded, "In any case, that's what we understood." The quality engineer, finding the situation quite ridiculous, remained silent for a few moments, before demanding, "Well, *is it* technically possible?" The meeting proceeded, but resentment and suspicion greatly impeded its efficiency.

The development engineers then requested standards to stipulate what they must do.

The quality engineers were puzzled by this collective tendency to interpret what had been said as an obligation, a requirement.

A technical director also present at the meeting later commented to the quality manager about the developers' request for standards. He observed that they had not wanted standards to comply with customer needs, but merely in order "not to have to think".

The overwhelming majority of the development engineers understood that the quality representative was acting in a normative fashion, whereas from the latter's point of view, he had intended to stimulate discussion and cooperation. Some development engineers evidently expect the quality function to act only in the normative and controlling mode.

6.3.7 Pathfinder role

Depth of technical knowledge is vital to competence in making decisions. Many small groups working on the development of a product simply have no time to explore the intricacies of a general area, especially if the time-to-market is to be minimized. Concrete examples of product assurance playing this role in an aerospace electronics company are establishing directives for PCB layouts, board and system testability, and environmental testing knowledge. It is difficult for a group to achieve depth in new areas or remain abreast of a technical field. A general rule for not being caught by surprise is to stay slightly ahead of the problem.[243] When the situation arrives, there should be someone who has the necessary knowledge to support the development group. This could shorten the reaction time of the company.

The advantageous situation of a central quality group may be the most appropriate to develop new areas of knowledge in order to increase the speed

[243] An old pilot's rule states "fly ahead of the aircraft". This is a variant of this rule.

of development. Again, through decentralization of activities, the quality group can complement the knowledge of the development teams while keeping the central group. This enables the quality group to continuously update their knowledge of general subjects, in a way that is not possible for decentralized, project-specific activities.

6.3.8 Development partner role

Sometimes, portions of a product development analysis can be done by the central quality function, which can act as a development partner for groups working on product development. Examples of such activities include evaluation of portions that require specific, extensive analysis , and development of test practices. Specific detailed knowledge of the quality area (such as statistical or reliability analyses) could be better acquired by means of a concentration of expertise. There are several successful examples of such cases.

The central quality function can maintain this development with slightly different goals than the product designers. Consequently they use a different yardstick (derived from Seghezzi, 1988, 358), which might be of interest for the entire company (e.g. increase of company base of knowledge).

6.3.9 Modern quality approaches

6.3.9.1 St. Gallen integrated quality management

The roles and issues discussed above can be addressed by application of the St. Gallen integrated management concept, stated generally by Bleicher (1992) and adapted for quality management by Seghezzi (1996). *Figure 20* depicts how the concept considers the reciprocal relations between its elements.

The rightmost column emphasizes the relevance of behavior for achieving quality. At this point, the necessity of the *organizational development* being covered by the quality department (or responsible persons), especially by means of quality leadership is emphasized. Organizations in which these processes are embedded feature a learning character that requires consideration of the quality of interaction among the participants. The issues faced today have their roots in the perception by participants that their environment does not foster the conditions necessary for the successful implementation of modern concepts such as this. Centuries-old mentalities must be changed in a learning process that, in the usual cases analyzed, have a long-term horizon. Experience shows that persons willing to change draw greater benefits from the learning process, although they too are aware of the predictable difficulties of an implementation.

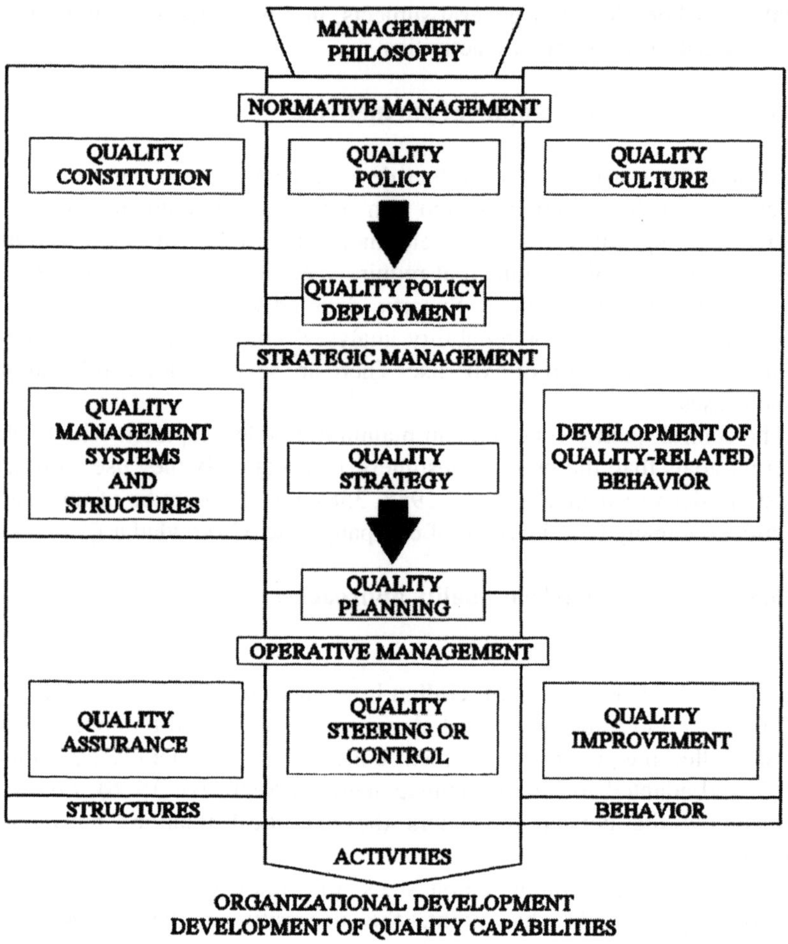

Figure 20. St. Gallen quality management concept (from: Seghezzi 1996, 50).

6.3.9.2 ISO 9000:2000

The evolution of the ISO 9000 series led to the 2000 version, which encourages the use of process orientation. "Encourages" is the term used by the International Standardization Organization to make explicit that the norm does not specify that all companies *must* use it in order to be certified.

The revised standard also makes use of the term "stages" instead of phases, maybe in an effort to preclude the implicit misunderstanding resulting from one model. Whether this semantical change was performed with this goal is not known by this author at the moment of publication, but readers are recommended not to merely rename phases into stages, but to take the opportunity to propose their own way to handle development activities. Moreover, the term "design control" is no longer used.

An auditor and instructor of a leading certifying organization declared that the changes were needed because of an observed exaggeration of formalism concerning quality systems.

The ISO 9004:2000 standard also mentions under "Use of Quality Management Principles" the topics listed in *Table 43*:

Table 43. Use of quality management principles in ISO 9004:2000.

	PRINCIPLE
a)	Customer focus
b)	**Leadership**
c)	**Involvement of people**
d)	Process approach
e)	System approach to management
f)	Continuous improvement
g)	Factual approach to decision making
h)	**Mutually beneficial supplier relationship**

Although ISO 9004 is likewise not obligatory for certification, note the principles highlighted in boldface in *Table 43*. Quality organizations with higher goals than basic certification are encouraged to give these non-obligatory issues their serious consideration.

6.3.9.3 EFQM

Another approach to quality was proposed by the European Foundation for Quality Management (EFQM) with the fundamental concepts listed in *Table 44*. The columns in the table refer to different organizational maturity stages.

This table is self-explanatory. In diagrammatic form, the model assumes the form depicted in *Figure 21*. The values indicate the weight for evaluation.

This model has a cyclic-dynamic character in that *innovation and learning* also serves to improve the *enablers*, thus leading to *results*. *Table 44* also reveals this explicitly dynamic character necessary for an organization to achieve business excellence.

Table 44. Fundamental concepts underlying the EFQM excellence model. (EFQM, 1999b, 8)[244]

	FUNDAMENTAL CONCEPT	START UP	ON THE WAY	MATURE
1.	Results orientation	All relevant stakeholders are identified	Stakeholder need are assessed in a structured way	Transparent mechanisms exist to balance stakeholder expectations
2.	Customer focus	Customer satisfaction assessed	Goals and targets are linked to customer needs and expectations. Loyalty issues are researched	Business drivers of customer satisfaction needs and loyalty issues are understood, measured and [acted upon]
3.	Leadership and constancy of purpose	Vision and Mission are defines	Policy, People and Processes are aligned. A leadership 'model' exists	Shared values and ethical role models exist at all organizational levels
4.	Management by processes and facts	Processes to achieve desired results are defined	Comparative data and information is used to set challenging goals	Process capability is fully understood and used to drive performance improvements
5.	People development & involvement	People accept ownership and responsibility to solve problems	People are innovative and creative in furthering organizational objectives	People are empowered to act and openly share knowledge and experience
6.	Continuous learning, innovation and improvement	Improvement opportunities are identified and acted on	Continuous improvement is accepted objective for every individual	Successful innovation and improvement is widespread and integrated
7.	Partnership development	A process exists for selecting and managing suppliers	Supplier improvement and achievements are recognized and key external partners have been identified	The organization and its key partners are interdependent. Plans and policies are co-developed on the basis of shared knowledge
8.	Public responsibility	Legal and regulatory requirements are	There is active involvement in	Societal expectations are measured and

[244] From: EFQM - European Foundation for Quality Management (1999). Eight Essentials of Excellence - The Fundamental Concepts and their Benefits, Brussels: EFQM. Reproduced here by kind permisssion of the European Foundation for Quality Management.

FUNDAMENTAL CONCEPT	START UP	ON THE WAY	MATURE
	understood and met	'society'	{acted upon]

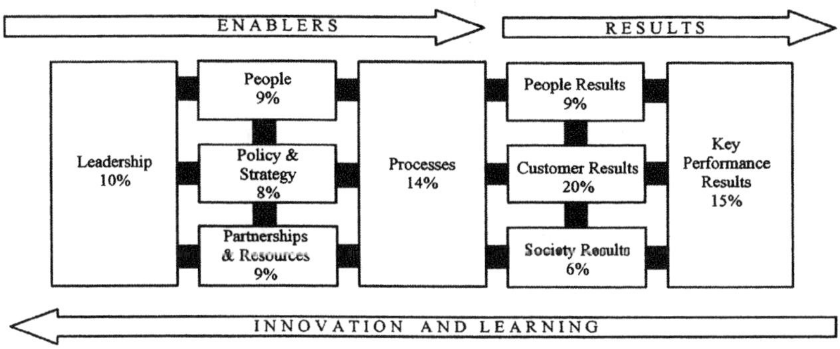

Figure 21. EFQM model. (EFQM, 1999a, 34)[245]

The principal characteristic of this model is its departure from the process/procedure-based approach still in use as the main quality approach in many industries. It is also required for companies applying to the EFQM. Attentive readers will note that the same approach is *possible* using the ISO 9004:2000 guidelines (*Table 43*).

Although these new quality models were proposed in a general sense, the elements contained therein strongly influence product development quality. They show the extent to which the quality function should be improved to encompass additional issues beyond controlling specifications.

In contrast to the ISO 9000:2000 model, the weighted importance of the non-system-related groups to the EFQM evaluation depicted in *Figure 21* shows a trend toward balancing the consideration of the non-system (or non-process) factors that determine quality or business excellence. This is especially true for the factors "people" and "people results", which when integrated make up 18% of the weight, hence surpassing the weight of the processes (14%). When they are added to leadership, they total 28% of the weight. This illustrates the enhanced importance of people issues in achieving quality—including product development quality—which is not necessarily covered by quality systems based on procedures.

[245] From: EFQM - European Foundation for Quality Management (1999). The EFQM Excellence Model, Brussels: EFQM. Reproduced here by kind permisssion of the European Foundation for Quality Management.

6.4 Redundancy in these two roles

There is an implicit redundancy in the roles of project manager and quality manager assigned to a project. This redundancy is associated with the quality leadership function discussed above. Recalling the need for integrative functions described on page 288, the quality department can actually introduce a redundancy to the role of the project/program manager.

There is a tendency of managers to regard this redundancy as unnecessary duplication, waste, or a potential source of a conflict of responsibility.

In addition to the sheer quantity of details to be considered, human reliability is not 100.0000%, even though in many cases it approaches 100%. In fact, it typically decreases in cases such as stress situations. This includes the project manager him/herself (or the quality manager). Hence a certain redundancy has many advantages of alleviating the load on a project manager and improving product characteristics. Similar to technical systems, redundancy is not a feature to be used indiscriminately. Redundancy can be used locally to improve the reliability of a specific critical point, or globally to improve the reliability of a particular function. In some cases, redundant systems can actually facilitate or accelerate the operation or provide pertinent information or functions by virtue of the additional equipment. A good example is the dual transceiver / dual navigator equipment found even in small aircraft. It provides not only a backup in case of failure but also improves communication with air traffic control. Moreover, it simplifies navigation and greatly reduces workload when flying single pilot IFR.

Hence, implementing an equipment redundancy can actually increase the probability of success of a system. Similarly, project manager—quality manager redundancy can actually increase the probability of project success, provided this redundancy accommodates the critical issues addressed in detail in this book.

6.5 Concluding Remarks

Developing the developers' job is not a trivial task. General elements of job design developed for the working world are presented briefly. The presentation of job design with a focus on manufacturing activities is intended to enhance the contrast between what has been improved in the blue-collar working world and what has been done or, in some cases, demolished in development activities. These considerations provide a starting point for evaluating job design; its form in terms of development is not yet final.

In this chapter, the need for two integrative functions was introduced. One is related to leadership aspects and the other is the partially complementary, partially redundant role of quality personnel.

It is emphasized that both should assume a leadership role, not in the sense of strong-handed leadership, but directed toward improving interactions.

Leader orientation and problem solving are related. This includes bringing expert knowledge to surface, e.g. letting the voice of a minority be heard. Leadership is also very important for avoiding the pitfalls of group work described in the previous chapter.

Quality has a long way to go before it can free itself from the stereotypes associated with control personnel and achieve an image of playing a proactive role. In spite of the quality systems in place, companies would be wise to invest in many other aspects that determine the quality of problem-solving processes. It is clear that *much more* than a system of rules or procedures is needed to satisfactorily accommodate these issues. Whether quality professionals and their management counterparts—either in aerospace or any other technical industry—are well prepared for this challenge remains to be investigated.

Chapter 7

GENERALIZATION OR PARTICULARIZATION?

Other technological sectors and Aerospace

Although there are significant differences between aerospace and other technical industries, the gap is closing for two main reasons. First, there has been extensive cross-fertilization between aerospace and other industries, and, second, aerospace is being subjected to the same cost-efficacy pressure as other industries.

Influence of aerospace on other industries

- Project management
- System engineering
- Skunk works
- Systematic approach to "controlling" development
- Engineering changes control
- Reliability & maintainability engineering
- Traceability
- Integrated 3-D CAD
- Integrated logistics

Influence of other industries on aerospace

- Self-control
- Outsourcing
- Lean management: The concept was tried in aerospace, but apparently without the advertised results

- Quality circles
- Group work / teamwork
- Benchmarking: Cessna, for example, used GM Saturn plant as a reference when restarting manufacturing of single-engine piston aircraft.

Examples of its influence on daily practices that were readopted in an improved form by aerospace are:

Design Review. Originated by the aerospace industry, this technique was tried in other branches and aerospace is now considering using design reviews internally and not only as the contractual requirements.

FMECA. Also originated by the aerospace industry, it was improved by the German Association of Automobile Industries[li] as an instrument to reduce the significant differences in prioritizing the myriad of issues that arise, assigning actors to the implementation in a product and comparing original and improved states. In this form it is being reintroduced into aerospace companies (especially in German-speaking countries).

Simultaneous engineering. This method originated in the aeronautical industry during World War II. It was widely applied outside of the aerospace industry before it started to be used again in its original sector. Some argue that the approach has been (and is) originated by every garage-style technical company. But many companies have tended to diverge from this approach as they grew in size, making the application of the simultaneous engineering approach an organizational goal, not a starting point. In this context, the specific advantages of a garage-style enterprise are meant.

Moreover, the "shift to commercial" in the aerospace industry has introduced elements from such areas as marketing, cost accounting, logistics, less emphasis on paper, etc.

Commercial products have evolved significantly since the introduction of many aerospace requirements that were necessary to enhance levels of quality and reliability. In the 1960s, for example, people were enraged and willing to speak up if they noticed that a billing invoice listed an exorbitantly high price for a bolt to be used in the Apollo program, compared to the price of a similar article at an ordinary hardware store. Thereby they typically overlooked the additional effort necessary to attain the desired reliability levels. During the same era, some engineers even made public statements to the effect that the computer used for navigation in the Apollo program was not even as powerful as one's then modern PDP-11 in the lab, yet the former was much more expensive. Even the legendary PDP-11 was perhaps not the most suitable solution to achieve the goal. Since then the gap has closed, but it would still not be admissible to use bolts from an ordinary hardware store for most aerospace applications. Most present-day personal computers would still not be suitable for flight guidance. Space agencies throughout the world are now analyzing the applicability of commercial technologies in space, but their use is still restricted.

This restructuring of industry, especially the aerospace industry, was intended to enhance efficiency, pushing it to new, unprecedented heights. But to avoid throwing the baby out with the bath water, it is necessary to consider the factors underlying various cases of success and failure.

According to Lambright (2001, 4) , NASA's faster – better – cheaper (FBC) paradigm was rooted in an analogy to the revolution in (non-aerospace) microelectronics. The concepts behind FBC worked well in electronics for the following reasons, among others.

- Improvements in photolithography and diffusion processes, which led to a better compaction and better yield in die manufacturing.
- Improvements in speed were possible as the limiting factors and the reduction in size became better understood.
- Interconnection and packaging was one of the most cost-intensive stages of production.
- Reducing an external component count generally improved reliability and reduced costs for the product manufacturer.
- Enhanced knowledge of failure mechanisms, such as the CMOS latch-up, hot carriers in MOSFETs, time-dependent dielectric breakdown due to high electric fields, electromigration due to high current density, soft errors in DRAMS, and others fostered improvements in size and speed (cf. Birolini, 1994, 100-102, 107-111).
- Improvements in packaging technology and in testing components of increasing complexity were possible at the same time.
- Reduction of required power levels were possible at the same time.

Implementing this in electronics requires a thorough understanding of the physical devices and their technology limitations. Otherwise, a component will fail due to overload or to internal weaknesses. Although the electronics industry is struggling today to uphold Moore's Law, products available on the consumer market are not necessarily becoming cheaper, reducing price, albeit faster and better. A closer look at the transfer of this paradigm to the product development process is required. This calls for a detailed study of the applicability of the concept to organizational design in order to avoid an overzealous pursuit of efficiency. Hence, in terms of the technical world:

- Ignoring the processes underlying the attainable size and production throughput would lead to an increase in defects;
- Overloads are possible but only for limited amounts of time;
- Requiring more from the same component may cause an overload of its weaknesses, leading to component failure;
- Failure mechanisms should be known;

- Reducing external component count is not an absolute rule and can potentially introduce unreliability issues;
- Improvements in the transmission channel alone do not lead to higher speeds; transmitters and receivers; encoders and decoders should also be considered;
- Lowering tension (to reduce power) has a very defined limitation to still explore the (nonlinear) effects of semiconductors. It is not possible to go from 5 to 0 V in 50% reduction steps (current low-power microprocessors run at 3 V), whereas bipolar silicon devices for analog electronics are still inherently limited to the 0.6 threshold voltage.

FBC-like initiatives should consider the equivalent of these limitations in the processes underlying the activities to be optimized instead of blindly cutting parameters in half in successive increments until the item catastrophically breaks down.

As Dr. Edward Weiler, NASA's associate administrator for space science, commented after the root cause of the loss of Mars Climate Orbiter (MCO) was disclosed: "the problem here was not the error, it was the failure of NASA's system engineering, and the checks and balances in our processes to detect the error" (Dornheim, 1999, 40). It is indeed easier to focus on the technical error and forget the *non-technical* organizational characteristics that would allow the early detection and correction of errors, whether as trivial as the latter or more complex.

Namely, the eight causes that contributed to the loss of MCO are as follows.

1. Undetected mismodeling of spacecraft velocity changes.
2. *Navigation team unfamiliar with spacecraft.*
3. Trajectory correction maneuver number 5 not performed.
4. *System engineering process did not adequately address transition from development to operations.*
5. *Inadequate communications between project elements.*
6. *Inadequate operations navigation team staffing.*
7. *Inadequate training.*
8. Verification and validation process did not adequately address ground software.

Five of these causes are essentially of non-technical nature.

Other industries besides aerospace are also realizing that some 60% of costs or more are associated with personnel. One of the differences between the so-called Third World and the industrialized nations is reflected in this percentage distribution: the former is characterized by materials being more costly than personnel; in the latter the situation is the opposite. Instead of concentrating on the advantages of improved employee education, training and experience, the recent trend is to regard these assets as a limiting factor to success that should be cut, not as a strategic advantage to be cultivated. Clearly the spreadsheet highlights the predominant cost component, but this does not necessarily mean that this is the thing to cut first. Without understanding what can enhance the efficiency of an R&D organization, looking solely at the numbers can lead to a loss of a company's most valuable assets. Blindly following the numbers instead of looking at what they actually mean is a behavior that Swiss entrepreneur Branco Weiss referred to by remarking that, "The world was not created using a spreadsheet" (in a lecture in 1989).

Hence, it is not surprising that the same four themes are mentioned most frequently in the failure investigations listed by the Stephenson Report, the investigation of project management at NASA after the loss of MCO. These four themes are (i) reviews, (ii) risk management/assessment, (iii) testing, simulation, verification/validation, and (iv) communications (Stephenson et al., 2000, 34). In its attempt to translate the economic *advantages* of the commercial world to space, the FBC seems to have brought the *problems* of the commercial world into the aerospace business.

At the same time, the Space Shuttle Independent Assessment Team (SIAT) revealed that "communication difficulties exist between parties particularly in accepting feedback from the workforce..." (McDonald et al., 2000, 15). Moreover, they stated that, due to downsizing and the attendant changes in work practices, "resources are stretched thin, and there are now situations in which technicians cannot rely on the same resources and infrastructure as before. In the spirit of open and honest communication, many of the negative effects of these changes might have been mitigated if the workforce were more appropriately prepared, retrained and, in some cases, retained. Because people are covering more jobs and more responsibilities and are working across different facilities, the training needs have changed and increased." (op. cit., p. 19). The foremost finding of the SIAT investigation of the Hypergols and auxiliary power unit stated that training is not conducted by the actual flight hardware vendor, "which removes technicians and engineers from a direct interface with the designers" (op. cit., p. 23). The assessment team contrasts their observations with the situation of technicians working in the self-contained atmospheric protective ensemble on the Titan 4B program. They praised the fact that courses provide "technicians and engineers a direct interface with the manufacturers/designers as opposed to relying only on manuals, drawings and maintenance procedures for education" (op. cit., p. 26).

Cutting activities and personnel, together with disrupting (apparently time-consuming) communication channels in an effort to improve company performance were measures taken previously in non-aerospace companies. The Wyatt Company (1993) published a study covering the period 1986-

1991 of 1005 organizations that had engaged in downsizing. Their findings revealed that only 46% succeeded in reducing expenses, only 22% increased productivity, and only 17% actually reduced bureaucracy. All of these goals had been the declared objective of the downsizing measures (cited in Cameron, 1994, 190). Complementing the results of the Wyatt survey and their own study, Hitt, Keats, Harback & Nixon (1994, 18) conclude that only 12% of downsized companies experienced increased market share, 9% improved quality and only 7% enhanced innovation (cf. Hamel & Prahalad, 1994). Other reported effects of downsizing included: the realization that the wrong professionals had been "outsourced", increased expenses for retraining, overtime, and health care, as well as impaired morale among the remaining work force. This in turn caused even more "desirable" professionals to leave, among other effects (cf. Downs, 1995 and Noer, 1993).

The SIAT report also indicated similar results from studies of non-aerospace companies. It suggests that the relevance of these trends to NASA organizations has to be researched, but also states explicitly that these indicators are "consistent with many of the emergent issues of the SIAT human factors observations and interviews" (McDonald et al., 2000, pp. 89-90). A further comparison of the recent results from the commercial sector with the findings by Fries on NASA in the post-Apollo era (cf. page 165ff.) suggests a similarity in the two situations that deserves further investigation.

As mentioned in the report submitted by the ISS Management and Cost Evaluation (IMCE) Task Force to the NASA Advisory Council regarding the cost overrun of the International Space Station, "At this phase of the ISS program, deleting more hardware saves very little money since the bulk of the expenditures are in the 'people' category" (Young et al., 2001, 5). The task force considered the currently planned ISS staffing levels unnecessary, but recognizes that it "will enable continued high levels of response".[246] It also suggests that flight anomalies can be addressed on the ground with technical expertise that is "on call" and not "on-tap" (op. cit., pp. 4,5). How this technical expertise could be updated to accommodate the current situation, or on which spreadsheet their (fixed) costs would appear is not mentioned. Note that some managers subscribe to the naïve notion that experts can not only be made to disappear from their cost spreadsheets on command, but will reappear when needed, of course in possession of all the current pertinent information.

Another perception reflecting this cross-learning between aerospace and non-aerospace industries is described by two complementary cases.

[246] The current approach was based on the tenant that technical excellence and human safety come first.

During the development of the Boeing 777, serious consideration was given to eliminating flight testing of a new power plant for the twin-engined airliner. The main argument was that the engine had been so thoroughly designed and tested on the ground that a flight test bed was unnecessary. But this was not the case. On its third flight, the engine surged[247] during take-off and had to be throttled back to idle. The surge was not disastrous, but it revealed problems that had been overlooked in the previous analyses and tests. As Mulally points out, "after we analyzed the data, we also realized that we had not run that condition during our test program, so we went back and ran that test condition and it also surged on the ground" (Sabbagh, 1996, 218). Mulally also states that 85 to 90% of the cases validated by engine flight data supported what they expected from their previous analysis of the engine, but flight tests provided additional valuable data. They could then introduce modifications to preclude undesirable performance. Mulally continues: "I think it's a perfect example of what you have to do—include all the lessons learned, do the best design you can, but then comprehensively test it ...and learn from that data and then be able to deliver an even better-quality engine" (op. cit., pp. 218-219).

The counterpart case from a non-aerospace company occurred with the compact A-Klasse car by the then Daimler-Benz. Testing of the first units marketed by an independent Swedish tester revealed the tendency of the car to overturn. The so-called "moose test" refers to a hypothetical situation of swerving abruptly at a speed of 60 km/h without braking to avoid hitting a moose unexpectedly crossing the road. This case, which happened to a very experienced and renowned automobile company, illustrates a potential flaw of modern product development despite improved simulations and application of computer technologies. Daimler-Benz used its 6-axis driving dynamics simulator—unique in the world—which has an impressive simulation record of six million kilometers. Without a doubt, extensive experiments can facilitate design choices, but relevant performance characteristics were revealed only by testing the real unit. The previously unknown moose test is now in almost every car test catalog. Nevertheless, a point revealed by J. Stockmar[248] does not seem to be on every development manager's checklist: that "the physical and psychological limits of many workers in the automobile industry—particularly in the development departments—was reached and exceeded. If one analyzes the reason for mishaps, one finds consistent evidence of overstretched workers in every case" (Fuths, 1997, 24).

---

247 A condition where the airflow over the engine blades separates.

248 Mr. Stockmar was then at the technical board of Opel.

The concept of product commonality[249] used, for example, in the XP-75 described on page 82, was adopted by non-aerospace industries, where it was widely used for a range of products from home appliances to cars. It later reappeared in aircraft design such as Boeing 757/767 family or the Embraer Regional Jets. Based on this concept, the inertial reference system (SRI) of Ariane 4 was used in Ariane 5, but the pyrotechnical results of insufficient testing have not been forgotten by many who watched the latter's first launch. The different trajectory of Ariane 5 caused an overflow in the data conversion from 64-bit floating point to a 16-bit signed integer value. The operand error was indicated by the SRI in the data bus, but was misinterpreted as flight data by the on-board computer. The computer then initiated a full deflection of the nozzles in the solid rocket boosters and main engine. The overflow was caused by a function not necessary to flight, but that had served to eliminate the need for a 45-minute SRI realignment of Ariane 4 when resuming an interrupted launch sequence (Lions et al., 1996, pp. 3-4). Three things should be remarked in the context of this chapter:

- The function for keeping SRI alignment was not necessary to Ariane 5 but was retained for commonality purposes (op. cit., p. 4).
- The prevailing idea that what already flew in 23 previous Ariane 4 missions is already space-qualified presumably fostered a [self-defeating] high degree of confidence (op. cit., handouts).
- The board recommendation R#14 states that "a more *transparent* organisation of the *cooperation* among the partners in the Ariane 5 programme must be considered. *Close engineering cooperation*, with clear cut authority and responsibility, *is needed to achieve system coherence*, with simple and clear interfaces between partners." (op. cit., p. 14, italics added).

The same kind of thinking is encountered in non-aerospace industries as well, and much cross-fertilization is expected to occur, hopefully with productive effects. Hence the answer to the following question perhaps posed by the reader is as follows: Is this a generalization from aerospace to other industries or is it a particularization to aerospace? In fact, it is both.

249 Different products using the same parts.

Chapter 8

CONCLUDING REMARKS

8.1 General conclusions

Following the definition of the research method as developed in the Appendix, seven broadly stated research propositions provided a baseline for this research.

Box 71. Initial research propositions for this work.

The main research propositions can be stated as follows:

1. A controlling mode of operation leads to limited results in the product development process and causes resistance from the developers' side.
2. A cooperative mode of improving quality in the product development can help solve resistance issues.
3. A participative mode of improving quality can improve results further by improving the dynamics of interaction between various different disciplines.

The secondary research propositions can be stated as follows.

4. There is an effect of the contrast between autonomy and control, with the usual situation that both of them are present.
5. Product development is essentially a group activity and as such depends on the pertinent social interactions.

6. Product development is also an individual activity in that the interactions with other participants is dependent on or influenced by designated individuals.
7. The process barriers are essentially of a social-science nature, not of a technological one.

The starting point for the main research propositions is leadership psychology. The starting point for the secondary research propositions lies in the socio-technical systems approach.

As to be expected from a grounded theory study, a *conceptually dense* work resulted from this research: there are many concepts involved and numerous connections among them. It also means that complex interpretations and the necessary data collection were guided by successively evolving interpretations made during the course of the study. Moreover, the reoccurrence of the coded events found throughout various different technical branches as well as across geographical borders provides a generalization of each point. Most of the cases cited represent a class of cases that is inherent to the research method (core categorization) and should not be implicitly understood as isolated events. Other cases serve to summarize sets of elements. In still other cases, the necessary partial isolation of a subset of the environmental characteristics was achieved. This allowed the differentiation among the various influencial factors.

The distribution of partial results and the conceptual density render the task of providing a *brief* synthetic conclusion of this research work difficult. However, the following general conclusions could be drawn.

The first proposition had to be modified to allow the general conclusion that the *form and intensity* of control impose a limitation on the exchange of information necessary to achieve a quality improvement during the product development process. It can be said that there is a relationship between the *perception* (by either side) of the controlling mode of conducting activities and the resistance from developers to those activities. The actual relationship, however, is complex. Merely changing the labeling of activities rarely solves a problem if the orientation remains the same.

There is a reciprocal relationship between cooperative behavior and way a situation is perceived. It is possible for a learning process to change defensive/non-cooperating behavior into coping/cooperating behavior. The cognitive, motivational and moral orientations result from and at the same time cause a person or group of persons to behave cooperatively or not. They are dependent on and influence the type of social relationship in question and, in the case of R&D professionals, are not only dependent on a person's technical proficiency.

The growing importance of the inherent interdisciplinarity or multidisciplinarity present in many product development processes requires an enhanced perception by the participants of the situation faced in the R&D environment and the capacity to trade with and influence it. This reinforces the importance of interaction theory for leadership, both for individual leaders and for quality leadership. In particular, the three categories (i) R&D environment, (ii) the individual and (iii) the group should not be considered isolated entities. They not only interpenetrate each other but have many reciprocal relationships. The complex nature of this interaction should be kept in mind when dealing with the product development environment.

An application of the results of this work is illustrated by the following real-life case.

Box 72. Example of possibilities of analysis for job design in product development.

Consider the use of different leadership styles for different phases of product development.

A "proposal" was repeatedly made that a development manager should adopt a participative leadership style until the detailed product specifications have been written down. Afterwards—according to this proposal—a different person should assume the role of the development manager and should adopt an authoritarian style.

This example shows the danger of indiscriminately and vaguely stating such a proposal. It presumably originated from the general statement that different tasks require different leadership styles. However, this application to product development will presumably fail to consider the difficulty of establishing a tight compartmentalization of development tasks in phases such that there are similar elements in all phases—albeit in different combinations, occurrences, intensities and importance. This will probably predispose such an approach to hinder or at least inhibit essential elements in either of the two phases proposed. The fact that some elements are less prevalent in the latter stages of product development than in earlier ones does not mean that they are not important. In fact, the possibility of exploring these essential elements often eventually leads to the implementation of a product characteristic associated with a competitive advantage or to the elimination of an embarrassing or even catastrophic flaw of the product conception. It might even be necessary to reconsider the original goals if market conditions, for example, have changed in the meantime. It also fails to consider the strong mutual dependence of the most appropriate leadership style and the individuals being led. It ignores the convergent – divergent nature of the problem-solving process, neglects the impossibility of ensuring the complete and unequivocal determination of every detail of product specifications, and it overlooks that the participants, not only the project manager, will have to be changed, but for other reasons.

Pertinent results are readily available in the open literature concerning the risks associated with this proposal. A report by Fiedler, Potter, Zais & Knowlton (1979) describes four studies in military organizations. By inferring the use of intelligence or experience by correlations of these variables with performance measures, the authors found consistent evidence of different uses of the intelligence and experience of military personnel in leadership and staff positions. These positions "presumably require the individual to abstract, conceptualize, and make complex judgments" (op. cit., p. 635). This study was not integrated into the main body of this work because of the limitations in generalizing its findings for civil organizations. However, in the context of this proposal, a limited amount of generalization is possible due to the similarity of leadership styles, i.e. authoritarian, and the similarity of the above-mentioned position requirements to those needed for development personnel, even those without leadership or staff functions. The authors' conclusion is that *individuals use their intelligence if their relationship with their superior is nonstressful*. Likewise, they fail to use their intelligence optimally or even misuse it if their relationship with their superior is stressful.[250]

One of the occurrences of this proposal can be traced back to the (even though partially valid) characterization of early development vs. main development. In the former phase, the task is to generate an idea to be realized. In the latter, the efficient implementation of the idea is the main goal. In the former, creativity and intuition play a major role because the goals are vague. In the latter, discipline and clearly defined decision rules are followed in a rigid time frame. Between the phases is a filter function that serves to separate the two worlds. One author states that at Honda, in the former phase, 99% error is permitted, whereas in the latter phase, 1% error is too much. No unequivocal association between this author and the proposal

[250] One of the investigated military organizations showed a considerable negative correspondence between intelligence and performance under stressful conditions. In this organization, individuals with a high score reported feeling "under considerable pressure but that their boss withheld guidance and information that would enable them to do the job" (op. cit., p. 639). This situation occasionally occurs in product development due to the many interfaces with other functions, to uncertainties in the choices to be made by the developers, to the prevalence of time pressure, or to the manager's characteristics. All four studies found a high correlation between experience and performance in stressful conditions and a low correlation in nonstressful situations. To preclude a possible misinterpretation of this study, it must be mentioned that previous studies found that the correlation between a leader's performance and intelligence or experience were very low.

could be established and this is the reason why this author will not be cited in this example.

If the connection does exist, however, then the following should be remarked as well: First of all, this description does not seem to apply to the *product specification* but to the issues of *idea generation* and *proposal selection* in that a few of the many project proposals generated by means of creativity and imagination will be selected for implementation. The filter function is not the milestone "product specification" but the "decision to go ahead" with the proposed idea or not. In many cases, once the project has been selected, a detailed specification will follow during which the actual development goals would be set.

Finally, it is necessary to specify what constitutes an "error". There is a difference between 99% *error* and 99% *failure*. It is possible to avoid failure in spite of the errors, and a failure could be avoided because "errors" were found and corrected.

It is conversely possible to experience failure even without making an error. Regarding the statement above concerning Honda, without going into detail as to whether the Japanese culture makes this semantic differentiation between *error* and *failure*, it does exist in western culture. Typical expectations of western managers varied from 1-out-of-3 successes for cases generated by means of creativity and imagination, to 1-out-of-10 successes,[251] where "success" is characterized by an idea selected for implementation. Apparently, once it has been selected, western management tends to have a much lower tolerance to failure than the 1-out-of-100 of Honda, where failure is understood as market failure.[252] Under these conditions, it might be more than reasonable to seriously consider the possibility of allowing a certain amount of freedom for making *and* correcting intermediate errors when implementing the desired product.

This example illustrates the danger of making a vague blanket proposal. On the other hand, comments regarding this example have not contained a counterproposal of "one best way" to conduct the product development process, as some of the cases mentioned in this book show that in certain situations, strong leadership (not to be confused with authoritarian leadership) might be an effective approach to achieve at least one result, not necessarily the most appropriate one. This includes the approach of "terrorizing" or even control (in the narrow sense) described above.

---

[251] These numbers are merely examples taken from a small sample. These numbers were provided by development executives in the electronics industry to represent the goals to be achieved by "creativity groups".

[252] It has been suggested by many development managers that, when asked "What if you do not succeed?" they answer "This is not an option!"

This is one of the possibilities for using the results contained in this work.

Another consequence of the inherent interdisciplinarity or multidisciplinarity in product development is that it requires that the interplay among the different disciplines and professionals involved in the process could be appropriately considered. As such, cooperation and participation in the forms described in this work assume an important role that deserves the consideration of learning processes (concerning the possible forms of interaction with the social environment) occurring inside individuals or groups. The simple removal of external barriers to cooperation (such as walls, cubicle partitions or organigrams) is unlikely to be sufficient to achieve effective changes. Having opportunities to make decisions regarding product implementations; to jointly develop products; or to elaborate concepts associated with development are some of the participatory possibilities that provide elements that increase the quality of the results. Merely having the opportunity to submit suggestions does not produce a self-sustainable approach. In any case, if somebody has previously been discouraged (by superiors, the organization or other development colleagues) from making suggestions, that person might be slower to react positively. Even when one does react promptly due, for example, to one's agility in grasping the issue or because one were just waiting for the chance, there is a learning process to be faced by all parties involved, including management, in getting the new work form adjusted and running. Of special consideration in multidisciplinary product development is the influence of the majority on an individual as well as the importance of the minority.

Motivation plays a very important role not only in solving (technical) problems but also in the enhanced perception of the situation needed to achieve the global optimization necessary in modern product development. Not only the job content or the achievement orientation of individuals plays an essential role in product development, but also a (complex) set of characteristics used by Maslow to describe self-actualizing persons. A general warning can be formulated in this context. With today's growing awareness of the importance of *social competence* in technical professionals, care should be taken not to overemphasize teamwork capabilities (or something labeled thus) *required by these individuals* such that they would become excessively relationship-oriented, resulting in a departure from task orientation. On the other hand, the importance of problem orientation does not mean that individuals or groups working in product development can neglect their main discipline. Leaving "discipline development" exclusively to a research "elite" can lead to worse consequences than those of the Tayloristic division of labor on production. This ambiguity in relationship – task – discipline orientations is a very delicate issue to be addressed by every professional who intends to stay abreast of his or her field of expertise as well as by organizations that expect valuable contributions from them. The point of departure for resolving this ambiguity is that no orientation should be pathologized. Optimal processes require healthy professionals (in Maslow's sense). Unhealthy professionals are expected to have difficulties coping with this issue, especially if they still do not perceive their own "pathologies".

Although the actual relationship is much more complex, many factors can be said to have an inverted-U relationship with R&D performance. This is the case for autonomy, time pressure, group pressure, disciplinary division of activities vs. interdisciplinary barrier elimination, degree of task division and control (in the narrow sense). They are also interrelated. Other factors affecting the performance of the product development process that are known to show an inverted-U shaped relation are: commitment of the professional to and familiarity with the subject.

There is an apparent ambiguity in the roles of the quality function in the product development process that poses a special challenge to quality organizations. Harmonizing these roles and the confusion that may arise from having to switch modes provide essential keys to the strategy to diffuse quality activities in product development. Fear of low tolerance to this ambiguity by the participants is justified. However, as quality organizations became accustomed to coping with the ambiguity of the roles of QC and QA in production more than a decade ago, it is possible to overcome this issue. Nevertheless it should not be disregarded that this issue may be much more complex to cope with than the former was.

A second example of an application of this study is given below:

Box 73. The minority faces a potential gremlin.

In a recent space project, a precision scientific instrument with superb resolution, tight tolerances, milling difficulties, excellent design features, fine operating details, and redundancies was achieved by means of direct, open communication between designers and prime experimenters. During vibration tests of a flight model, the quality engineer noticed a slight slippage of one of the contact fingers. He showed it to the designer, who showed it to the others. The entire group became concerned about what would happen to the fingers during launch acceleration. Complete slippage would result in the failure of the entire instrument. After a while, this concern started to fade, but was re-stimulated by one or other member of the group. Asked whether there was a requirement about this in the specification, the quality engineer answered, "This is not a *specifications* problem, it's a *physics* problem". The experienced structure engineer, who had requested the spec replied, "Well, the (dynamic) vibration acceleration levels are greater than the launch static acceleration." The quality engineer answered, "Yes, *but* if you have sinusoidal acceleration, you integrate it over time twice and also obtain a sinusoidal displacement, which averages to zero. But with constant acceleration you have a non-zero average displacement." The contact finger was held in place by the cable bundle, so there was a compensation of forces that would keep everything in place, so his point was not necessarily applicable. The discussion continued and at some point, the quality engineer withdrew his misgivings to alleviate pressure and allow the designer to ponder the issue. The designer continued to entertain constructive questioning without becoming defensive. The project manager also participated in the discussion, and clearly did not force his own view. They did not need to arrive at a conclusion in this round. A couple of days later, the quality engineer approached the designer again. After all, it would be quite embarrassing to all involved if such a fine instrument were crippled by a flaw as basic as a slipping contact during launch. The designer reported he had *just* re-discussed his concerns with the project manager. In the end, even though the technical results of the vibration tests resulted in sufficient confidence that there was no significant risk, it was recommended nevertheless that the contact finger to the pin be affixed with a suitable adhesive.

In this case, the designer, the project manager, the structure engineer and the quality engineer all compensated for the tendency of the majority to overlook a possible design weakness. The improvement made was the result of allowing the doubt expressed by a minority (the designer and the QA alternated in the role of the minority voice) to be channeled and addressed. The quality of group interaction was the main reason that this simple point was discovered at the end of a development program, but prior to deployment. Many would have expected the quality engineer to assume a posture that said, "You are obliged to do such-and-such," or "it's in the requirements" (i.e. "it should have been" because, in fact, it was not), or to insist on having the final word on whether the group would accept the flaw, because after all, that is what product assurance officers "traditionally" do. This individual, however, refused to do so in the majority of situations.

Now the reader should try to imagine this kind of interaction occurring in the definition and redefinition process of the myriad of details that make the difference between a good idea and a well-functioning product.

Micro-interactions such as these occur throughout a project, whether in a dyadic group, in plenum, in formal events or at the coffee machine. Innumerable discussions take place concerning the possible forms of a given implementation. Every detail cannot be scheduled; every issue cannot be adequately foreseen and fixed, no matter how thorough the specification. Problems can appear at any time, at any level of detail, in any circumstance and—in keeping with Murphy's law—sometimes at the most inconvenient moment.

This case serves as an example of how to cope effectively with development issues using the concepts presented in this book. Such ways of dealing with quality issues can substantially increase the likelihood of correcting errors, such as those that crippled two recent Mars missions (cf. page 198 and page 218).

It is of utmost importance to consider the "big picture" when designing the job of the designer. This apparently trivial statement must be made, however, because development managers can be found time and again to insist on the need to apply pressure on developers without really justifying the criteria for doing so. Schedule compression is increasingly being indiscriminately used without regard to mid- and long-term consequences, many times with overly optimistic assumptions. The same trivial statement is applicable to the potentially "false" expectations of the key role played by the "bridger", "gatekeeper", or "product champion" to communicate relevant development information without considering, for example, the whole communication process or other factors affecting the diffusion of relevant information.

Analogously, although project management is a necessary function, it becomes ineffective if conducted to an excessive level of detail, i.e. micromanagement, or if conducted with unrealistic confidence in its capacity to reduce uncertainty. The complementary function of quality management can significantly contribute to overcoming these limitations. Likewise, design review is a very useful technique that has tremendous potential to improve the quality characteristics of a product *during* development. Yet it becomes ineffective if, depending on the type of review, any approach other than a problem-oriented one prevails, e.g. discipline orientation, favoring of political preferences, exploitation of personal differences, creation of defensiveness, or an emphasis on the "show". These approaches render the entire exercise merely a formal event. Quality systems make important contributions to product quality, but without considering their limitations or of the latent implementation errors they harbor, they may also become not only ineffective but even obstructive. It is therefore emphasized here that these techniques are still important and useful, but should not be expected to fulfill exaggerated expectations owing to the limiting or even destructive consequences to the entire process if not appropriately applied.

8.2 Work-oriented versus technology-oriented

Clearly, the utilization of simulation and visualization tools, CAD/CAM/CAE, schedule networking, etc. can facilitate the product development process. The issue to be examined here is the emphasis on buying (or advocating) such tools as *the* solution to a given problem. Tools should not be placed at the center of such issues as how to develop products or how to organize better products development. Instead, they should be used to address the characteristics of a given task in order to serve the persons using them. In other words, they are a means to an end, not an end in themselves. It is commonly acknowledged that *people* solve problems, not the quality system, computers, or other equipment. Especially in cases where this is not only lip service by management, it should be natural to put people at the center of an issue, and to put the technical equipment and tools around them to support their work. Yet this is unfortunately not always done. Often, a company acquires a technological system that determines the work form or the organizational structure, not the contrary. Insisting that "there is no system that offers what you want," or "everybody uses it, so it must be good," or "everybody does it like that" does not satisfy innovation-oriented companies interested in obtaining the best from their employees.

The potential advantages of quality systems are widely recognized and accepted. Burns & Stalker (1994) remind us, however, that there are basic differences between organic and mechanistic forms of organization (cf. page 41). If companies that implement a quality system also impose, say, a mechanistic view of product development activities, they should not be surprised to find that what they end up creating does not help personnel produce higher-quality products. Instead they may well discover that they have defined a structure that actually works against the personnel who are supposed to use it. At best, some people may merely tolerate such an imposed structure. These are the honest ones, because they do not buy into the illusion that they are achieving quality by these means. Similarly, if the good ideas behind quality systems are overshadowed by bureaucratic implementations, then the entire benefit may be lost, despite the certificates, manuals, and other evidence meant to demonstrate "effectiveness".

The situation with project management systems is no different. Thirty years ago, management systems proliferated in the American Department of Defense Programs. At some point, contractors earnestly began to question whether the added effort to provide all the documentation required by the management systems was consistent with one of the "most vaunted management tools of Defense Secretary Robert S. McNamara himself—cost effectiveness" (anon., 1967, 198). In 1966, the DoD-CODSIA Advisory Committee for Management Systems Control started to investigate this issue. By March of 1968, it had identified a staggering total of 934 management control systems impacting the defense industry (Rothstein, 1968, 10), many of them of "dubious value" (Sapolsky, 1972, 248). Indeed, many vendors of project management software boast that the original deployment of a product contributed to the success of the Polaris project and various NASA programs. Even users often subscribe to this belief.

Software suppliers today promote networked project management tools that supposedly solve the (technical) problems of existing programs. The organizations that acquire such systems must take special care in analyzing the work situations they have or wish to have. Tying an engineer with a fear of monolithic systems to systems that were conceived only around technical considerations may result in a repetition of such errors as the Swiss CIM-Ruins (Strohm, Kuark & Schilling, 1993), even if management attempts to stem fear by means of a drag-and-drop-operated Gantt chart.

In the CIM-Ruins[253] case, following an excessively technical-oriented implementation of Computer Integrated Manufacturing led to catastrophic performance. The investigated companies from the capital goods industry had set as goals:[254] (i) increase in delivery punctuality (68%), (ii) reduction of processing time (65%), (iii) increase in market flexibility (51%), (iv) reduction in inventory (48%), (v) quality improvement (42%), among other goals. However, among the companies that declared these goals as very important, only the following percentage recognized the goals as achieved: (i) increase in delivery punctuality (29%), (ii) reduction of processing time (27%), (iii) increase in market flexibility (27%), (iv) reduction in inventory (14%), (v) quality improvement (36%) (op. cit., p. 138). Reasons for failures are associated with the technology itself, but are mainly caused by the work structures that are imposed or created by the technology-oriented approach. Only 4% of the companies followed a work-oriented approach.

Similarly, this technology orientation can cause the errors committed by companies that discovered that the production planning and control software to which they had bound themselves forced them to "please the software, not the actual need" and sometimes had to generate a set of documents known to be flawed or ineffective. Or they found themselves entirely dependent on software that precluded them from introducing the form of organization they desired, e.g. partially autonomous working groups. There was a case of a company in the space industry whose purchasing software forced them to

---

[253] The term was coined by the Upper House of the Swiss parliament to describe the results of excessively technical implementations, where work structures are left out or are treated as a problem derived from technology issues. (Schweizerischer Bundesrat, 1989, 125, cited in Ulich, 2001, 345, 618).

[254] Numbers refer to the percentage of companies that declared the goals as very important. N= 434-450.

declare certain procedures as "released", even though they did not need those procedures at that moment. Otherwise the software would not issue the purchase orders that were actually needed urgently in order to adhere to the tight schedule established (case #55).

Another issue to consider is that the opportunities to reduce errors during product development are created by the opportunities to *make errors* by experimenting, learning, and correcting problems as they are encountered. Simulations, for instance, can mitigate this situation by allowing various different possibilities to be examined before implementing a physical prototype. If, however, developers are constrained to feeding data into a simulation engine and simply forwarding the results of the simulation, even when they are aware of the device's limitations in predicting reality, then one can obviously expect unsatisfactory results.

Schmelzer states that companies often try to improve their efficiency with regard to the time required to develop products by using technological tools such as CAD and CASE.[255] He argues that in many industries, the use of these tools is standard and that a competitive advantage can only be achieved with these tools if they are used efficiently. This is less a technical question than one that is much more dependent on the organizational context and the motivation and qualification of the development personnel (Schmelzer 1990, 35).

In view of all the advances made in technical methods for product development, especially with regard to computational power, we should keep in mind that overemphasizing the technological aspects of engineering led to numerous incidents in the past, some of which had dire consequences for those involved. In still more cases, the exact extent of the damage could not even be assessed at all. In 1912, the sinking of the Titanic sent a shock through a society that had placed unrealistic faith in technology's victory over the elements. One of the many documentaries released after the wreck was located in 1986 pointed out that the company operating the ship did not coin the attribute "unsinkable". Instead, it was the press that had interpreted the redundant features as such. Society demanded it. As Davie (1986, 215) stated, "Edward Kamuda of the Titanic Historical Society had been searching for evidence of this myth [that the Titanic had been hailed by its builders and owners as unsinkable] for thirty years, but failed to find it." The journal *Engineering* used the words "*practically* unsinkable" (op. cit., p. 215, italics added), and it is easy to imagine how an overeager society would be quick to overlook the word "practically". A comparable case occurred when

[255] Computer-Aided Software Engineering.

a manager at NASA estimated that the confidence in the Space Shuttle main engines was "100 percent" (Feynman, 1988, 181-183).[256]

Since then, lessons have fortunately been learned. Today, for instance, instrument flight rules for training aircraft pilots explicitly warn students that, in spite of the navigation technology, "if they think that instrument flight rating allows them to fly *independently* of the weather, they are making a *deadly* mistake." In another case examined in this book (page 83), the *concern* raised by the failure of simulation analyses to predict the performance of modifications to the F/A-18 E/F aircraft suggest the belief in the victory of technology over uncertainty. The problems faced by the development teams of this aircraft are *to be expected* in any development. Technology *can* help developers achieve better technical and commercial results by providing a means for gaining knowledge about a given product configuration or its implementation alternatives. Thus product qualities can be improved such that, for example, customers can use the product to better cope with their pertinent needs. At the same time, technology *exposes* developers to new problems and new interactions between the problems and their associated uncertainty. Uncertainty is also introduced by technology itself due to its imperfections. It is not possible for technology alone to achieve the ultimate victory over uncertainty.

Turning back to the beginning of the previous century, statistics show that the first-class passengers of the Titanic had a much higher survival rate than the third-class passengers. The American Commission led by Senator William Alden Smith[257] investigating the accident made intensive efforts to determine whether lower-class passengers had been barred by force from entering the few available lifeboats, but found no *conclusive* evidence of this (from a judicial point of view). It has been speculated that an "invisible" class barrier existed in the prevailing mentality (Büscher 1984, 1:45').[258] More applicable to the scope of this book is that, today, breaking down the barriers between departments as proposed long ago by Deming (page 221) reveals the existence of similar invisible barriers. This book has attempted to

[256] In the aftermath of the Challenger incident, Feynman asked three engineers and their boss, Mr. Lovingood (who also had been trained as an engineer), to write down on a piece of paper their estimates of the probability that a "flight would be uncompleted due to a failure in this engine". The engineers' estimates ranged from 1 in 200 to 1 in 300. After trying to "weasel out" of the situation, the manager was asked again what his confidence on the engines was. Feynman reports: "He says, '100 percent'—the engineers' jaws drops, my jaw drops; I look at him, everybody looks at him—'uh, uh, minus epsilon!' " (op. cit., p. 183).

[257] Senator Smith was also a lawyer. He used the pertinent skills to conduct investigations.

[258] If they were systematically barred by physical means to remain to be verified in the wreckage (cf. Davie, 1986, 63). Its historical verification had a low priority due to its low relevance to the context of this work.

provide evidence of this phenomenon. Technology can *help* overcome these barriers, but only within the context of the social aspects *within* and *amongst* its users, and provided that the organizational context supports them. A work-oriented approach can ensure the essential difference.

8.3 Evaluation systems

A willingness to admit errors advocated by the concept of *critical testing* mentioned above can be jeopardized if the mutual respect based on individuals' competence and the cooperation amongst them is threatened by individuals oriented strongly toward ambition and money. As many personal career systems still value "aggressiveness" as a trait necessary for assuming a managerial function, those interested in exploring the flaws of a product in order to deal with them might be implicitly threatened by an organizational structure of this kind. Nevertheless, most personnel evaluation processes are built around successes of the past, not on failures nor on what was learned from them, nor on how failures were handled in order to foster future success. It is assumed that successes of the past are repeatable, even if a success was a matter of sheer luck. Thus, success is not a determinant of future success, whereas failures are frequently detrimental to careers, even when much was learned and could be successfully applied to future programs.

A system that tolerates errors made while trying to achieve innovative solutions is necessary. A good evaluation system should therefore avoid promoting achievements made by omitting failures. Evaluation of past performance as a criterion for professional progress should emphasize a person's potential for solving *future* problems in the company. Not only the present knowledge of an employee should be evaluated, but also the individual's potential to learn from mistakes and acquire knowledge.

Superior evaluation and reward systems should place a premium on cooperation or cooperative behavior and on participation or participative behavior. Many current systems, however, are based solely on an individual's isolated performance. This can prove in turn to impede cooperative and participative behavior.

Professionals with multiple biographies have a greater potential to understand the diverging perspectives of different professionals. This is beneficial to successful product development. Professionals with dual backgrounds in such fields as production and development, marketing and development, hardware and software design, quality and development, production and quality can make extremely valuable contributions. Their dual insight enables them to better communicate with both sides while simultaneously providing complementary experience and knowledge. It is the belief of the author that the concepts of *retrievable qualifications*[lii] and *polyvalence* deserve serious consideration in the product development environment (cf. Discussion of *qualification in partially autonomous manufacturing groups* by Ulich, 2001, 227-229; discussion of *salary concepts effecting qualification either negatively or positively*, op. cit. pp. 515-521; and description of *experiences with learning-oriented salary concepts*, op. cit. pp. 521-525).

The capacity to propagate useful information to others should also be rewarded in an evaluation system.

8.4 Two different time perspectives

Market dynamics requires that companies shorten the so-called time-to-market in order to realize a competitive advantage. The extent to which this time-to-market orientation really applies is beyond the scope of this work. Nevertheless it should be explicitly acknowledged that lost time—unlike lost money—is not a recoverable quantity. It is important to solve the immediate issues faced by a company, perhaps even associated with its survival. However, care should be taken not to restrict one's perspective to extremely short time horizons. The nature of efforts to increase one's knowledge of an issue or set of problems is long term. This contradicts the short-term loop required by modern product development. Both are important to a company's long-term survival.

In many circumstances, this long-term time horizon also has a slow—sometimes very slow—response character. In order to be certain about the consequences of a new technology or procedure, it is necessary to invest in this gradual process.

Reliability studies, for instance, are implicitly slow because of the order of magnitude of failure rates and because of the difficulty to accelerate the testing of hypotheses.

Box 74. Cases illustrating the short-term and long-term time perspectives.

In SMT (surface mount technology)-printed circuit boards, the components leads are soldered to small openings in a protective mask called pads. To protect the exposed copper layer from oxidation, which would render the metallurgical bond impossible, a passivation process called hot-air leveling was used in the industry to manufacture the bare boards. This process entails applying a layer of solder to the board and removing the excess by blowing hot air over the board, thus producing a level surface. This process worked very well, but with decreasing lead pitches down to 0.3

mm, the dimensions of the pads and the spaces between them caused problems to appear. Owing to surface tension effects, the surface of the pads is not perfectly flat, thus creating production problems when soldering the components. There are other solutions than the pad passivation method, but the industry tended to favor this conventional process.

The issue was perceived as a potential problem as early as 1991, but it was not sufficiently well understood to declare it a problem until 1994 when its characterization had become clear (cf. Grossmann, 1994 and Grossmann, 1995).

In 1997, industry was beginning to have significant quality problems with this effect and it became clear that the existing understanding of the problem was insufficient to cope with it. If investments had been made in determining the effects, it would have taken two years to completely characterize the reliability problem because of the long and slow nature of the tests required. It was not possible to exploit the lead of six years of having perceived the problem to analyze it thoroughly because industry could not quantify it. Thus it could not justify the necessary investments. The reason for not investing in this endeavor was caused to a considerable extent by the conventional practice of not taking action until a problem is measurable. It was not possible to predict the cause of the problem in quantitative terms, although they perceived the problem itself. The main difficulty faced by the technical personnel in proceeding with their research of the problem was that it was difficult to justify the necessary investment when no concrete, hard data was available predicting what would occur in the future (case #56).

The case of potentially problematic transfer described on page 127 complements these two time perspectives with a consumer industry that was already using the components in question while the long-term characterization took place.

In one company in the electronics industry, there was a general perception by the technical management that a need for product standards existed. They intended to foster the achievement of homogeneity in the type of solutions that the various development teams were adopting for a complex product. After some evaluation, they arrived at four main subjects to be focused upon by the standardization teams. These teams were multifunctional and included at least one member each of the development, production and quality engineering teams. They were given a couple of weeks to elaborate the required standards. The best work was performed by the group working out the guidelines for printed circuit board design. The least satisfactory work was performed by the group working on surface treatment for materials. High-level management, which evaluated only the final product, expressed particular satisfaction with the best work and praised the group that had produced it. What they did not recognize was that that group had actually started working on related issues many months in advance. Starting with the perception of the issue by a quality engineer who enjoyed an excellent rapport with highly competent and experienced development engineers, this group conducted numerous discussions and succeeded in listing many points of interest prior to the management decision to implement the results. There had also been some past interaction between the quality engineer and production engineers on this same subject. This was possible in part because the quality engineer's supervisor granted him this freedom. The engineer had also discussed his perception of the problem briefly and informally with his supervisor, who supported him in this endeavor.

The other group working on the surface treatment standards faced many problems, starting with the perception of the problem itself. The mechanical engineer representing the development group agreed that the problem was relevant but declared that he was participating only because his boss had assigned it to him. The discussions conducted by this group were not well focused, and the standard they arrived at contained only a minimum of items. It was a starting point, but it was not a high-quality product. If they had been given more time to interact and search for information, based on their competence and experience (especially that of the development and production engineers), they could have achieved a much better result. Moreover, group members who contribute readily and enthusiastically are more likely to achieve better results than someone "forced" to participate, like the mechanical engineer. Additional interaction time would also be necessary to win this person over as a valuable group member. In these two contrasting situations, which occurred within the same company, the importance of the amount of previous interaction and work that directly and indirectly contributed to success was overlooked by top management (case #57).

These examples illustrate the importance of previous work in terms of the speed with which one can react to a given issue, which is particularly crucial within the short-term time horizon. When starting from scratch, a considerable amount of luck is necessary to achieve success quickly, despite the fact that no professional with a solid background truly starts at zero, of course. The point is that if professionals are continuously overloaded with short-term projects, it becomes virtually impossible for them to renew and cultivate the internal resources that allow them to react promptly to urgent problems.

Again, this issue is not new. Goldstein[259] (1976, 22) referred to so-called "tiger teams... with little or no planning other than dictated by the immediate emergency." It must be emphasized that whereas speed is indeed essential, this should not become an excuse for not investing in the long-term development of personnel. Crucial expertise will be lacking even in the mid-term if a company or society neglects the importance of these two time perspectives.

When new technologies are applied, there is rarely sufficient hard data—if any at all—concerning suspected or anticipated problems, meaning that problem characterization proceeds slowly. When a problem characterization is clear from the use of technology, the speed of response depends on the amount of research work done previously.

Moreover, to ensure the reliability and quality of a product in development, it is necessary to enhance one's knowledge of the actual performance of the equipment. Releasing a product with unknown characteristics due to the insufficient time available to test it and modify it can have highly unpleasant consequences.

A related problem of interest was reported by Allen (1986). To improve the productivity and performance of R&D, it is necessary to maintain two levels of information. One is required to "coordinate the many tasks and subsystem interrelations that exist". The other is required to ensure "that the technical staff of the project remain in close contact with the developments in their technical activities." Allen found that different types of organizational structures support these layers: Project organization supports the former, whereas functional organization the latter. When Allen mentioned this issue at the 1985 European Industrial Research Management Association Conference, an interesting comment was made by Mr. U. Creutzburg of Dornier Systems during the panel discussion. Creutzburg said, "We frequently have the situation where the project manager is calling for additional people to be put on the project, whereas the functional departments claim that they will not be able to operate if the people are moved. I am under the impression that the decision is made based on who screams the loudest." (paraphrased from Ruiter et al. 1986, 78).

Different organizational forms support different time horizons, and a means has to be devised to support them all. Considering only the short term and the continuous reduction of the time-to-market has certain shortcomings. This was analyzed by Backhaus & Gruner (1994) and supported by simulation data from C.–F. von Braun (1994). As Ulich (1996, 195) notes, C.–F. von Braun's simulation shows that "many of the so-called growth markets are in reality acceleration markets,[liii] and they are about to decay".

---

[259] R. Goldstein was with Litton Data Systems.

The scenarios simulated by C.–F. von Braun for turnover in the post-acceleration phase have alarming consequences for enterprises as described in the cited publications. Exclusive consideration of the short term leads, for example, to a planning blind spot in that the decay in turnover is not anticipated by the planners. Backhaus & Gruner mention that the corresponding shorter product lifecycle is also associated with a narrower time window to amortize the rising costs of R&D. This in turn increases the entrepreneurial risks involved (op. cit. p. 29).[260] The entire situation of concentrating only on the short term does not appear to be sustainable, in response to which the expression *acceleration trap* was coined. Ulich quotes the warning by Tom Sommerlatte, senior vice president of A. D. Little, who said that, he who uses "the shortening of the product lifecycle as a strategic instrument is placing himself in grave danger" (paraphrased from op. cit. p. 195).[261]

Similarly, the study of downsizing by Cascio & Morris (Cascio 1995, 3-6) confirms this transitory effect. These authors investigated the financial performance of 25 large firms (target companies) over a seven-year period. These companies, which cut their work force by an average of 28%, were analyzed three years before and three years after the downsizing took place. The results were compared with those for a corresponding group of 91 firms that had cut less than 15% of their work force over the same period. The data indicate that, although the target companies had succeeded in limiting the decline in earnings in the short term, after the three-year period, their earnings had recovered by 183.4%, compared to a recovery of 422.5% by the comparison group. The authors conclude that "there was virtually no link between the magnitude of the employment cuts by target companies and their subsequent earnings". The authors also indicate that, in the target companies, an investment made at the beginning of the downsizing year produced a cumulative return on investment of 4.7% after the third year following downsizing, whereas the comparison group reached a level of 34.3%.

Concentrating solely on the short term leads to *Brownian management.* This term stems from the so-called Brownian motion first observed by the botanist Robert Brown, describing the phenomenon that particles of pollen immersed in a solution exhibit erratic movements. One explanation for Brownian motion is that high-energy electrons randomly collide with the lightweight pollen particles, inducing their chaotic movement. The term Brownian manager was coined to describe an actual manager who reacted only to the immediate problems of an emergency nature or that were

---

260 Reference to Bullinger (1990).

261 Ulich quotes here from Deutsch (1995, 83).

requested by other managers above him or at the same level, i.e. with "high-energy" content. Lower-energy requests (his or others' subordinates) could not budge him, regardless of whether they concerned significant issues with consequences for the entire factory. As a consequence, his group appeared to jump from one direction to the other in a seemingly erratic manner, sometimes even without have properly completing previous tasks. Their activities were dictated solely by these high-energy impulses. They had no overall concept of their activities. Like the pollen particles, their actions were focused solely on the short term. Recalling a basic premise of stochastic phenomena, however, we know that *Brownian motion has an average displacement of zero*. In other words, people under this style of management may travel very, very great distances, but the entire system goes nowhere.

What is necessary, then, to achieve a "sustainable" development? Let us look for analogies in agriculture, which is also concerned with growth and cultivation. If soil is overused, it becomes exhausted and loses its richness. Or let us imagine a field invaded by locusts: stripped bare of every last stalk of growth. In agriculture, several approaches to this situation have been taken. Nomadic agricultural cultures deal with this issue simply by moving to new land. Other cultures use massive amounts of chemicals to extract at least a minimum of yield from what once was fertile land (with the implicit market limitations for the product). An alternative is to allow the land to rest after a period of intensive use. Yet another is to switch to a different product, called crop rotation, in which essential elements in the soil are biologically restored by complementary crops.

This analogy has obvious counterparts in industry, but with one crucial difference. The role played by soil in agriculture is that played in industry by human beings, which are not an "expendable" resource.

It is possible to overcome these discrepancies between the two time horizons, as Wernher von Braun did (indirectly) with his "dirty-hands" philosophy. Other approaches foresee continued education or cyclical education with alternative periods of work and study, technical sabbaticals, exchange of professionals, involvement with issues between functional and project organizations, to name a few.

8.5 No one best way

It is not the purpose of this book to establish general rules. This research orientation was necessary because of the complexity of the relationships that lead to success or failure. For example a strong leader with above-average knowledge, a stimulating, participative leader can succeed or fail. However, there are basic differences associated with certain styles described in this work. Neither the controlling nor the participative quality leadership is the *sole* correct approach to pursue. There are situations in which a cooperative mode would be advisable as well as situations in which a control form of action would lead to better results.

The purpose of developing and buttressing this framework with quantitative and qualitative data was to provide insight into their combined effects on the product development process. Practitioners should find here a basis for orientation in dealing with the factors that influence the processes they conduct. Finding a balanced combination of characteristics appropriate to the given situation is perhaps the greatest challenge in improving development results.

One recommendation cannot be omitted, however, and that is to regard the term "control" in the appropriate context. In this work, the term was necessary to emphasize essential differences between behaviors and methods. In practice, it is important not to merely replace this term with a trendy euphemism but to actually exhibit appropriate, coherent behavior. Success is not achieved by putting fancy labels on impediments.

8.6 Principles to guide product development activities

As described in the Chapter 2, significant progress has been made since Taylor's *scientific management*, passing through Max Weber's *bureaucracy*, Follett's *law of the situation* and her concept of *power-with*, as well as the advent of *human relations*, to today's *socio-technical systems* and the *organic* vs. *mechanistic* view of enterprises. The results of applying, not labeling, the concepts defended by these schools of thought regarding product development processes were covered in this work.

Very basic scientific principles expounded by Descartes, Popper, Kuhn, Maslow, and Churchman, as well as the concept of serendipity were discussed as different forms of perceiving science. Product development rightly claims to use scientific methods. It uses the Cartesian method, but includes Popperian concepts of *falsification*, not of demonstration. It does not necessarily evolve in a stepwise progression. It has a strong intrinsic-motivational component and a relation to the object of study, which affects the way of perceiving it. It has a formal, deductive component and an experiential, inductive, consensual component. It comprises a crucial integrative aspect to "reconcile its parts". It has a conflictual, antagonistic component and sometimes a synthetic, interdisciplinary and holistic component.

The challenge of product development lies in conceiving the development activities such that they accommodate these general science concepts appropriately. There are numerous lessons to be learned from science in order to (i) grow, (ii) achieve better products, and (iii) satisfy its participants. The latter three objectives are strongly interrelated.

Box 75. Engineers coping with the social world

The human processes behind product development are highly nonlinear and strongly dependent on interactions of work content as well as interactions among participants. They are also dependent on previous "inputs to the system", must be matched to the situation and considered in context. Effectively coping with its challenges goes far beyond the limits of technocratic thinking.

I hope that my engineering colleagues have thus become acquainted with some of the limitations of our art or, in other words: why technology alone is not enough. Moreover, the need is emphasized here to incorporate humanities *effectively* into engineering training. This is essential for the field of engineering because the technical and social subsystems exert reciprocal influence. Certain engineering schools, such as M.I.T., ETHZ, and ITA, do so as a didactical concept. Far from applauding their approach to the concept, however, my intention is actually to call the attention—both that of students and of lecturers—to the far-reaching implications of *integrating* social science fields into the engineering sciences. To these and other institutions I emphasize that this is intended as constructive criticism.

Joseph Campbell's work *The inner reaches of outer space* approached the effect of the early space explorations on humankind and its philosophy. To paraphrase Campbell, I close this cycle in the following sense: to reach outer space (as well as many things on earth), we have to care for our inner space—in all its splendor.

Appendix

SCIENTIFIC APPROACH

used to build up this work

A.1 Need for a scientific approach

This work emerged from the practice of product development and associated quality activities by this author. In the practical world, the influence of non-technical disciplines on the life and work of engineering professionals has become evident. Thus, based on engineering experience, the challenge was to dive into the academic world to integrate the three areas of work psychology, organizational theories and (technical) engineering to describe the manifold interactions that occur among the technical, social and organizational aspects of product developments. These factors cannot be examined independently and simply summed up.

A scientific approach to the issues of interest was needed in order to achieve a generalization. This author's opinion alone is clearly insufficient to propose adequate solution approaches to engineering professionals, at least in terms of a generalization of observed cases. When is an example mere anecdotal evidence, and when is it a representative case?

To support the generalization of the issues explored here and to filter out geographic-cultural factors, a situation assessment was conducted in 1993 within the framework of the research underlying this book. The assessment included informal contacts with local Swiss and international experts, and attempts to appreciate issues examined by other authors in the specialized literature. This evolved in turn into an extension of the data collection process.

A.2 Establishing the research method

A.2.1 False expectations and assumptions

One of the difficulties encountered at the beginning of this work was to fulfill the expectations placed on the method used to demonstrate or contradict the concepts proposed. An example of false expectations was encountered during the first contacts, in which an interested practitioner willing to contribute to this work said: "We have the cases and you have the statistics." Up to that point, nothing had been said regarding the use of statistics. Yet somehow an implicit expectation arose that statistics would or should be used. In another discussion with social scientists, similar false expectations resulted from the desire to be "scientific" and therefore to be *objective* instead of *subjective*. An experienced psychologist mentioned the need to quantify findings about the real world. Although the conversation was very productive otherwise, the suggestion prompted a misguided search for the appropriate "scientific methods", which would have given the research an inclination toward being a quantitative study.

The issue of subjectivity versus objectivity is very broad in social research. As Muckler & Seven, (1992, 444) state, "The degree to which data that are considered objective contain subjective aspects varies from one research area to another. ...In some areas of research, objectivity seems to mean consensus of subjective opinion."

Generally speaking, there is no absolute requirement in science that quantitative measures scientifically "prove" a proposition or theory. Sometimes one can find "objectively assessed" but nevertheless misleading data. Moreover, objective study artifacts can block entire areas of knowledge. Of course, the demonstration of concepts and conclusions must be scientific, but this does not necessarily imply hypothesis testing by performing quantitative measurements of a population or by means of experiments. There are indeed other ways to scientifically demonstrate concepts or theories.

A.2.2 Data collection

Most of the start-up data was collected on-the-job as a participant in and observer of processes that occur during product development. The author participated in some of these processes, sometimes in an active and sometimes in a passive capacity. In some cases he was directly involved in the development; in other cases he was engaged in other departments working on related issues.

As the research evolved from this start-up data into a search for confirmation data, a wealth of complementary data, contradicting cases, and existing approaches to these issues was found. In this search, the author's "insider" perspective proved very useful in finding and selecting cases, and filtering the abundance of situation reports described in publications. This facilitated the understanding of the development situation, and also provided an integration basis owing to actual experience with the object under study.

There is, however, a potential bias of an inside observer that is not easily compensated. This is a known source of potential "deficiencies", or is at best a weakness in a social sciences research project. An insider tends to impose his or her own perception on the issues under discussion. It takes additional effort on the part of the researcher to question one's bias repeatedly—a difficult task to accomplish alone. Replacing an insider by an outsider to the field, however, is clearly not the way to achieve freedom from bias. Any researcher, whether an insider or an outsider to the fields under analysis, has to state and verify the many explicit and implicit assumptions.[262] Furthermore, the researcher also affects the measurements. Consider the classic case reported in the Hawthorne studies that the manifold informal interactions among workers, and among workers, superiors and researchers necessary for the study actually had the effect of improving productivity (Ulich, 2001, 40), more or less independently of whatever approach was being tried.

In addition, specifically to R&D, there is evidently a certain degree of discomfort perceived by development personnel in the presence of external observers. This affects both the insiders and external personnel, and constitutes an obstacle that is difficult to circumvent. To expand the data set during the course of this research, several companies were contacted. One of the interested companies tried to involve this researcher in their activities, but found they could not "sell" the idea to their development leaders, who rejected the idea of an external observer, even one from the same technical discipline. It was clear that this had nothing to do with the company's information security.

[262] The diversity of these assumptions is perhaps one of the keys to interdisciplinary R&D. It is not impossible to accommodate it, but it takes time and intensive interactions between different researchers, if a multiple-researcher approach is followed.

A.2.3 Empiricism, plausibility and theories "in theory"

The origins of this work go back to the early 1980s. The basic approach was to examine product development issues in practice in order to appreciate them empirically. Then theoretical considerations were applied to summarize and integrate the issues experienced. Hence, a pragmatic approach was taken, starting with practice, then a theory was formulated, the results of which were reapplied to practice. This approach imposed stringent requirements regarding the scientific methodology to support this endeavor.

It is not sufficient that explanations be merely plausible, because this would introduce the danger of making invalid generalizations. In many cases of success or failure, the plausible explanations presented by the media, sometimes even by the specialized press, often omit the factors that in fact were the most crucial. Instead, frequently those factors are listed that reflect the momentary context in which the event took place. An example of this is the case of project management techniques. The success of selected development cases was attributed primarily to the use of PERT/CPM networks, and the reasons cited to confirm their influence on project performance were entirely plausible. Yet the reader will see that, in many cases of the successful application of that technique, there were other hidden success factors that played a more significant role in the project's success. The context in which it had originated in the late 1950s called for a way to get matters "under control", "managed", "organized", "disciplined". None of these notions are "bad", but the approaches based on these objectives are dependent on the interpretation valid at that time. As a result of the emphasis placed on the effectiveness of this method, there was a tendency to exaggerate the results obtained. People simply tended to believe it was *the* way. Not surprisingly, many managers today still exhibit this tendency, and again, company context has to be considered to filter this effect out. Similarly, the defenders of bureaucracy in the 1930s and 1940s had no lack of context-dependent reasoning, and their "success" stories were analogous. Further discussion of these aspects can be found in Chapter 2. Modern contextual factors that affect the attainment of results can be hypothesized to exist as a product of exaggerating the role of information technologies (computer hardware and software) for achieving success in product development.

Returning to project scheduling, there is a considerable difference between theory (e.g. of PERT/CPM) and what is actually revealed when these techniques are applied in an engineering environment. Significant issues arise that were not foreseen by the theoretical model. Objective measurements of these observations, which could be used in a formal (statistical) demonstration, are unfortunately not available, perhaps because practitioners were not interested in appreciating the use of a model in a systematic and unbiased manner, but only in achieving practical results with them. Laboratory measurements of the effects of the use of PERT/CPM (here used as an example) are not known and are thought to be inconclusive because of the complexity of the environment necessary to simulate it. However, experiential data about its use do exist, the analysis of which reinforced the empiricism of this research.

A.1.1.1 Measurement variables and surveys

Measurement variables are crucial in the analysis process and can provide an indication of the influence factors. Nevertheless, their relative importance in a complex, real environment is not easy to assess.

Project success can be measured in diverse forms. Depending on how "success" is measured, various results can be obtained. Some of the measurement variables in product development are of a conflicting nature, as illustrated by the following examples: execution time vs. quality of work as perceived by the customer; time-to-market vs. product competitive advantage; development cost vs. development time, etc.

A major difficulty associated with measuring the success of product development is that most of the variables are only partial factors of success. Unfortunately, achieving success in terms of one parameter does not necessarily produce overall success, and vice versa.

Taking again project scheduling methods as an example, it is measurable that a tool to reduce uncertainty may facilitate the achievement of success, but its application can exhibit limiting factors that had not been known previously or that are masked by the measurements. Nevertheless, even though project scheduling has already been used for nearly forty years, there is still an expectation that these techniques will reduce uncertainty in project implementation. Experience shows that its use is not as straightforward as it might seem. Many users attach an exaggerated importance to it, perhaps wishfully expecting results that might never appear or because they have to "sell" it to others inside or outside the company. This was not found in many of the quantitative studies. For example, only an attribution on a Likert scale of the perceived reason for success would not compensate for it. Special care was taken in attempting to filter out this effect.

In addition, when assessing the development process, instead of examining only the development *results*, a certain resistance to and skepticism toward quantitative surveys have been reported. One example is cited by Tompkins (1977) regarding an evaluation by the Marshall Space Flight Center (MSFC) of the National Aeronautics and Space Administration (NASA) in Huntsville, AL. The researcher had to rely on qualitative methods because MSFC personnel demonstrated such an aversion to quantitative research. It is not known whether this aversion still exists among development groups.

Other associated issues are due foremost to the fact that the formal personnel evaluation and progression systems are still based primarily on past performance, not on the potential of the individual or group. Hence, openly admitting errors from the past is tied to a fear of negative consequences. Even if much was learned from a "failure" and the possibility of improvements in the future is considerable, there is an understandable reluctance to disclose information concerning the process of developing products. Second, in an environment like product development where purely rational arguments are predominantly used, there is a tendency to avoid displaying emotion or other supposedly "irrational" behavior. A former development director described a case where his personnel had acted contrary to "reasonable" and "rational" arguments. He then, however, asked this author not to cite the case, despite the assurance that the persons involved would not be named. He feared that the case could cast a negative light on their professionalism. This situation illustrates that although it is not entirely impossible to cross the barrier between rational and irrational actions, doing so either quantitatively or qualitatively poses a challenge to the researcher.

One of the possible artifacts that distort a given assessment is that the research population might be inclined to believe one or the other technique or method, depending on the trends of the moment, thus potentially obscuring the results of quantitative research. On the other hand, contextual situations of specific cases analyzed by qualitative methods should be considered in order to determine the possible effect of such methods on those situations. As these aspects are believed to play a major role in indicating the influential factors, special care was been taken to critically appraise considerations of an exclusively numerical nature as well as qualitative descriptions.

A.1.1.2Personal opinions of the author come into play

Issues that affect the scientific validity of the researcher's statements or personal opinions are evident in qualitative studies. Hence this author was initially reluctant to use qualitative results. Some were nevertheless included if they constituted a confirmation of results obtained by other means, or if a certain pattern was experienced. An author's opinion can be valid if based on a class of or a group of behaviors.

In the social sciences, the issue of introspection is very often considered biased and is hence eschewed. As an engineer focusing on engineering activities, this author is not free of bias. Consequently, experimental data were used in order to identify which issues could be confirmed by other sources. Narratives of true cases were analyzed to confirm or reject these issues. Cases described in the open literature on technology, including biographies or memoirs, had to be used selectively, not only in terms of their technical suitability but also their applicability within the context in which they occurred. Special care was taken not to project a certain expectation into a case by first filtering a given aspect out the data and then trying to prove the contrary. The validity of a description was questioned rigorously, causing cases that did not stand up to scrutiny to be eliminated. In some cases, the bias of the person describing the situation was obvious, which led this author to use such examples with extreme care if for some reason they could not be eliminated entirely. In this way, data of "dubious" applicability was removed from the main concept and, if used, is accompanied by an explicit mention of its potentially distorted nature.

A.2.4 History of technology

The difficulty of including historical cases is associated with the issue of unbiasing the sources. Some historians say that history is implicitly biased by the persons who write it and that it would be impossible to write a completely impartial account, even when only mentioning concrete facts. This is no easy task for historians to solve and involves detailed knowledge of the context in which the described situation occurred and, of course, depends on the quality and reliability of the sources chosen.

Unfortunately, some good work on technical systems is based on unreliable sources for citing facts or behavior exhibited during actual events, thus potentially casting doubt on the validity of such studies. In this work, an attempt was made to compensate for this effect by comparing different texts, especially those that had originally served different purposes or reflect different perspectives. Even sources originating from program or political advertising were analyzed, albeit very carefully, if material of sufficient quality was available. Moreover, historical texts were scrutinized for inconsistencies and compared with other cases. Once again, the mere plausibility of a description was not considered sufficient for the purposes of this work, and an attempt was made to filter out occurrences of (supposedly) distorted information, even when this meant omitting seemingly reasonable descriptions. A few exceptions were made in cases of particularly compelling anecdotes, but these are described explicitly as potentially fictive. Examples of such cases are:

- The rejection by the U.S. Navy of the "Sea Shadow" Stealth ship concept proposed by Lockheed, on the grounds that the small number of ships required to fulfill their function would not further an officer's career, or because it was unattractive to officers to command a ship with a crew of only four (including the commander), compared with commanding a frigate with a crew of more than 300 (Rich & Janos, 1994, 277-278).

- The Stealth submarine concept was rejected by the Navy because they would "never [have] built a modern submarine that looked like that" (op. cit., p. 273). Although there are many repetitions of such cases of rejecting promising innovations, these two cases were omitted due to insufficient records available..
- The description by Knopp (1997) of the designers of the Hindenburg airship as "negligent"[liv] for not having considered galvanic continuity in the airship structure. This would have caused sparks to appear in adverse weather conditions, eventually igniting either the aluminum-based reflective paint and its fabric support or the hydrogen that escaped from a cell. Although the oversight did occur, there is no conclusive evidence of negligence. Therefore, Knopp's article was not incorporated into this work.
- The case of the Grumman manager who, after the service module of the Apollo 13 mission was disabled, was asked whether the lunar module were able to push the entire composition (command and service modules plus lunar module). He refused to answer, insisting only that the lunar module had not been constructed for this purpose. The situation was depicted in the motion picture *Apollo 13* directed by Ron Howard, but no description of this event was found in the book *Lost Moon* by the Apollo 13 commander (Lovell & Kluger, 1994) on which the screenplay is based. This discrepancy was also mentioned by Chaikin (1995). The technical possibility of using the lunar module as a *lifeboat* had in reality already been verified by the Apollo 9 mission during the earth orbit trials, which however were not as complex as the situation encountered by Apollo 13. Thus, the incident involving the manager may never have happened. Hence, although there are scores of real cases in which managers of contractors or suppliers have behaved *exactly the same way*, this case was rejected as a possible dramaturgical artifact of the motion picture.
- The SR-71 spy plane built by Lockheed featured many major innovations. Lockheed claims to have introduced the fabrication in halves to reduce production time (Rich & Janos, 1994, p. 216). This was also claimed by North American more than two decades earlier regarding the P-51 Mustang fighter.
- One author speculated that the P-51 Mustang must have been conceived using simultaneous engineering techniques merely because it was flown within an amazingly short time. Yet there is insufficient evidence to support this assumption. The main reason for its short development time is more likely due to the fact that this aircraft was based on a well-proved

layout configuration of an earlier fighter featuring the innovation of the laminar flow wing profile and the position of the radiator scoop.

It must be said that most of the doubtful cases could indeed be proved or disproved historically, but do so in detail would have constituted a major departure from the objective of the present work. If the supporting documentation was considered insufficient, the example was not used.

If these "distortions" can be corrected or avoided, the history of technology offers a good instrument to confirm or reject propositions concerning the influence factors that affect the product development process. A detailed discussion of the sources used is beyond the scope of this book.

Coincidentally during the course of this research, the two motion pictures *Apollo 13* and *Titanic* were released. Although of course these are films for entertainment purposes, they were both fairly well researched. In any case, they recalled these two historic catastrophes to the public mind. Certain aspects of these two cases are pertinent to this work, and indeed it was particularly enjoyable to watch their dramatic interpretation. They did not, however, serve here as case studies.

Readers acquainted with historical research will note the absence of an abridging short essay about the sources used. Unfortunately, this had to be omitted due to space constraints.

A.3 Guidelines for the scientific-theoretical approach

Published studies on product development and related areas offer a broad range of data. They support the generalization of individual cases, provided the connection between the various findings could be sufficiently supported in scientific terms. How these connections are established is crucial in any scientific work and deserves careful examination and definition.

Initial experimental data and compiled cases were complemented by similar descriptions of documented cases from the history of technology, as well as by studies of cognition theory, problem-solving processes, R&D activities, creativity, leadership, and organizational forms and techniques. This allowed a generalization of the experienced cases.

A.3.1 Deductive vs. inductive connection

The deductive connection of the various findings is complicated by the fact that most of the studies establish or try to establish a unidirectional causal relationship between dependent and independent variables. Consider, for example, the results of two studies concerning the effect of the independent variables *A* and *B* on the dependent variable *Z* in two-valued symbolic logic:

- Study 1 concluded that: $A \rightarrow Z$.
- Study 2 concluded that: $B \rightarrow Z$.
- Hence it follows *only* that: $(A \text{ or } B) \rightarrow Z$.

Let us express this as an illustrative (hypothetical) example:

- Study 1: Greater freedom-of-action leads to a better solution of technical problems.
- Study 2: A cooperative strategy for conflict handling leads to a better problem solution in the development.
- Hence it follows that greater freedom *or* a cooperative conflict handling strategy lead to better results in product development.

It is considered insufficient for connecting the studies in a framework.

On the other hand, the inductive connection of partial results harbors the (almost) implicit problem of generalizability (cf. Popper 1959, 27 ff.). What is valid for one group or branch of engineering sciences is not necessarily valid for other groups or branches in engineering; what is obtained in one case cannot be generalized without many detailed considerations of the situation in question.

A.3.2 Narrative reviews and meta-analysis

Merely compiling studies and cases in a narrative review was not pursued in the research underlying this book. A condensation of the existing studies would produce an overview of what has been done in this field and would serve as an orientation point for further research. A good overview of the published works on product development is necessary to appreciate the findings and the limitations of the conclusions or methods. It also helps avoid conceptual replications of studies or experiments, and indicates better future approaches to a given issue. Although that kind of work is not to be undervalued, it was not the goal of this work.

Connecting the various studies within a narrative review alone cannot compensate for the bias issue. In some cases, maybe because of the subjectivity of the procedure, it is common to find that different narrative reviews of the same subject draw different conclusions.

A meta-analysis might objectively connect the results of the quantitative research on the subject. However, it was neither the central point in this work, nor the declared goal. The success of this work depends on arriving at practical, usable concepts. A meta-analysis, if used, should contribute to this, instead of being the main objective. The main goals of a meta-analytical procedure are the cumulation of the findings of various studies and an estimate of the variations among studies due to artifacts. A new theory or a new approach does not directly follow.

The true core of the work is the connection of concepts, theories, and objectively evaluated studies of research and development and related approaches, such as quality management and problem-solving processes. Subjective data arising from interviews or personal opinions of experts, also deserved consideration.

As with science itself, the "meaning is not in the text" but in the "context" of science, of development practices. Many studies cover only a certain aspect of product development. Integration of all these aspects is only possible by considering the real environment in which they occurred The dynamics of complex groups—composed of complex persons or characterized by complex interactions—can only be appreciated within the context in which they operate.

Furthermore, meta-analytical reviewers are plagued by the accusation of "comparing apples with oranges". This is the basic difficulty of pursuing a meta-analytical approach. As the evolution of the meta-analysis methodology has shown, the operationalization of variables to assess the dependent and independent variables are often quite different. This would cause potential problems in establishing the connection between different studies in a measurable (i.e. quantitative) fashion.

In addition, the objective measures of the success of such a complex activity are sometimes contradictory. The same program could be evaluated as successful or unsuccessful, depending on the definition of "success". Is success measured by adherence to the planning schedule, or by the efficacy of the achieved product characteristics on the market? For example, do the cost overruns in the development of the C-5A transport aircraft due to a structural problem with the wing render it a failure, or does the reduction of transport cost per unit weight by the end user render it a success? On the other hand, some of the measurement variables are not applicable to the job design because they are dependent on too many external factors beyond the control of the participants. They are indeed useful for planning purposes, but not necessarily as conclusive success factors. For example, defining a minimum the number of changes as a performance measure (i.e. how efficiently can a group provide a product definition on the first try) can be misleading if modifications become necessary to accommodate a changed market situation, or of changes are made in the production environment, or if the availability of resources changes. In many cases, groups that can cope with a greater number of modifications in a given time period are often sufficiently nimble to react to changes vital to product success. This kind of flexibility is in fact a measure of efficiency, not an indication of inefficiency.

Blindly measuring success in terms of arbitrary criteria can even lead to failure.

All these factors should be filtered out. This requires the critical analysis of an expert in the field of analysis. At the same time, it should be done with care because an "expert judgment call" can be both a solution and a source of problems. On the one hand, it enables the selection of applicable studies and the appropriate reduction of data, i.e. the filtering of "bad data". On the other hand, it might potentially introduce bias to the results.

The meta-analytical procedure attempts to solve the problem of bias in subjectivity by connecting objective values. Nevertheless, it does not completely resolve the issue. There are also cases of independently assessed meta-analytical studies that do not agree with each other. Very often a narrative review is required in connection with this method of analysis in order to minimize this effect.

A.4 Grounded theory and framework

The methodical approach known as the *grounded theory* can resolve all these issues. The following text translated from Breuer (1996, 23-24) is the most compact form found to describe the method.

Box 76. Description of grounded theory by Breuer (1996, 23-24).[263]

The relationship between insight[lv] (knowledge, theory) and experience[lvi] (perception, observation) is, we believe, mutually dependent. When experience leads to knowledge, then knowledge in turn leads us to search the world of experience; this structures our perception. 'Thinking ... imposes order upon a topic to the same extent to which the order inherent in the topic influences our thinking' (Goodman & Elgin, 1993, 203). A corresponding assessment can be found in psychological models of perception based on iterative-cyclical processes, such as those described by Neisser (1976) (cf. page 238ff.). Cognitive schemes of the subject influence his exploration within his field of perception and guide his selection of the available information. This in turn alters his cognitive schemes—and so on.[264]

In our opinion, the grounded theory concept operates in this framework with a sensible idea concerning the role of deductive (tautological) and inductive (creative, order-creating) aspects of the research process. On the one hand, it does so by means of reflective and self-critical use of previous

[263] Excerpt from Breuer, F. (1996). Theoretische und methodologische Grundlinien unseres Forschungsstils, in: F. Breuer (ed.), Qualitative Psychologie - Grundlagen, Methoden und Anwendungen eines Forschungsstils, pp. 14-40, Opladen, Germany: Westdeutscher. Reproduced here by kind permission of Westdeutscher Verlag GmbH.

[264] A detailed description of Neisser's concept is provided on page 238ff.

knowledge and of the knowledge obtained in the course of the research on a given subject as well as a continuous willingness to revise. On the other hand, it does so with openness, sensitivity, thoroughness, and attention to detail during the analysis of the empirical phenomena of interest and their structures. The *methodical explication* of both components in a transparent manner—especially the potential action for realizing the inductive part—and the resultant ability to learn the process makes this approach especially attractive.

The *research logic* of the *grounded theory* approach consists basically of the following:

Starting with the object under study, the researcher collects *field data* in a broad sense. These are *coded* in a detailed, interpretative analysis according to their conceptual content (first step: *open coding*). Ideally, this is done word by word or line by line of the transcriptions. The resulting knowledge gained determines which empirical phenomenon (which case?) of theoretical interest is chosen next (*theoretical sampling*). In this process, special importance is attached and attention given to establishing *contrasts* and *comparisons*. The coding procedure is then repeated with the new material and, as the initial concepts become somewhat stabilized, continued in a more structured form (*axial coding, selective coding*). This determines the further data collection process (additional cases of interest, contrasting phenomena, etc.) This procedure creates abstracting, structuring categories or constructs, which are incrementally specified, theoretically densified and finally interconnected in a relationship system (possibly with a core category to give coherence and perspective). Strauss[265] proposes a process-oriented *coding-paradigm* as background heuristics: the phenomena referred to by a category are differentiated according to their *conditions*, the *interactions* among the actors, their *strategies* and *tactics*, as well as their *implications*.

During the entire coding process, the researcher writes *theoretical memos*, in which the idea fragments that arise (relating to the various categories, relationships, methodical procedures, etc.) are captured. In the process of the model development, the *memos* are then ordered, completed, specified, etc. and integrated and extended into a *grounded theory*.

This iterative-cumulative process (data collection and processing are repeated cyclically, in a spiral-like fashion) until the point of *theoretical saturation* is reached, at which the process can be stopped. This point is the state at which the researcher can no longer extract any essential contribution to the conceptual extension of the resulting theory. In principle, any combination of the research phases *data collection, category building, and theoretical modeling* are possible and meaningful: from memos to further

[265] Anselm L. Strauss, one of the creators of the method.

> data collection (after theoretical sampling considerations); from memos to further coding runs; from coding to further data collection. Inductive and deductive aspects are productive in different steps. The emphasis of the analysis lies not...in that masses of data are collected and ordered, but instead that the great variety of thoughts occurring to the researcher can be organized.

The grounded theory approach offers additional advantages.

The experiential data of this researcher with product development and quality management in the product development is not only a supply of concrete, real cases but constitute an essential foundation that features:

- it amplifies the researcher's *theoretical sensitivity,* i.e. the ability to regard data in theoretical terms, especially including the nuances of the processes that would be less accessible to researchers without experiential knowledge;
- it suggests a rich quantity of comparisons, useful for finding *variations* and to support the *wide sampling on theoretical grounds.*

All these features support the formulation of a carefully ordered and *conceptually dense* theory.

This approach is in apparent contradiction to the usual conventions regulating scientific research, which state that personal experience leads to biased results. The effects of bias in experiential data are minimized by a thorough execution of the triad*: data collection, coding, and memoing*. This triad provides explicit control over the bias of the researcher (Strauss, 1987, 11).

The construction of a grounded theory proceeds similarly to the human perception process, in a repetitive, iterative manner with a very strong connection to practice. Not only new knowledge[lvii] is attained with this method but the risks of straying too far from the reality of product development is minimized.

Historical cases can also be considered adequately and integrated in the study, even though they are usually not measured quantitatively and are not necessarily free of distortion.

The method was developed by Glaser and Strauss in the early 1960s and is based on the American pragmatism of John Dewey, George H. Mead and Charles Peirce as well as on the tradition of sociology at the University of Chicago, "which extensively used field observations and intensive interviews as data-collecting techniques, and furthered much research on the sociology of work." (Strauss 1987, 6).

It was proposed as an approach to integrate qualitative data on social research into a theory. The pragmatic approach and the integration method allow the resulting theory to be *grounded* on the practical data. The originators of the method argue that in 1967, it was necessary to emphasize the term "*grounding* of the generated theories". By the end of the 1980s, this was no longer the case (op. cit., pp. 5-6).

One source of data used in this work that is not prescribed by the grounded theory method is the utilization of quantitative studies. To improve the generalization, quantitative studies on the R&D processes were included in the integration. This is not a necessary part of the grounded theory method, but it adds to the framework described herein. Use of historical (technology) cases is not prescribed by the method either. This could be done only because of the many accounts provided by persons who had previously interviewed participants, thus providing the necessary documentation to allow its integration. Similarly, descriptions by the participants also supplied a richness of detail concerning the processes taking place during the product development.

Hence, the grounded theory method has been chosen to integrate the diverse findings.

The research method used having been defined, one point that warrants explicit mention is that the initial cases analyzed were previously "open*ly* coded" with a host of questions and provisional answers. This raised issues pertaining to the: (i) organizational framework in which the situation occurred; (ii) technical facts that were the subject of or affected by the situation; (iii) social aspects characteristic of the situation, (iv) behavior of the participants and (v) consequences for the product, the participants, and the organization. Fortunately, this corresponded to the method defined by Strauss (1987, 28), leading to the intermediate results concerning conditions, strategies, interactions, and consequences. These issues, as well as the concepts and dimensions arising from this phase, are of a *provisional* nature. Therefore it is natural that they are not likely to appear in the case descriptions.

Other studies related to R&D using this integration technique are presented by Thomas (1994), who examines the political dynamics surrounding the design and implementation of new production technology. It provides insight into the complex dynamics of technological change and power in industry. Whyte (1994, 297) establishes a parallel between the work of Thomas and a study by Cole (1989) of the American, Japanese and Swedish automobile industries.[266]

---

[266] Whyte relates a situation in which Cole was "describing an ambitious research program involving a team of U.S. and Japanese auto manufacturers" to a group of sociological researchers. Cole's talk, Whyte says, "fell flat. In various ways, several people wanted to know what hypothesis Cole was testing. When he could not answer that question, some people seemed to be assuming that the program was not really scientific." This reflects one

A.5 Main research fields and extent of research

The research fields chosen were mainly the aerospace, electronics and computer industries. There are many reasons for this choice:

- The complexity of the products developed by these organizations is sufficiently high to yield intriguing issues to occur.
- They possess a short—sometimes, a extremely short—product lifecycle, thus allowing learning effects to be applied quickly to the issues encountered.
- There is experiential data available to the author based on his background and professional activity in these sectors. This justifies the use of the grounded theory approach and allows a relatively fast processing of the cases.
- There is a thorough documentation system based in the industries and certifying organs; and analyses by organizational scientists, technology historians and accounts by involved persons are available.
- In the case of the aerospace industry, there is the additional advantage that established quality assurance systems and management control systems have been in place longer than in other industries.

There are isolated cases taken from other industrial sectors.

Some indicators of the extent researched are given by the following numbers:

A detailed transcription of 36 critical incidents was obtained at the beginning of this research using the questioning prescribed by the grounded theory method. The cases described in the published literature were not protocolled in the form of critical incidents, but the research questioning still followed the same pattern. The references were processed in the form of "reading notes" to provide the necessary traceability of the coded elements to the original works and their exact location therein to allow their subsequent integration into the body of the framework. The literature search comprised qualitative and quantitative studies of research and development, quality management, and problem solving processes, among others. It included reference works in the many pertinent fields for this research. More than 1,600 works were collected for this purpose, including books and articles that were evaluated during the course of the dissertation research. Further research in the subsequent four years increased the number of

---

of the myths of science that research should be done exclusively by means of hypothesis formulation and testing (cf. Chapter 2).

sources used to more than 2,130 works. The processing of the literature was first done in the form of the previously mentioned "reading notes". This resulted in 151 word processor files comprising 4 Mbytes of (formatted) documents. For rough comparison, the five files containing the initial report comprise 1.2 Mbytes without figures. During the evolution of the work, another four critical incidents were protocolled. The general rule for the grounded theory of *protocolling only as needed instead of writing everything down* was followed. Prior to beginning the research manuscript, a total of 755 items were coded and grouped in 40 categories. This was performed after numerous passes through the database and after many tentative integrative sessions. After beginning the manuscript, the writing process served to integrate the components, and further coding and protocolling was done directly in the document.

No further time could be allocated to filtering the database to count the relevant works. A rough, worse-case estimate is that from the 2,130 works originally recorded, filtering out the irrelevant ones, double entries due, for example, to translations of a work, reprints that were entered separately, as well as the database entry points for edited volumes, some 1,500 references remained. Not counting the general reference works, some 1,100 works are estimated to have contributed significantly to this study. Not all of them have been integrated explicitly in this work due to time and size constraints.

A.6 Further nomenclature issues

During the development of this work, it was found that some words have several meanings, giving rise to a certain ambiguity. In an effort to maintain a satisfactory degree of neutrality in describing various forms of dealing with quality management issues, it is necessary to define these words for the sake of clarity. Basically, neutral terms were sought to describe the essential ideas.

A.6.1 Cooperation and participation

A.1.1.3Participatory design

The so-called "participatory design" approach (cf. Schuler & Namioka, Eds. 1993) to software development is a very interesting and promising concept. It says that customers' real needs can be assessed with the help of the customer. Many examples of software systems that have considered only the developers' side in their conception reinforce the need to pursue concepts in this manner. From nuclear power plants to flight cockpits to office programs, the world is full of good examples of this problem. The influence of human factors in the man-machine interface is taken seriously into consideration in high-risk or high-reliability systems. Nevertheless, the frequent repetition of cases where programmers have very limited or sometimes even no idea of how the user will work with the final program suggests that many lessons are still being unlearned. They often remain forgotten until a complete or near product failure prompts countermeasures to be taken to correct what could have been avoided from the outset. There are many cases of man-machine interfaces that turned out to be inappropriate for use, caused by the isolation of the programmer from the end user. There is some evidence that this isolation was intentionally sought in the past to ensure that the programmer would not be distracted by interaction with the user. Participatory design is an attempt to highlight the role of the user in achieving better software.

Some innovations have been made in software engineering with this approach, one of which is the possibility of building software "mock-ups" or building an evaluation prototype to discuss with customers.

In spite of the adjective "participatory", the approach is actually predominantly a cooperative one, as the user's participation is limited to being observed and, in some cases, making suggestions.

Contrary to this approach, customers have been known to approach a with a fixed list of requirements that a software *mus*t fulfill. The developer then has to implement them without question, thus going to the other extreme and curtailing the capabilities of software specialists to proposing an innovative tool. Both customers and entrepreneurs lose in such cases.

A.1.1.4Inter-entrepreneurial cooperation

The following definition of a new form of relationship among companies is given by Hoffmann (1996, 53), based on cooperation among German middle-sized companies: "A cooperation is a longer-term joint effort of two or more legally independent enterprises. *Intensive exchange of experience, task forces or joint ventures* are examples of cooperation. Excluded from this definition are: relationships with suppliers, sales representatives, or the sale/purchase of licenses.[lviii] Based on this definition, the meaning of cooperative ventures is actually participative in the sense described above.

In financial jargon, the term *participation in the share or in the profits* means being a part of the venture, but does not necessarily mean taking an active part in the activities.

A.6.2 Design

The word *design* is used to describe both:

- the activity of product development as a technical concept and an implementation of product functions, as well as
- the activity of conceiving ergonomic and esthetically pleasing aspects of form and function in a product, exemplified by names such as Bauhaus, Ital Design (G. Giugiaro) or Porsche Design, among others.

As both these activities endeavor to produce the conception of a device, both could be referred to as design. However, as they are based on different premises, namely the former on engineering and the latter on architecture and anthropometry, they differ in the way they approach the task. Therefore the word *design* will be used as follows:

1. Industrial design refers to the architecture approach.
2. The conception of manufacturing plants, for example for the activity of industrial engineers, is *not* referred to as industrial design.
3. The word *design* refers to the technical conception even if this is performed by an architect or an artist, provided the work was based on technical considerations (e.g. ergonomics). Whenever a distinction was necessary, it was only used in the sense of *industrial design* as defined above.

A.7 Limitations and further research

The first observation regarding the scope of this research refers to the extent to which the examples presented constitute isolated events or typical occurrences in the product development process that have an influence on quality. There is an implicitly generalizing character in the main research method chosen that ensures the generalization required to characterize the *existence* of the issues described and to compensate for the researcher's bias. The major concern addressed in this work is not primarily the frequency of occurrences or elements described herein but the characterization of their existence and their relevance to the achievement of better results. However, information on the frequency of occurrences or a numerical weighting of their significance is provided by the many quantitative studies published in the open literature.

The theoretical saturation required by the grounded theory approach was achieved. However, owing to the same constraints that determined this work, it cannot cover everything in this extensive field. Even with the saturation reached, there was a limit to its integration, i.e. to the consideration of the aspects of conflict management that could be elaborated and improved.

Moreover, the emotion-related aspects of product development could also be covered in greater detail.

Personality is believed to have an influence on product development processes. This was purposely not pursued in this work, as this would require a further qualification of this researcher in psychology in order to avoid the attendant risk of misinterpreting certain concepts. It remains a promising direction of research.

Other suggestions for research would be to elaborate the investigation of how quality and reliability organizations effectively cope with new challenges presented by the evolving product development process. A specific study of the inconsistencies of performance measurement variables could be implemented to improve the knowledge of the global (as opposed to local) optimization of the product development process. A general model description was not the basic intention of this research work. Research work to achieve a better descriptive model of the process considering the interaction between the participants and the internal mental processes constitute another intriguing research topic. This work aimed at proposing *solution approaches*, not actual solutions. Elaboration of a good set of rules for guiding the job design of product developers as well as of a measurement instrument may provide useful tools for practice as well as theory. Owing to its limited scope, issues pertaining to organizational development were not covered in this work.

This research work is an attempt not only to combine the advantages but also to compensate for the disadvantages of both the qualitative and the quantitative techniques. The result is a scientific step toward a better understanding of the complexity of the product development environment. This venture is not free of risk, however, and a critical appraisal of the content of this work would presumably be one of the next steps to be pursued in future research.

In order to protect the identity of professionals who shared touchy information as well as to respect proprietary technical information including confidential content, the technical descriptions are intentionally limited. A balance between presenting the necessary information to describe the situation of interest and not divulging too much information was pursued with special care.

A gender-neutral style was sought but could not be implemented in all cases, mainly because of the original texts cited. In order to adhere as closely as possible to the original sources, no attempt was made to rectify the style in this sense.

Last but not least, some evaluations presented in the text could potentially lead to the misunderstanding that it harbors negative opinions toward such conventional techniques as project management, quality systems, design reviews and others. This was a byproduct of critical appraisal necessary to lead to a better understanding of the potentialities and limitations of any technique or method. As this work was not intended as a specific book about any of the described techniques, the basic descriptions concentrated on describing the findings of *this* study and the results of its integration. All these techniques offer some potential for dealing with various issues in the research and development environment. This was not handled explicitly in the present text because almost every fundamental book on such techniques can provide this information. As mentioned above, it was not the purpose of this research to rewrite the many fine fundamental books on these techniques. This side-effect was perceived to occur during the draft stages of the research manuscript, and an effort was made to compensate for it in the final text. If this was not sufficiently done, it must be said first that the position of this researcher regarding the cited techniques is as follows. If working in product development, he would use the techniques of PERT/CPM, computer simulations, quality systems, reliability programs, design reviews, etc., but he would not apply them uncritically.

Finally, following the notion of critical testing, this author manifests his desire to "[know] the weaknesses and difficulties of [these] thoughts and solutions, [... to] entertain counterarguments, and [... to] welcome alternative arguments in order that he be given the opportunity to compare, modify and revise his position." (page 46)

I thus submit this book to critical testing by the reader—in the sense defended by this work for product development, for which this author invites readers to provide their insight and experience for further development of this work.

References

Abelson, R. P. (1975). Representing Mundane Reality in Plans, in: D. G. Bobrow & A. M. Collins (Eds.), *Representation and Understanding*, New York: Academic Press.
Abelson, R. P. (1976). Script Processing in Attitude Formation and Decision-Making, in: J. S. Carroll & J. W. Payne (Eds.), *Cognition and Social Behavior*, Hillsdale, NJ: Lawrence Erlbaum.
AGREE - Advisory Group of Electronic Equipment (1958). Reliability of Military Electronic Equipment, Washington DC: Office of Assistant Secretary of Defense.
Albaugh, J. (2000). How Aerospace can Compete in a Dot.com Economy, *Aviation Week & Space Technology*, May 22, 2000, p. 70.
Albert, H. (1971). Plädoyer einer Kritischen Rationalismus, München: R. Piper.
Alexander, T. (1969). The Unexpected Payoff of Project Apollo, *Fortune*, July 1969, pp. 114-117, 150-156.
Allen, T. J. (1970). Communication Networks in R&D Laboratories, *R&D Management*, vol. 1, no. 1, pp. 14-21.
Allen, T. J. (1986). Organizational Structure, Information Technology, and R&D Productivity, *IEEE Transactions on Engineering Management*, Vol. 33, No. 3, pp. 212-217.
Amabile, T. M. (1983). The Social Psychology of Creativity, New York: Springer.
Amabile, T. M. (1988). A Model of Creativity and Innovation in Organizations, in: B. M. Staw & L. L. Cummings (Eds.), *Research in Organizational Behavior*, Vol. 10, pp. 123-167, Greenwich, CT: Jai Press.
Amabile, T. M. & Gryskiewicz, S. S. (1987). Creativity in the Research Laboratory, Technical Report No. 30, Greensboro, NC: Center for Creative Leadership.
Ames, C. & Archer, J. (1988). Achievement Goals in the Classroom: Students' Learning Strategies and Motivation Processes, *Journal of Educational Psychology*, Vol. 80, pp. 260-267.
Andrews, F. M. & Farris, G. F. (1972). Time Pressure and Performance of Scientists and Engineers: A Five-Year Panel Study, *Organizational Behavior and Human Performance*, Vol. 8, No. 2.
anon. (1963). Industry's Ethics Could Be Better..., *Space/Aeronautics*, Vol. 39, No. 3, March 1963, pp. 11-17.
anon. (1967). The Pentagon builds a Monster, *Business Week*, 18 Feb 1967, pp. 198-199.
anon. (1997). Boeing Struggles with Super Hornet Wing-Drop Problem, *Flight International*, Oct 8-14, 1997, p. 18.
Argyris, C. (1964). Integrating the Individual and the Organization, New York: Wiley.
Argyris, C. (1982). The Executive Mind and Double-Loop Learning, *organizational Dynamics*, Autumn 1982, pp. 5-22.
Asch, S. E. (1951). Effects of Group Pressure on the Modification and Distortion of Judgements, in: H. Guetzkow (Ed.), *Groups, Leadership and Men*, Pittsburgh: Carnegie.
Asch, S. E. (1956). Studies of Independence and Conformity: A Minority of One Against a Unanimous Majority, *Psychological Monographs*, Vol. 70, No. 9, whole no. 416.
Backhaus, K. & Gruner, K. (1994). Epidemie des Zeitwettbewerbs, in: K. Backhaus & H. Bonus (Hrsg.), *Die Beschleunigung oder der Triumph der Schildkröte*, pp. 19-46, Stuttgart: Schäffer-Poeschel.
Bailyn, L. (1985). Autonomy in the Industrial R&D Laboratory, *Human Resource Management*, Vol. 24, No. 2 Summer, pp. 129-146.
Bednarek, E. (1985). Veränderungen der Arbeitsmotivation durch Qualitätszirkel und Lernstatt, *Dissertation, TU München*.
Bergaust, E. (1976). Wernher von Braun, Washington, D.C.: National Space Institute.

Bernstein, D. A.; Roy, E. J.; Srull, T. K. & Wickens, C. D. (1988). Psychology, Boston, etc.: Houghton Mifflin Company.

Bilstein, R. E. (1980). Stages to Saturn: A Technological History of the Apollo/Saturn Launch Vehicles, NASA-SP-4206, Washington DC: NASA.

Bilstein, R. E. (1994). Flight in America - From the Wrights to the Astronauts (Revised Edition), Baltimore, London: John Hopkins University Press.

Birdsall, S. (1970). Flak Bait and True Grit, *Air Classics*, Vol. 6, no. 5, June 1970, pp. 22-33.

Birkan, M. A. (1992). Laser Propulsion: Research Status and Needs, *Journal of Propulsion and Power*, Vol. 8, No. 2, pp. 354-360.

Birolini, A. (1994). Quality and Reliability of Technical systems: Theory - Practice - Management, Berlin: Springer.

Bleicher, K. (1992). Das Konzept Integriertes Management, 2. Aufl., Frankfurt: Campus.

Bradley, M. E. (1974). The P-51 over Berlin, *Aerospace Historian*, Vol. 21, no. 3, Sept. 1974.

Bravo, E. (1993). The Hazards of Leaving Out the Users, in: D. Schuler & A. Namioka (Eds.), *Participatory Design: Principles and Practices*, pp. 3-11, Hillsdale, NJ: Lawrence Erlbaum Associates.

Breuer, F. (1996). Theoretische und methodologische Grundlinien unseres Forschungsstils, in: F. Breuer (ed.), *Qualitative Psychologie - Grundlagen, Methoden und Anwendungen eines Forschungsstils*, pp. 14-40, Opladen, Germany: Westdeutscher.

Briggs, J. E. (1960-61). Training the Aerospace Force, *Air University Quarterly Review*, Vol. XII, No. 3 and 4 - Winter/Spring 1960-1961, pp. 186-194.

Brooks, F. P., Jr. (1982). The Mythical Man-Month - Essays on Software Engineering, Reading, Mass: Addison-Wesley, reprint with corrections of 1972 edition

Bullinger, H.-J. (1990). Integrierte Produktentwicklung, *Handelsblatt Dokumentation*, 3-5.

Burns, T. (1971). Mechanistic and Organismic Structures, in: D. S. Pugh (Ed.), *Organization Theory*, New York: Penguin Books.

Burns, T. & Stalker, G. M. (1994). The Management of Innovation (3rd ed.), Oxford, New York, etc.: Oxford University Press.

Büscher, L. (Dir.) (1984). Titanic - Ein Untergang zieht Kreise - Nachspiel einer Katastrophe, Video: Auftrag von ZDF und SRG.

Butler, R. (1987). Task-involving and Ego-involving Properties of Evaluation: Effects of Different Feedback Conditions on Motivational Perceptions, Interest, and Performance, *Journal of Educational Psychology*, Vol. 79, pp. 474-482.

Cameron, K. S. (1994). Strategies for Successful Organizational Dowsizing, *Human Resource Management*, Vol. 33, No. 2, summer, pp. 189-211.

Casani, J. (Chairman); Albee, A.; Leising, C; Battel, S.; MacPHerson, D.; Brace, R.; Menard, W.; Burdick, G.; Rose, R.; Burr, P.; Sackheim, R.; Schallenmuller, A.; Dipprey, D.; Whetsel, C. & Lavell, J. (2000). Report on the Loss of the Mars Polar Lander and Deep Space 2 Missions - JPL Special Review Board, JPL D-18709, Pasadena, CA: Jet Propulsion Laboratory.

Cascio, W. (1995). Guide to Responsible Restructuring - U.S. Dept. of Labor's Office of the American Workplace, Washington, DC: U.S. Government Printing Office.

Chaikin, A. (1995). Lunar Odissey, *Scientific American*, Vol. 273, No. 4, Oct 1995, pp. 162-166.

Chaisson, E. J. (1994). The Hubble Wars: Astrophysics Meets Astropolitics in the Two-Billion-Dollar Struggle over the Hubble Space Telescope, New York NY: HarperCollins.

Churchman, C. W. (1971). The Design of Inquiry Systems, New York: Basic Books.

Coad, P., Jr. (1988). DOD-STD-2167, Defense System Software Development: Point, Counterpoint, and Revision A, *Proceedings of the Computer Standards Conference 1988,*

Computer Standards Evolution: Impact and Imperatives, 21-23 March 1988, Washington DC, pp. 47-50, Los Alamitos, CA: IEEE Computer Society.
Cohen, A. & Bradford, D. (1985). Managing for Excellence, New York: Wiley.
Cole, R. E. (1989). Strategies for Learning - Small-Group Activities in American, Japanese, and Swedish Industry, Berkeley, Los Angeles: University of California Press.
Collins, E. G. & Guetzkow, H. (1964). A Social Psychology of Group Processes for Decision-Making, New York: Wiley.
Cooke, R. A. & Kernaghan, J. A. (1987). Estimating the Difference Between Group Versus Individual Performance on Problem-Solving Tasks, *Group and Organization Studies*, vol. 12, no. 3, pp. 319-342.
Dalton, G. W. & Perry, L. T. (1987). Organizing for Innovation, in: D. Christiansen (Hrsg.), *Engineering Excellence - Cultural and Organizational Factors*, pp. 101-107, New York: IEEE Press.
Davie, M. (1986). The Titanic - The Full Story of a Tragedy, London: The Bodley Head.
Davis, J. H. & Restle, F. (1966). The Analysis of Problems and Prediction of Group Problem Solving, *Journal of Abnormal and Social Psychology*, Vol. 66, No. 2, pp. 103-116.
De George, R. T. (1991). Ethical Responsibilities of Engineers in Large Organizations: The Pinto Case, in: D. G. Johnson (Ed.), *Ethical Issues in Engineering*, pp. 175-186, Englewook Cliffs, NJ: Prentice-Hall, reprint from Business and Professional Ethics Journal, vol. 1, no. 1 (1981), pp.1-4
Deitz, D. (1996). A Re-Engineered Strike Fighter Takes Off, *Mechanical Engineering*, Nov. 1996, pp. 70-73.
Demarco, T. (1996). Personal Approach to Improving Software Development, *IEEE Software*, November 1996, pp. 116-117.
Deming, W. E. (1982). Out of the Crisis, Cambridge, MA: Massachussets Institute of Technology, Center for Advanced Engineering Study, 9th print 1989
Descartes, R. (1969). Discours de la Méthode (Pour Bien Conduire la Raison, & Chercher la Verité dans les Sciences) - Von der Methode des richtigen Vernunftsgebrauchs und der wissenschaftlichen Forschung, in: Übersetzt und herausgegeben von Lüder Gäbe, Hamburg: Felix Meiner.
Descartes, R. (1994). Discours de la Méthode - Discourse on the Method -- A Bilingual Edition with an Interpretive Essay, in: Edited, translated, and introduced by George Heffernan, Notre Dame, IN: University of Notre Dame Press.
Deutsch, M. (1982). Interdependence and Psychological Orientation, in: V. J. Derlega & J. Grzelak (Eds.), *Cooperation and Helping Behavior - Theory and Research*, pp. 15-42, New York NY, etc.:Academic Press.
Deutsch, C. (1995). In der Falle, *Wirtschaftswoche*, Vol. 49, No. 23, pp. 83-86.
DGQ / NTG - Deutsche Gesellschaft für Qualität / Nachrichtentechnischen Gesellschaft - Arbeitsgruppe 143 (1986). Software Qualitätssicherung - Aufgaben, Möglichkeiten, Lösung, 1. Aufl., DGQ-NTG-Schrift Nr. 12-51, Berlin: Beuth; Berlin:VDE Verlag.
DoD - Department of Defense (1980). MIL-STD-785B - Reliability Program for Systems and Equipment Development and Production, Washington DC: Department of Defense.
DoD - Department of Defense (1981a). Quality Program Requirements, MIL-Q-9858A, Amd. 2, Washington DC: Department of Defense.
DoD - Department of Defense (1981b). Technical Reviews and Audits for Systems, Equipments, and Computer Programs, MIL-STD-1521A, Nt.2, Washington DC: Department of Defense.
DoD - Department of Defense (1985). Defense System Software Development, DOD-STD-2167, Washington: Department of Defense.

DoD - Department of Defense (1996). Update on the New "Dual Use" Handbook on Maintainability - MIL-HDBK-470A.

Dooley, A. R. (1964). Interpretations of PERT, *Harvard Business Review*, Vol. 42, No. 2, pp. 160-172.

Dornheim, M. A. (1999). Faulty Thruster Tale Led to Mars Mishap, *Aviation Week & Space Technology*, Oct 4, 1999, pp. 40-41.

Dornheim, M. A. (2000). Space Industry Dogged by Lack of Engineers, *Aviation Week & Space Technology*, Sep. 25, 2000, p. 64.

Downs, A. (1967). Inside Bureaucracy, Boston: Little, Brown.

Downs, A. (1995). Corporate Executions: The Ugly Truth About Layoffs - How Corporate Greed is Shattering Lives, Companies, and Communities, New York: American Management Association.

Driskell, J. E. & Salas, E. (1992). Collective Behavior and Team Performance, *Human Factors*, Vol. 34, no. 3, pp. 277-288, Santa Monica, CA: The Human Factors Society.

Drucker, P. F. (1974). Management, New York: Harper and Row.

Drucker, P. F. (1979). Adventures of a Bystander, New York: Harper & Row.

Drucker, P. F. (1995). Introduction - Mary Parker Follett: Prophet of Management, in: P. Graham (Ed.), *Mary Parker Follett - Prophet of Management: A Celebration of Writings from the 1920s*, pp. 1-9, Boston, MA: Harvard Business School Press.

Duda, J. L. & Nicholls, J. G. (1992). Dimensions of Achievement Motivation in Schoolwork and Sport, *Journal of Educational Psychology*, Vol. 84, pp. 290-299.

Dweck, C. S. (1989). Motivation, in: A. Lesgold & R. Glaser (Eds.), *Foundations for a Psychology of Education*, pp. 87-136, Hillsdale, NJ: Lawrence Erlbaum.

Eastman, R. M. (1980). Engineering Information Release Prior to Final Design Freeze, *IEEE Transactions on Engineering Management*, Vol. EM-27, No. 2, pp. 37-42.

Eddy, P.; Potter, E. & Page, B. (1992). Fat, Dumb, and Happy: The Failure of the FAA, in: J. H. Fielder & D. Birsch (Eds.), *The DC-10 Case. A Study in Applied Ethics, Technology and Society*, pp. 109-122, Albany, NY: State of New York University Press, reprint from "Destination Disaster: From the Trimotor to the DC-10", 1976

EFQM - European Foundation for Quality Management (1999a). Eight Essentials of Excellence - The Fundamental Concepts and their Benefits, Brussels: EFQM.

EFQM - European Foundation for Quality Management (1999b). The EFQM Excellence Model, Brussels: EFQM.

Elliott, E. S. & Dweck, C. S. (1981). Children's Achievement Goals as Determinants of Learned Helplessness and Mastery-oriented Achievement Patterns: An Experimental Analysis, Unpublished Manuscript, Harvard University.

Elmes, M. & Wilemon, D. (1992). Determinants of Cross-Functional Cooperation in Technology-Based Organizations, *IEEE Engineering Management Review*, Vol. 20. No. 1, pp. 40-46.

Emery, F. E. (1959). Characteristics of Socio-Technical Systems, Document No. 527, London: Tavistock Institute of Human Relations.

Emery, F. E. & Trist, E. L. (1960). Socio-Technical Systems, in: F. E. Emery (Ed.), *Systems Thinking*, Harmondsworth: Penguin Books.

Epton, S. R. (1972). The Underestimation of Project Duration: An Explanation in Terms of a Time-Horizon, *R&D Management*, Vol. 2, No. 3, pp. 141-142.

ESRO - European Space Research Organisation (1973). Design Review Programme, QRA-05, Issue No. 2.

Ewing, D. W. (1964). Tension Can Be an Asset, *Harvard Business Review*, Vol. 42, No. 5, pp. 71-78.

Fazio, R. H. & Williams, C. J. (1986). Attitude Accessibility as a Moderator of the Attitude Perception and Attitude-Behavior Relations: An Investigation of the 1984 Presidential Election, *Journal of Personality and Social Psychology*, Vol. 51, pp. 505-514.

Festinger, L. (1957). A Theory of Cognitive Dissonance, Stanford: Stanford University Press.

Festinger, L. (1964). Conflict, Decision, and Dissonance, Stanford, CA: Stanford University Press.

Feynman, R. P. (1988). What Do You Care What Other People Think? - Further Adventures of a Curious Character, New York, etc.: W. W. Norton & Co.

Fiedler, F. E.; Potter, E. H., III; Zais, M. M. & Knowlton, W. A., Jr. (1979). Organizational Stress and the Use and Misuse of Managerial Intelligence and Experience, *Journal of Applied Psychology*, Vol. 64, No. 6, pp. 635-647.

Fielder, J. (1992). Floors, Doors, Latches, and Locks, in: J. H. Fielder & D. Birsch (Eds.), *The DC-10 Case. A Study in Applied Ethics, Technology and Society*, pp. 69-81, Albany, NY: State of New York University Press.

Fielder, J. H. & Birsch, D. (Eds.) (1992). The DC-10 Case. A Study in Applied Ethics, Technology and Society, Albany, NY: State of New York University Press.

Fisch, R. & Daniel, H. D. (1983). Determinanten des Erfolgs universitärer Forschungsprojekte - eine empirische Untersuchung forschungshemmender und -fördernder Faktoren. 1. Die Form der Projektorganisation als Determinante für Leistung und Zufriedenheit der Mitarbeiter, Universität Konstanz , Zwischenbericht.

Fleeter, R. (1996). Reducing Spacecraft Cost, in: J. R. Wertz & W. J. Larson (Eds.), *Reducing Space Mission Cost*, pp. 161-191, Dordrecht: Kluwer.

Fleischer, M. & Liker, J. K. (1992). The Hidden Professionals: Product Designers and Their Impact on Design Quality, *IEEE Transactions on Engineering Management*, vol. 39, no. 3, pp. 254-264.

Follett, M. P. (1926). The Giving of Orders, in: H. Metcalf (Ed.), *Scientific Foundations of Business Administration*, Baltimore: Williams & Wilkins, reprint.

Follett, M. P. (1949a). The Illusion of Final Authority, in: L. Urwick (Ed.), *Freedom and Coordination: Lectures in Business Organisation by Mary Parker Follett*, London: Management Publications Trust.

Follett, M. P. (1949b). The Process of Control, in: L. Urwick (Ed.), *Freedom and Coordination: Lectures in Business Organisation by Mary Parker Follett*, pp. 77-89, London: Management Publications Trust.

Fraisse, P. (1964). The Psychology of Time, London: Eyre and Spottiswoode.

French, P. (1992). What is Hamlet to McDonnell Douglas or McDonnell Douglas to Hamlet?:DC-10, in: J. H. Fielder & D. Birsch (Eds.), *The DC-10 Case. A Study in Applied Ethics, Technology and Society*, pp. 177-192, Albany, NY: State of New York University Press, reprint (1982)

Frey, D. (1986). Recent Research on Selective Exposure to Information, in: L. Berkowitz (Ed.), *Advances in Experimental Social Psychology*, Vol. 19, Orlando, FL: Academic Press.

Fries, S. D. (1992). NASA Engineers and the Age of Apollo, NASA-SP-4104, Washington DC: NASA.

Fromm, E. (1956). Die Kunst des Liebens, Frankfurt a.M., Berlin: Ullstein, Neue Übersetzung von Mickel, L. und Mickel, E. 1989

Frosch, R. A. (1969). A New Look at Systems Engineering, *IEEE Spectrum*, Vol. VI, No. 9, pp. 24-28.

Frühauf, K. (1990). Die Ältesten Fehler Sind die Teuersten, *Technische Rundschau*, Nr. 28/90, pp. 32-36.

Fuths, T. (1997). Sonnenfinsternis, *mot*, Heft 26, pp. 16-24.

Gibb, C. A. (1969). Leadership, in: G. Lindzey & E. Aronson (Eds.), *The Handbook of Social Psychology, 2nd. ed.*, Vol. 4 - Group Psychology and Phenomena of Interaction, pp. 205-282, Reading, MA: Addison-Wesley.

Glumb, R. J. & Krier, H. (1984). Concepts and Status of Laser-Supported Rocket Propulsion, *Journal of Spacecraft and Rockets*, Vol 21, No. 1, pp. 60-79.

Goldstein, R. (1976). Failure Review Board - A Management Tool for Producibility and Profit, *Quality Progress*, June 1976, pp. 22-25.

Goltz, G. E. (1986). A Guide to Development, *R&D Management*, Vol. 16, No. 3, pp. 243-249.

Goodman, N. & Elgin, C. Z. (1993). Revisionen. Philosophie und andere Künste und Wissenschaften, Frankfurt: Suhrkamp.

Gray, M. (1992). Angle of Attack: Harrison Storms and the Race to the Moon, New York, London: W. W. Norton & Co.

Greenwood, J. T. (1974). The Air Force Ballistic Missile and Space Program (1954-1974), *Aerospace Historian*, Vol. 21, no. 4, Dec. 1974, pp. 190-205.

Grote, G. (1997). Autonomie und Kontrolle. Zur Gestaltung automatisierter und risikoreicher Arbeitssysteme, Zürich: vdf.

Grzelak, J. (1988). Conflict and Cooperation, in: M. Hewstone; W. Stroebe; J.-P. Codol & G.M. Stephenson (Eds.), *Introduction to Social Psychology - A European Perspective*, Oxford: Basil Blackwell.

Hacker, W. (1978). Allgemeine Arbeits- und Ingenieurpsychologie, 2. Aufl., in: E. Ulich (Hrsg.), *Schriften zur Arbeitspsychologie*, Band 20, Bern: Huber.

Hacker, W. (1986). Arbeitspsychologie: Psychische Regulation von Arbeitstätigkeiten, *Schriften zur Arbeitspsychologie*, Nr. 41, Bern, etc.: Hans Huber.

Hacker, W. (1987). Software-Ergonomie: Gestalten rechnergestützer Arbeit?, in: W. Schönpflug & M. Wittstock (Hrsg.), *Software Ergonomie '87: Nützen Informationsysteme dem Benutzer?*, pp. 31-54, Stuttgart: Teubner.

Hackman, J. R. (1968). Effects of Task Characteristics on Group Products, *Journal of Experimental Social Psychology*, Vol. 4, pp. 162-187.

Hackman, J. R. & Morris, C. G. (1975). Group Tasks, Group Interaction Process, and Group Performance Effectiveness: A Review and Proposed Integration, *Advances in Experimental Social Psychology*, Vol. 8, pp. 45-99, New York, London: Academic Press.

Hackman, J. R. & Oldham, G. R. (1975). Development of the Job Diagnostic Survey, *Journal of Applied Psychology*, Vol. 60, pp. 158-170.

Hackman, J. R. & Oldham, G. R. (1976). Motivation Through the Design of Work: Test of a Theory, *Organizational Behavior and Human Performance*, Vol. 16, pp. 250-279.

Hagemann, M. (1996). Globales Collaborative Working, in: H. J. Bullinger (Hrsg.), *Kooperation im Engineering - Tagungsunterlagen IAO- Forum, 19. Juni 1996, Stuttgart*, pp. 101-118, Stuttgart: Fraunhofer IRB Verlag.

Hall, D. T. & Lawler, E. E., III (1971). Job Pressures and Research Performance, *American Scientist*, Vol. 59, Jan-Feb, pp. 64-73.

Hall, J. & Williams, M. S. (1970). Group Dynamics Training and Improved Decision Making, *Journal of Applied Behavioral Science*, Vol. 6, pp. 27-32.

Hamel, G. & Prahalad, C. K. (1994). Competing for the Future, *Harvard Business Review*, July-August, pp. 122-128.

Hellpach, W. (1922). Sozialpsychologische Analyse des betriebstechnischen Tatbestandes 'Gruppenfabrikation', in: R. Lang & W. Hellpach, *Gruppenfabrikation*, pp. 5-186, Berlin: Springer.

Heppenheimer, T. A. (1997). Countdown - A History of Space Flight, New York: John Wiley & Sons.

Herzberg, F. (1968). Work and the Nature of Man, London: Staple Press.

Herzberg, F. (1987). One More Time: How do You Motivate Employees?, *Harvard Business Review*, Sep-Oct 1987, pp. 109-120, reprint of 1968 article

Herzberg, F.; Mausner, B. & Snyderman, B. (1959). The Motivation to Work, New York: Wiley.

Hesse, W.; Merbeth, G. & Fröhlich, R. (1992). Software-Entwicklung, *Handbuch der Informatik*, Band 5.3, München: Oldenburg.

Hill, G. W. (1982). Group Versus Individual Performance: Are N+1 Heads Better Than One?, *Psychological Bulletin*, Vol. 91, No. 3, pp. 517-539.

Hill, W.; Fehlbaum, R. & Ulrich, P. (1992). Organisationslehre 2 - Theoretische Ansätze und praktische Methoden der Organisation sozialer Systeme, 4. Aufl., Bern: Verlag Paul Haupt.

Hirschhorn, L. & Gilmore, T. (1992). The New Boundaries of the "Boundaryless" Company, *Harvard Business Review*, May-June 1992, pp. 104-115.

Hitt, M.; Keats, B.; Harback, H. & Nixon, R. (1994). Rightsizing: Building and Maintaining Strategic Leadership and Long-Term Competitiveness, *Organizational Dynamics*, Vol. 23, No. 2, pp. 18-32.

Hoffmann, D. A. (1993). The Influence of Goal Orientation on Task Performance: A Substantively Meaningful Suppressor Variable, *Journal of Applied Social Psychology*, Vol. 23, pp. 1827-1846.

Hoffmann, M. (1996). Kooperation von mittelständischen Unternehmen - Erste Ergebnisse einer gemeinsamen Umfrage des Fraunhofer-IAO und der VDI-Nachrichten, in: H.-J. Bullinger (Hrsg.), *Kooperation im Engineering*, pp. 49-61.

Hoffmann, D. A. & Strickland, O. J. (1995). Task Performance and Satisfaction: Evidence for a Task- by Ego-Orientation Interaction, *Journal of Applied Social Psychology*, Vol. 25, No.6, pp. 495-511.

Holley, I. B., Jr. (1987). A Detroit Dream of Mass-produced Fighter Aircraft: The XP-75 Fiasco, *Technology and Culture*, Vol. 27, July 1987, pp. 578-593.

Hornby, A. S. with Cowie, A. P. & Gimson, A. C. (1974). Oxford Advanced Learner's Dictionary of Current English, 3rd ed., Oxford: Oxford Univeristy Press.

Humphrey, W. S. (1995). A Discipline for Software Engineering, Reading, MA: Addison Wesley.

Huzel, D. (1962). Peenemünde to Canaveral, Englewood Cliffs, NJ: Prentice-Hall.

Hygate, B. (1990). British Experimental Jet Aircraft, Hertforfshire, UK: Argus Books.

Irving, C. (1993). Wide Body: The Making of the 747, London, etc.: Hodder & Stoughton.

ISO - International Organisation for Standardization (1987a). ISO 9001 - Quality Systems - Model for Quality Assurance in Design/Development, Production, Installation and Servicing, 1987-03-15.

ISO - International Organisation for Standardization (1987b). ISO 9004 - Quality Management and Quality Systems Elements - Guidelines, 1987-03-15.

Jagacinsky, C. M. & Nicholls, J. G. (1984). Conceptions of Ability and Related Affects in Task Involvement and Ego Involvement, *Journal of Educational Psychology*, pp. Vol. 76, 909-919.

Janis, I. L. (1972). Victims of Groupthink: A Psychological Study of Foreign-Policy Decisions and Fiascoes, Boston: Houghton Mifflin.

Janis, I. L. & Mann, L. (1977). Decision Making - A Psychological Analysis of Conflict, Choice, and Commitment, New York, NY: The Free Press.

Judd, C. M. & Kulik, J. A. (1980). Schematic Effects of Social Attitudes on Information Processing and Recall, *Journal of Personality and Social Psychology*, Vol. 38, pp. 569-578.

Juran, J. M. (1992). Juran on Quality by Design - The New Steps for Planning Quality into Goods and Services, New York, etc.: The Free Press.

Juran, J. M. & Gryna, F. M., Jr. (1980). Quality Planning and Analysis - From Product Development Through Use (2nd ed.), New York, etc.: McGraw-Hill.

Kanter, R. M. (1989). When Giants Learn To Dance - Mastering the Challenges of Strategy, Management, and Careers in the 1990s, London: Routledge.

Kanter, R. M. (1991). Improving the Development, Acceptance, and Use of New Technology: Organizational and Interorganizational Challenges, in: National Academy of Engineering & Commission on Behavioral and Social Sciences and Education, National Research Council, *People and Technology in the Workplace*, pp. 15-56, Washington DC: National Academy Press.

Katz, R. & Allen, T. J. (1982). Investigating the Not Invented Here (NIH) Syndrome: a Look at the Performance, Tenure and Communication Patterns of 50 R&D Project Groups, *R&D Management*, Vol. 12, No. 1, pp. 7-19.

Katz, D. & Kahn, R. L. (1978). The Social Psychology of Organizations, 2nd. Ed., New York: John Wiley & Sons.

Knauer, M. (1990). Konrad Zuse - Porträt des Computerpioniers und seiner Maschinen, Zürich: Filmkollektiv Zürich.

Knopp, G. (1997). Eine 'silberne Zigarre' fängt Feuer, *Damals*, 5/97, pp. 8-11.

Kohn, A (1986). No Contest: The Case Against Competition, Boston: Houghton Mifflin.

Kolleck, K. D. (1995). Entscheidend ist der Mensch, *Qualität und Zuverlässigkeit*, Band 40, Heft 1, pp. 45-48.

Krassowski, G. & Treffert, C. (1991). Fachchinesisch, *Qualität und Zuverlässigkeit*, Band 36, Nr. 10, p. 561.

Krubasik, E. G. (1989). Der Königsweg zum Neuen Produkt, *HarvardManager*, (3)1993, pp. 25-31.

Kuhn, T. S. (1962). The Structure of Scientific Revolutions, Chicago: University of Chicago Press.

Kutzer, A. (1989). Herausforderung an ein Praxisgerechtes Projektmanagement, in: H.-J. Bullinger (Hrsg.), *Forum für Management in Forschung und Technologie, Tagungsunterlagen, 16-17 Okt 1989, Frankfurt am Main*, pp. 365-395, München: GfMT--Gesellschaft für Management und Technologie- Verlags.

Lambright, W. H. (1995). Powering Apollo - James E. Webb of NASA, Baltimore, London: John Hopkins University Press.

Lambright, L. H. (2001). Transforming Government: Dan Goldin and the Remaking of NASA, Arlington, VA: PricewaterhouseCoopers Endowment.

Larson, C. E. & LaFasto, F. M. J. (1989). TeamWork - What Must Go Right / What Can Go Wrong, Newbury Park, CA: Sage.

Launius, R. D. (1994). NASA: A History of the U.S. Civil Space Program, Malabar, FL: Krieger.

Lawson, M. & Karandikar, H. M. (1994). A Survey of Concurrent Engineering, *Concurrent Engineering Research and Applications*, no. 2 (1994), pp. 1-6.

Lazard, J. Y. (1987). Accélération de Développement des Prototypes et Réduction des Coûts par Intégration Pluridisciplinaire, *Flight Vehicle Development Time and Cost Reduction*, AGARD-CP-424, pp. 26F-1 -- 26F-9, Neully sur Seine, France: NATO - North Atlantic Treaty Organization.

Lazarus, R. S. & Eriksen, C. W. (1952). Effects of Failure Stress Upon Skilled Performance, *Journal of Experimental Psychology*, Vol. 43, pp. 100-105.
Lazarus, R. S. & Folkman, S. (1984). Stress, Appraisal, and Coping, New York: Springer.
Lerner, F. (1988). Geschichte der Qualitätssicherung, in: W. Masing (Hrsg.) , *Handbuch der Qualitätssicherung, 2. Aufl.*, pp. 19-32, München: Carl Hanser.
Lester, J. T. (1983). Organizational Stress: A Literature Review: 1974-1982, Boston: Office of Naval Research.
Levine, A. S. (1982). Managing NASA in the Apollo Era, NASA-SP-4102, Washington DC: NASA.
Lewin, K. (1982). Werkausgabe, in: C. F. Graumann (Ed.), Vol. 4: Feldtheorie, Bern: Huber.
Lewis, R. S. (1988). Challenger: the Final Voyage, New York: Columbia University Press.
Lions, J. L. (chairman) et al. (1996). Ariane 501 Failure - Report by the Inquiry Board.
Lippisch, A. (1982). Alexander Lippisch - Erinnerungen,, Steinebach-Wörthsee: Luftfahrtverlag Axel Zuerl.
Lovell, J. & Kluger, J. (1994). Lost Moon: The Perilous Voyage of Apollo 13, Boston, New York: Houghton Mifflin Company.
Lowe, V. & Schneewind, J. B. (1990). Alfred North Whitehead - The Man and His Work, Vol. II: 1910-1947, Baltimore, ML: John Hopkins University Press, (the unfinished manuscript by Lowe was edited by Schneewind).
Luce, D. R. & Raiffa, H. (1967). Games and Decisions - Introduction and Critical Survey, New York: John Wiley & Sons.
MacCrimmon, K. R. & Ryavec, C. A. (1962). An Analytical Study of the PERT Assumptions, *RAND Memorandum*, RM-3408-PR, December 1962, Santa Monica, CA: The RAND Corporation.
Maderthaner, R. (1989). Kommunikationsprozesse, in: E. Roth (Hrsg.), *Enzyklopädie der Psychologie*, Themenbereich D: Praxisgebiete; Serie III: Wirtschafts-, Organisations- und Arbeitspsychologie; Band 3 Organisationspsychologie, pp. 487-504.
Maguire, S. (1994). Du Code et des Hommes - Stratégies de Suivi de Project, Les Ulis Cedex: Microsoft Press.
Maier, N. R. F. & Hoffman, L. R. (1965). Acceptance and Quality of Solution as Related to Leader's Attitudes, *Journal of Applied Behavioral Science*, Vol. 1, No. 4, pp. 373-386.
Maier, N. R. F. & Solem, A. R. (1952). The Contribution of a Discussion Leader to the Quality of the Group Thinking: The Effective Use of Minority Opinions, *Human Relations*, vol. V, no. 3, pp. 277-288, London: Tavistock.
Mainsfield, E.; Rappaport, J.; Schnee, J.; Wagner, S. & Hamburger, N. (1981). Research and Innovation in the Modern Corporation, in: R. R. Rothberg (Ed.), *Corporate Strategy and Product Innovation*, pp. 416-427.
Maitra, K. & Whittaker, J. (1993). A Survey of Engineers' Attitudes and Careers Expectations, *Leadership in Technology, Creativity, and Innovation: An International Competitive Imperative, 14th Annual Conference of the American Society for Engineering Management, Dallas, Texas, Oct. 24-26, 1993*, pp. 255-261, American Society for Engineering Management.
Mankin, D. (1978). Towards a Post-Industrial Psychology, New York: Wiley.
March, J. G. & Simon, H. A. (1958). Organizations, New York: Wiley, 9th printing 1967.
Mark, H. & Levine, A. (1984). The Management of Research Institutions - a Look at Government Laboratories, NASA-SP-481, Washington, DC: NASA.
Maslow, A. H. (1954). Motivation and Personality, New York: Harper & Row.
Maslow, A. H. (1962). Toward a Psychology of Being, New York: Van Nostrand.

Maslow, A. H. (1966). The Psychology of Science - A Reconnaissance, New York, etc.: Harper & Row.

Mason, F. K. (1986). Tornado: Full Development Story of the World's Foremost Combat Aircraft, Wellingborough, UK: Patrick Stephens.

Mayo, E. (1930). The Human Effect of Mechanization, *Papers and Proceedings of the 42th Annual Meeting of the American Economic Association*, Vol. XX, No. 1, pp. 156-176.

Mayo, E. (1933). Human Problems of an Industrial Civilization, New York: Macmillan.

McCarthy, J. (1995). Dynamics of Software Development: "Don't Flip the Bozo Bit" and 53 More Rules for Delivering Great Software on Time, Redmont, WA: Microsoft Press.

McConnell, S. (1993). Programmation Professionelle, Les Ulis Cedex: Microsoft Press.

McCurdy, H. E. (2001). Faster, Better, Cheaper - Low Cost Innovation in the U.S. Space Program, Baltimore, ML: John Hopkins.

McDonald, H. (Chairman); Eaton, D.; Ernst, R.; Hopson, G.; Kanki, B.; Lahoff, J.; Conahan, M.; McKeown, J.; Newman, J.; Sackheim, R.; Slenski, G.; Strauss, R. & Young, J. (2000). Space Shuttle Independent Assessment Team - Report to Associate Administrator Office of Space Flight, Washington, DC: NASA.

McGill, M. E.; Slocum, J. W., Jr. & Lei, D. (1992). Management Practices in Learning Organizations, *Organizational Dynamics*, Vol. 21, pp. 5-16.

McGrath, J. E. (1976). Stress and Behavior in Organizations, in: M. D. Dunnette (Ed.), *Handbook of Industrial and Organizational Psychology*, pp. 1351-1395, Chicago: Rand McNally.

McGrath, M. E.; Anthony, M. T. & Shapiro, A. R. (1992). Product Development: Success Through Product and Cycle-Time Excellence, Boston, etc.: Butterworth-Heinemann.

McGregor. D. (1957). The Human Side of the Enterprise, *Management Review*, Nov. 1957, pp. 22-28, 88-92.

Medaris, J. B. & Gordon, A. (1961). Die Zukunft wird heute entschieden - Raketen, Flugkörper und Satelliten, Köln: Verlag Wissenschaft und Politik.

Meinefeld, W. (1988). Einstellung, in: R. Asanger; G. Wenninger (Hrsg.), *Handwörterbuch der Psychologie, 4. Aufl.*, pp. 120-126, München; Weinheim: Psychologie Verlag Union.

Merton, R. K. (1940). Bureaucratic Structure and Personality, *Social Forces*, Vol. 18, pp. 560-568.

H. C. Metcalf & L. F. Urwick (Eds.) (1941). Dynamic Administration - The Collected Papers of Mary Parker Follett, Bath: Management Publications Trust.

Michaelsen, L. K.; Watson, W. E. & Black, R. H. (1989). A Realistic Test of Individual Versus Group Decision Making, *Journal of Applied Psychology*, vol. 74, no. 5, pp. 834-839.

Miller, J. G. (1960). Information Input, Overload, and Psychopatology, *American Journal of Psychiatry*, Vol. 116, pp. 695-704.

Mitroff, I. I. (1974). The Subjective Side of Science - A Philosophical Inquiry into the Psychology of the Apollo Moon Scientists, Amsterdam: Elsevier.

Mitroff, I. I. & Mason, R. O. (1974). On Evaluating the Scientific Contribution of the Apollo Moon Missions via Information Theory: A Study of the Scientist-Scientist Relationship, *Management Science*, Vol. 20, No. 12, August 1974, pp. 1501-1513.

Moorhead, G.; Ference, R. & Neck, C. P. (1991). Group Decision Fiascoes Continue: Space Shuttle Challenger and a Revised Groupthink Framework, *Human Relations*, vol. 44 no. 6, pp. 539-550, Tavistock Institute of Human Relations.

Morton, D. O. (1987). How to Attract and Retain Innovators, in: D. Christiansen (Hrsg.), *Engineering Excellence - Cultural and Organizational Factors*, pp. 17-21, New York: IEEE Press.

Moscovici, S. (1976). Social Influence and Social Change, London: Academic Press.

Moscovici, S.; Lage, E. & Naffrechoux, M. (1969). Influence of a Consistent Minority on the Responses of a Majority in a Color Perception Task, *Sociometry*, 32, pp. 365-380.

Muckler, F. A. & Seven, S. A. (1992). Selecting Performance Measures: "Objective" versus "Subjective" Measurement, *Human Factors*, vol. 34, no. 4, pp. 441-455, Santa Monica, CA: The Human Factors Society.

Muirhead, B. K. & Simon, W. L. (1999). High Velocity Leadership - The Mars Pathfinder Approach to Faster, Better, Cheaper, New York: HarperCollins.

Müller, D. (1989). Methoden der Ablauf- und Terminplanung von Projekten, in: H. Reschke; H. Schelle & R. Schnopp (Hrsg.), *Handbuch Projektmanagement*, Band 1, pp. 263-311, Köln: Verlag TÜV Rheinland.

Murray, C. & Cox, C. B. (1989). Apollo - The Race to the Moon, London: Secker & Warburg.

Neisser, U. (1976). Cognition and Reality: Principles and Implications of Cognitive Psychology, San Francisco: W. H. Freeman.

Nemeroff, P. M.; Passmore, W. A. & Ford, D. L. (1976). The Effects of Two Normative Structural Interventions on Established and Ad Hoc Groups: Implications for Improving Decision Making Effectiveness, *Decision Sciences*, Vol. 7, pp. 841-855.

Nemeth, C. J. (1994). The Value of Minority Dissent, in: S. Moscovici; A. Mucchi-Faina & A. Maass (Eds.), *Minority Influence*, pp. 3-15, Chicago: Nelson-Hall.

Nemeth, C. & Kwan, J. (1987). Minority Influence, Divergent Thinking and Detection of Correct Solutions, *Journal of Applied Social Psychology*, 17, pp. 786-797.

Nemeth, C. J. & Wachtler, J. (1983). Creative Problem Solving as a Result of Majority vs. Minority Influence, *European Journal of Social Psychology*, Vol 13, pp. 45-55.

Nemeth, C.; Swedlund. M. & Kanki, B. (1974). Patterning of the minority's responses and their influence on the majority, *European Journal of Social Psychology*, 4-I, pp. 53-64.

Nicholls, J. G. (1984). Achievement Motivation: Conceptions of Ability, Subjective Experience, Task Choice, and Performance, *Psychological Review*, pp. Vol. 91, 328-346.

Noer, D. (1993). Healing the Wounds - Overcoming the Trauma of Layoffs and Revitalizing Downsized Organizations, San Francisco: Jossey-Bass.

Nohria, N. (1995). Mary Parker' Follett's View on Power, the Giving of Orders, and Authority: An Alternative to Hierarchy or a Utopian Ideology?, in: P. Graham (Ed.), *Mary Parker Follett - Prophet of Management: A Celebration of Writings from the 1920s*, pp. 154-162, Boston,MA: Harvard Business School Press.

Nolen, S. B. (1988). Reasons for Studying: Motivational Orientations and Study Strategies, *Cognition and Instruction*, pp. Vol. 5, 267-287.

Norris, K. P. (1971). The Accuracy of Project Cost and Duration Estimates in Industrial R & D, *R&D Management*, Vol. 2, No. 1, pp. 25-36.

Noyce, R. N. (1987). The Importance of the Individual's Contribution to Industrial Progress, in: D. Christiansen (Hrsg.), *Engineering Excellence - Cultural and Organizational Factors*, pp. 23-29, New York: IEEE Press.

O'Connor, P. D. T. (1991). Practical Reliability Engineering (3rd ed.), Chichester: Wyley.

Pelz, D. (1956). Some Social Factors Related to the Performance in a Research Organization, *Administrative Science Quarterly*, Vol. 1, pp. 310-325.

Pelz, D. C. (1967). Creative Tensions in the Research and Development Climate, *Science*, vol. 157. no. 3785, pp. 160-165.

Pelz, D. C. & Andrews, F. M. (1976). Scientists in Organizations - Productive Climates for Research and Development, revised edition, Ann Arbor, MI: Institute for Social Research.

Perez, M. (1995). Interdisziplinarität - eine Buchse der Pandora?, *SIDOS News*, 1995, No. 2, pp. 5-9.
Perrott, R. H. & Allison, D. C. S. (1984). Pascal for Fortran Programmers, Rockville, MD: Computer Science Press.
Piechowiak, R. (1989). Griff nach den Sterne - Der Wettlauf zum Mond, ZDF.
Pill, J. (1971). Technical Management and Control of Large Urban Studies, A Comparative Analysis of Two Cases, Ph.D. Dissertation, Case Western University.
Poletti, G. (1964). The Diffusion of Network Techniques Throughout Government Publications, *IEEE Transactions on Engineering Management*, Vol. EM-11, March 1964, pp. 43-50.
Popper, K. R. (1959). The Logic of Scientific Discovery, London: Hutchinson.
Proctor, P. (1997). Public Opinion, Not Safety, Shaping FAA Enforcement, *Aviation Week and Space Technology*, Aug. 18, 1987, p. 48.
Quinn, J. B. (1985). Managing Innovation: Controlled Chaos, *Harvard Business Review*, May-June 1985, pp. 73-84.
Ranftl, R. M. (1977). A Study Program and Its Application, *Research Management*, Vol. XX, No. 1, pp. 25-29.
Ranftl, R. M. (1978). R&D Productivity - An Investigation of Ways to Improve Productivity in Technology-Based Organizations - Study Report, 2nd Ed.
Rauterberg, M.; Spinas, P.; Strohm, O.; Ulich, E. & Waeber, D. (1994). Benutzerorientierte Software-Entwicklung - Konzepte, Methoden und Vorgehen zur Benutzerbeteiligung, Zürich: vdf Hochschulverlag an der ETH Zürich; Stuttgart: Teubner.
Reiser, G. (1981). Einstellungen (Meinungen), in: P. G. v. Beckerath; P. Sauermann & G. Wiswede (Hrsg.), *Handwörterbuch der Betriebspsychologie und Betriebssoziologie*, pp. 143-147, Stuttgart: Ferdinand Enke.
Rice, A. K. (1958). Productivity and Social Organizations: The Ahmedabad Experiment, London: Tavistock.
Rich, B. R. & Janos, L. (1994). Skunk Works - A Personal Memoir of My Years at Lockheed, Boston: Little, Brown & Co.
Ritland, O. J. (1960-61). Concurrency, *Air University Quarterly Review*, Vol. XII, No. 3 and 4 - Winter/Spring 1960-1961, pp. 237-250.
Rivers, J. & Thompson, M. (1998). People Management in Aerospace - Report Summary, London: The Society of British Aerospace Companies.
Roberts, J. V. (1985). The Attitude-Memory Relationship After 40 Years: A Meta-Analysis of the Literature, *Basic and Applied Social Psychology*, Vol. 6, pp. 221-241.
Roethlisberger, F. & Dickson, W. (1939). Management and the Worker, Cambridge, MA: Harvard University Press.
Rogers, E. M. (1962). Diffusion of Innovations, New York: The Free Press.
Roozenburg, N. F. M. & Eekels, J. (1995). Product Design: Fundamentals and Methods, Chicester, etc.: John Wiley & Sons.
Rothstein, A. J. (1968). DOD-Industry Efforts Pass Milestone Toward Management Systems Control, *Aerospace Management*, Spring-Summer, 1968, pp. 5-20.
Rubenstein, A. H.; Chakrabarti, A. K. & O'Keefe, R. D. (1974). Field Studies of the Technological Innovation Process, PB 243 403, 15 Sep 1974, Washington, DC: National Science Foundation.
Ruiter, L. H.; Tubbs, M. R.; Randall, G. L. P.; Hoogeveen, L. P. J.; Allen, T. J. & other Panel Participants (1986). Panel Discussion on Theme II - Improving Productivity in the Laboratory, in: European Industrial Research Management Association, *Productivity and Effectiveness of R&D*, EIRMA Conference Papers, Vol. XXXII, pp. 76-87, Paris: EIRMA.

Saad, K. N.; Bohlin, N. H. & van Oene, F. (1992). R&D de Troisième Génération - La Gestion en Partenariat, Paris: Les Editions d'Organisation.
Saad, K. N.; Roussel, P. A. & Tiby, C. (1991). Management der F&E-Strategie, in: A. D. Little (Hrsg.), Wiesbaden: Gabler.
Sabbagh, K. (1995). Twenty-First Century Jet - The Building of the Boeing 777, Episode I: To Design a Plane, Seattle: KCTS; London: Channel Four.
Sabbagh, K. (1996). Twenty-First Century Jet - The Making and Marketing of the Boeing 777, New York: Scribner.
Sachse, P. (1995). Entwicklung und Bewertung einer computergestützten Entscheidungshilfe, Frankfurt a.M: Peter Lang, Dissertation Technische Universität Dresden 1994.
Sadow, R. M. (1964). How PERT Was Used in Managing the X-20 (Dyna-Soar) Program, *IEEE Transactions on Engineering Management*, Vol. EM-11, December 1964, pp. 138-154.
Samaras, T. T. & Czerwinski, F. L. (1971). Fundamentals of Configuration Management, New York: Wiley.
Sande, G. N. & Zanna, M. P. (1987). Cognitive Dissonance Theory: Collective Actions and Individual Reactions, in: B. Mullen & G. R. Goethals (eds.), *Theories of Group Behavior*, pp. 49-70, New York: Springer.
Sapienza, A. M. (1995). Managing Scientists - Leadership Strategies in Research and Development, New York: Wiley.
Sapolsky, H. M. (1972). The Polaris System Development: Bureaucratic and Programmatic Success in Government, Massachussets: Harvard University Press.
Saranson, I. G. (1975). Anxiety and Self-Preoccupation, in: I. G. Saranson & C. D. Spielberger (Eds.), *Stress and Anxiety*, Vol. 2, pp. 27-44, Washington DC: Wiley-Hemisphere.
Schank, R. C. & Abelson, R. P. (1977). Scripts, Plans, Goals, and Understanding: An Inquiry into Human Knowledge Structures, Hillsdale, NJ: Lawrence Erlbaum.
Schlotterbeck, D. (Dir.) (1984). Titanic - Mythos und Wirklichkeit eines Traumschiffes, Video: ARD.
Schmelzer, H. J. (1990). Steigerung der Effektivität und Effizienz Durch Verkürzung von Entwicklungszeiten, in: R. Reichwald & H. J. Schmelzer (Hrsg.), *Durchlaufzeiten in der Entwicklung - Praxis des Industriellen F+E Managements*, pp. 27-63, München, etc.: R. Oldenburg.
Schmienz, B. (1979). Kybernetik, in: W. Kern (Hrsg.), *Handwörterbuch der Produktionswissenschaft*, pp. 1022-1028, Stuttgart: Poeschel.
Schoderbek, P. P. (1965). A Study of the Applications of PERT, *Academy of Management Journal*, Vol. 8, No. 3, pp. 199-210.
Schuler, D. & Namioka, A. (Eds.) (1993). Participatory Design: Principles and Practices, in: D. Schuler & A. Namioka (Eds.), *Participatory Design: Principles and Practices*, Hillsdale, NJ: Lawrence Erlbaum Associates.
Schwarz, K. (1995). Osprey auf dem Weg zur Serie, *Flugrevue*, Apr.95, pp. 53-57.
Schwarz, K. (1997). Neuer Start für Osprey, *Flug Revue*, April 1997, pp. 15-17.
Schweizerischer Bundesrat (1989). Botschaft über Sondermassnahmen zugunsten der beruflichen und akademischen Weiterbildung und zur Förderung neuer Technologien im Fertigungsbereich, Bern: Eidgenössische Drucksachen- und Materialzentrale.
Scott, W. B. (1999a). Industry's Hire-and-Fire Paradigm Is Obsolete, *Aviation Week & Space Technology*.
Scott, W. B. (1999b). New Management Incentives are Key to Change, *Aviation Week & Space Technology*.

Scott, W. B. (1999c). Panel Links Launch Failures to Systemic Ills, *Aviation Week & Space Technology*, Sep. 13, 1999, pp. 24, 26.
Scott, W. B. (1999d). 'People' Issues are Cracks in Aero Industry Foundation, *Aviation Week & Space Technology*, Jun. 21, 1999, pp. 63-66.
Scott, W. B. (2000). Industry's Loss of Expertise Spurs Counterattack, *Aviation Week & Space Technology*, Mar 13, 2000, pp. 60-61.
Scott, W. B. (2001). Worries Deepen Over Dearth of Technical Talent, *Aviation Week & Space Technology*, Apr 23, 2001, pp. 32-33.
Seghezzi, H.-D. (1988). Qualitätssicherung von Neuen Produkten, in: W. Masing (Hrsg.) , *Handbuch der Qualitätssicherung, 2. Aufl.*, pp. 331-359, München: Carl Hanser.
Seghezzi, H.-D. (1992a). Bewirtschaftung der Qualität - eine Betriebswirtschaftliche Aufgabe, *Wf-Dokumentation Betriebswirtschaft*, Nr. 1/92, Zürich: Gesellschaft zur Förderung der schweizerischen Wirtschaft.
Seghezzi, H.-D. (1992b). Qualitäts-Management - Quo Vadis?, *SAQ Bulletin ASPQ*, 9/92.
Seghezzi, H.-D. (1996). Integriertes Qualitätsmanagement - Das St. Galler Konzept, München, Wien: Carl Hanser.
Selznick, P. (1949). TVA and the Grassroots, Berkeley, Cal.
Selznick, P. (1952). Cooptation: A Mechanism for Organization Stability, in: R. Merton et al. (Eds.), *Reader in Bureaucracy*, Glencoe, Ill: Free Press.
Shackle, G. L. S. (1961). Decision, Order and Time in HUman Affairs, Cambridge: Cambridge University Press.
Shaw, M. E. (1932). A Comparison of Individuals and Small Groups in the Rational Solution of Complex Problems, *American Journal of Psychology*, Vol. 44, pp. 491-504.
Shaw, M. E. (1976). Group Dynamics - The Psychology of Small Group Behavior, 2nd. Ed., New York: McGraw-Hill.
Signol, C. (1997). Influence Sociale et Dissonance Cognitive: Contribution a la Théorie de l'Élaboration de Conflit, Université Toulouse-Le-Mirail, Thèse de Doctorat Nouveau Régime, Octobre 1997.
Smith, C. G. (1970). Age of R and D Groups: A Reconsideration, *Human Relations*, Vol. 23, No. 2, pp. 81-96, London: Plenum Press.
Spangenburg, R. & Moser, D. K. (1995). Wernher von Braun - Space Visionary and Rocket Engineer, New York: Facts On File.
Specht, G. & Schmelzer, H. J. (1991). Qualitätsmanagement in der Produktentwicklung, Stuttgart: C. E. Poeschel.
Speiser, A. P. (1981). Die Z4 an der ETH Zürich - Ein Stück Technik- und Mathematikgeschichte, *Elemente der Mathematik* , Vol. 36, No. 6, 1981 - Separatum, pp. 145-153, Basel: Birkhäuser.
Stahlberg, D. & Frey, D. (1988a). Attitudes I: Structure, Measurements and Functions, in: M. Hewstone; W. Stroebe; J.-P. Codol & G.M. Stephenson (Eds.), *Introduction to Social Psychology - A European Perspective*, pp. 142-166, Oxford: Basil Blackwell.
Stahlberg, D. & Frey, D. (1988b). Einstellung I: Struktur, Messung und Funktionen, in: W. Stroebe, M. Hewstone, J. P. Codol & G. M. Stephenson (Hrsg.), *Sozialpsychologie - Eine Einführung*, pp. 144-170, Berlin: Springer.
Steiner, I. D. (1972). Groups Processes and Productivity, New York: Academic Press.
Stephenson, A. G. (Chairman); LaPiana, L. S.; Mulville, D. R.; Rutledge, P.J.; Bauer, F. H.; Folta, D.; Dukeman, G. A.; Sackheim, R. & Norvig, P. (2000). Report on Project Management in NASA by the Mars Climate Orbiter Mishap Investigation Board, Mar 13, 2000, Washington, DC: NASA.

Stine, G. H. (1991). ICBM - The Making of the Weapon that Changed the World, New York: Orion Books.

Stoff, W.-D. (1990). Hindernisse im Innovationsprozess, in: H.-J. Bullinger (Hrsg.), *Forum für Management in Forschung, Entwicklung und Technologie, Tagungsunterlagen 1990, Zürich, 3. - 4. Juli 1990*, pp. 21-25.

Strauss, A. L. (1987). Qualitative Analysis for Social Scientists, Cambridge: Cambridge University.

Strohm, O.; Kuark, J. K. & Schilling, A. (1993). Integrierte Produktion: Arbeitspsychologishce Konzepte und empirische Befunde, in: G. Cyranek & E. Ulich (Hrsg.), *CIM- Herasuforderung an Mensch, Technik und Organisation*, pp. 129-140, Zürich: Verlag der Fachvereine, Stuttgart: B. G. Teubner.

Stuhlinger, E. & Orway, F. I., III (1994). Wernher von Braun - Crusader for Space - A Biographical Memoir, Malabar, FL: Krieger.

Surber, H. (1997). Technologie-Transfer aus den Erfahrungen der SF, Schweizerische Unternehmung für Flugzeuge und Systeme, anhand von Raumfahrt-Projekten und dem F/A-18 Projekt, *Tagungsunterlagen - Technologie Transfer Seminar, Swiss Space Days SPADA 97, Luzern, 5. Mai 1997.*

Susman, G. I. (1976). Autonomy at Work: A Sociotechnical Analysis of Participative Management, New York: Praeger.

Takeuchi, H. & Nonaka, I. (1986). The New Product Development Game, *Harvard Business Review*, Jan-Feb 1986, pp. 137-146.

Taylor, F. W. (1911). The Principles of Scientific Management, New York: Harper & Row.

Thamhain, H. J. (1983). Managing Engineers Effectively, *IEEE Transactions on Engineering Management*, vol. EM-30, no. 4, pp. 231-237.

Thamhain, H. J. & Wilemon, D. L. (1987). Building High Performing Engineering Project Teams, *IEEE Transactions on Engineering Management*, Vol. 34, No. 3, pp. 130-137.

The Wyatt Company (1993). Best Practices in Corporate Restructuring: Wyatt's 1993 Survey of Corporate Restructuring, Washington, D.C.: The Wyatt Company.

Thomas, R. J. (1994). What Machines Can't Do: Politics and Technology in the Industrial Enterprise, Berkeley, CA:University of California Press.

Thompson, M. (2000). The UK People Management Audit 2000 - Final Report: The Bottom Line Benefits of Strategic Human Resource Management, London: Society of British Aerospace Companies.

Tomaszewski, T. (1981). Struktur, Funkjtion und Steuerungsmechanismen menschlicher Tätigkeit, in: T. Tomaszewski (Hrsg.), *Zur Psychologie der Tätigkeit*, pp. 11-33, Berlin: Deutscher Verlag der Wissenschaften.

Tompkins, P. K. (1977). Management Qua Communication in Rocket Research and Development, *Communication Monographs*, Vol. 44, No. 1, pp. 1-26.

Trist, E. L. & Bamforth, K. (1951). Some Social and Psychological Consequences of the Longwall Method of Coalgetting, *Human Relations*, Vol 4, pp. 3-38.

U. S. Air Force, PERT Orientation and Training Center (1963). Bibliography: PERT and other Management Systems and Techniques, Washington, D.C.: Bolling Air Force Base.

Udris, I. & Alioth, A. (1980). Fragebogen zur <subjektiven Arbeitsanalyse> (SAA), in: E. Martin; I. Udris; U. Ackermann & K. Oergerli (Eds.), *Monotonie in der Industrie*, pp. 67-68, 204-207, Bern: Huber.

Ulich, E. (1978). Über mögliche Zusammenhänge zwischen Arbeitstätigkeit und Persönlichkeitsentwicklung, *Psychosozial*, Band 1, pp. 44-63.

Ulich, E. (1996). Neue Produktionskonzepte - Ein kritischer Diskussionsbeitrag, *Zeitschrift für Arbeitswissenschaft*, 50 (22F), 1996/4, pp. 193-198.

Ulich, E. (2001). Arbeitspsychologie (5. Aufl.), Zürich: vdf; Stuttgart: Schäffer Poeschel.
Ulich, E.; Baitsch, C. & Alioth, A. (1987). Führung und Organisation (2. Aufl.), *Die Orientierung*, Nr. 81, Bern: Schweizerische Volksbank.
van Avermaet, E. (1988). Social Influence in Small Groups, in: M. Hewstone; W. Stroebe; J.-P. Codol & G.M. Stephenson (Eds.), *Introduction to Social Psychology - A European Perspective*, pp. 350-380, Oxford: Basil Blackwell.
van Ishoven, A. (1975). Messerschmitt: Aircraft Designer, London: Gentry Books.
Vandivier, K. (1972). Why Should My Conscience Bother Me? The Aircraft Brake Scandal, in: R. L. Heilbroner et. al. (Ed.), *In the Name of Profit*, pp. 3-31, Garden City NY:Doubleday & Co.
Vann, F. (1993). Willy Messerschmitt: First Full Biography of an Aeronautical Genius, Sparkford: Haynes.
VDI - Verein Deutscher Ingenieure (1987). Systematic Approach to the Design of Technical Systems and Products, Berlin: Beuth; Düsseldorf: VDI Verlag.
Velocci, A. L., Jr. (2000a). Brain Drain Threatens Aerospace Vitality, *Aviation Week & Space Technology*, Apr. 24, 2000, pp. 24-26.
Velocci, A. L., Jr. (2000b). Loss of Intellectual Capital Challenges Company Ingenuity, *Aviation Week & Space Technology*, Apr. 24, 2000, p. 26.
Vincenti, W. G. (1990). What Engineers Know and How They Know It - Analytical Studies from Aeronautical Story, Baltimore: John Hopkins.
Volpert, W. (1979). Der Zusammenhang zwischen Arbeit und Persönlichkeit aus handlungspsychologischer Sicht, in: P. Groskurth (Hrsg.), *Arbeit und Persönlichkeit*, pp. 21-46, Reinbek: Rowohlt.
Volpert, W. (1987). Psychische Regulation von Arbeitstätigkeiten, in: U. Kleinbeck & J. Rutengfranz (Hrsg.), *Arbeitspsychologie*, pp. 1-42, Enzyklopädie der Psychologie, Themebereich D, Serie III, Band 1, Göttingen: Hogrefe.
von Braun, W. (1956). Teamwork: Key to Success in Guided Missiles, *Missiles and Rockets*, Vol. 1, No.1, Oct. 1956, pp. 38-42.
von Braun, C. -F. (1994). Der Innovationskrieg, München: Hanser.
von Diemer, R. (1991). Qualität der Kommunikation und Kooperation, *SAQ-Bulletin-ASPQ*, 11+12/91, pp. 9-12.
von Diemer, R. (1994). Motivation, in: W. Masing (Hsrg.), *Handbuch Qualitätsmanagement, 3. Aufl.*, pp. 1061-1074, München: Hanser.
von Neumann, J. & Morgenstern, O. (1966). Theory of Games and Economic Behavior, Princeton NJ: Princeton University Press.
Voss, T & Abraham, M. (2000). Rational Choice Theory in Sociology, in: S. R. Quah & A. Sales (Eds.), *The International Handbook od Sociology*, pp. 50-83, London: Sage.
Walker, C. R. (1950). The Problem of the Repetitive Job, *Harvard Business Review*, 28, pp. 54-58.
Warwick, G. (1997). Lockheed Martin Aims to Solve C-130J Hercules Stall problem, *Flight International*, 7-13 May 1997.
Warwick, G. (1998a). Boeing and the US Navy Identify F/A-18E/F Wing-Drop Solutions, *Flight International*, 7-13 Jan 1998, p. 13.
Warwick, G. (1998b). Wingdrop is a 'National Issue', says Panel, *Flight International*, 21-27 Jan 1998, p. 24.
Weaver, W. (1959). Dither, *Science*, Vol. 130, No. 3371, p. 301.
Weber, W. G. (1997). Analyse von Gruppenarbeit - Kollektive handlungsregulation in soziotechnischen Systemen, Bern: Hans Huber.

Weinberg, G. M. (1971). The Psychology of Computer Programming, New York: Van Nostrand Reinhold.

Wells, W. P. & Pelz, D. C. (1976). Groups, in: Pelz, D. C. & Andrews, F. M., *Scientists in Organizations - Productive Climates for Research and Development, revised edition*, Ann Arbor, MI: Institute for Social Research.

Wertz, J. R. (1996). Process Changes to Reduce Cost - Radical Cost Reduction Methods, in: J. R. Wertz & W. J. Larson (Eds.), *Reducing Space Mission Cost*, pp. 34-53, Dordrecht: Kluwer.

Weyrich, C. (1998). Industrieforschung vor globalen Herausforderungen, in: W. Fricke (Hrsg.), *Innovationen in Technik, Wissenschaft und Gesellschaft*, Forum Humane Technikgestaltung, Band 19, pp. 55-74, Bonn. Friedrich-Ebert Stiftung.

Whitehead, A. N. (1917). The Organization of Thought, Educational and Scientific, London: Williams and Norgate.

Whitehead, A. N. (1967). Science and the Modern World: 1925 Lowell Lectures, New York: Free Press.

R. R. Whyte (Ed.) (1975). Engineering Process Through Trouble, London: The Institution of Mechanical Engineers.

Whyte, R. R. (1978). Development Costs - A Law of Overspend?, in: R. R. Whyte (Ed.), *Engineering Progress Through Development*, pp. 4-7, London: Mechanical Engineering Publications.

Whyte, W. F. (1994). What Machines Can't Do, *Qualitative Sociology*, Vol. 17, No. 3, pp. 293-298.

Wiebecke, G.; Tschirky, H. & Ulich, E. (1987). Cultural Differences at the R&D-Marketing Interface: Explaining Interdivisional Communication Barriers, *Proceedings of the 1987 IEEE-Conference on Management and Technology, Atlanta*, pp. 94-102.

Wilke, H. & van Knippenberg, A. (1988). Group Performance, in: M. Hewstone; W. Stroebe; J.-P. Codol & G.M. Stephenson (Eds.), *Introduction to Social Psychology - A European Perspective*, pp. 313-349, Oxford: Basil Blackwell.

Wischnewski, E. (1996). Modernes Projektmanagement - PC-gestützte Planung, Durchführung und Steuerung von Projekten, 5. Aufl., Braunschweig: Friedr. Vieweg & Sohn.

Wolf, G. (1992). Wege zum Effektiven Motivations-Management, *Qualität und Zuverlässigkeit*, Band 37, Heft 3, pp. 133-137.

Wood, R. & Bandura, A. (1989). Impact of Conceptions of Ability of Self-Regulatory Mechanisms and Complex Decision Making, *Journal of Personality and Social Psychology*, Vol. 56, pp. 407-415.

Yerkes, R. M. & Dodson, J. D. (1908). The Relation of Strength and Stimulus to Rapidity of Habit-Formation, *Journal of Comparative Neurology and Psychology*, Vol. 18, pp. 459-482.

Young, T. A. (Chairman) et al. (2001). Report by the International Space Station (ISS) Management and Cost Evaluation (IMCE) Task Force to the NASA Advisory Council, Nov 1, 2001, Washington,DC: NASA.

Zerega, J. E. (1976). Down and Up with PERT at Goddard, *Astronautics and Aeronautics*, Vol. 14, Feb 1976, p. 65.

Ziemke, M. C. & Spann, M. S. (1991). Warning: Don't Be Half-Hearted in Your Efforts to Employ Concurrent Engineering, *Industrial Engineering*, Vol. 23, No. 2, pp. 45-49.

Persons index

Index

Original terms and citations

i German: *Pflichtenheft*

ii German: *Um den künftigen Bedarf von Forschung und Entwicklung nach Interdisziplinarität und Innovationsfähigkeit der Mitarbeiter decken zu können, müssen schon in der Ausbildung nichttechnische Inhalte, Bereitschaft zu Teamarbeit und zu mehr Risiko vermittelt werden. Die Persönlichkeitsbildung des Ingenieurs und Naturwissenschaftlers muss in der Zukunft verstärkt werden, indem frühzeitig Verantwortung übertragen, aber auch eingefordert wird*

iii German: *Später zeigte es sich, wie wertvoll dieser Umweg gewesen war, die Erfahrungen, die wir mit der Z4 gewonnen haben waren für unsere spätere elektronische Maschine vom grössten Wert. Viele der betrieblichen Beobachtungen, die wir gemacht haben haben wir einbeziehen können.*

iv German: *Kundenanteil*

v German: *Menschenbild*

vi German: *Fertigkeit*

vii German: *Versachlichung*

viii German: *Herrschaft*

ix German: *fachliche*

x German: *Aktenmässigkeit*

xi German: *überwunden*

xii French: *ne recevoir jamais aucune chose pour vraie, que je ne la connusse évidemment être telle: c'est-à-dire d'éviter soigneusement la précipitation et la prévention; et de ne comprendre rien de plus en mes jugements, que ce qui se présenterait si clairement et si distinctement à mon esprit, que je n'eusse aucune occasion de le mettre en doute*

xiii French: *diviser chacune des difficultés que j'examinerais, en autant de parcelles qu'il se pourrait, et qu'il serait requis pour les mieux résoudre* xiv French: *conduire par ordre mes pensées, en commençant par les objets plus simples et les plus aisées à connaître, pour monter peu à peu, comme par degrés, jusques à la connaissance des plus composés; et supposant même de l'ordre entre ceux qui ne se précèdent point naturellement les uns les autres*

xv French: *faire partout des dénombrements si entiers, et des revues si générales, que je fusse assuré de ne rien omettre*

xvi German: *Die Idee der kritischen Prüfung ist eine methodische Idee, die darauf zurückgeht, dass unser Denken und Handeln der Irrtumsmöglichkeit unterworfen ist, so dass derjenige, der ein echtes Interesse an der Wahrheit hat, daran interessiert sein muss, die Schwächen und Schwierigkeiten seiner Denkresultate und Problemlösungen kennenzulernen, Gegenargumente zu hören und seine Ideen mit Alternativen konfrontiert zu sehen, um sie zu vergleichen, modifizieren und revidieren zu können. Nur Anschauungen, die kritischen Argumenten ausgesetzt werden, können sich bewähren. Nur auf dem Hintergrund alternativer Auffassungen lassen sich die Vorzüge und Nachteile bestimmter Konzeptionen beurteilen.*

xvii German: *Wir Deutschen, die damals hierüber kamen, wir waren technisch nicht besser als die Amerikanischen Kollegen. Die waren ihrer Ausbildung nach, ihrer inneren Beteiligung nach, ihrer Interesse nach, so gut wie wir waren. Aber wir in Deutschland haben zwölf Jahre mehr Zeit [gehabt], Fehler zu machen und von den Fehlern zu lernen.*

xviii German: *Erfahrungswissen*

xix German: *Nachrichtentechnischen Gesellschaft – NTG im Verein Deutscher Elektrotechniker – VDE*

xx German: *DGQ – Deutsche Gesellschaft für Qualität e.V.*

xxi German: *Fehler, die nicht gemacht werden können, brauchen auch nicht behoben zu werden*

xxii German: *Ablauforganization*

xxiii German: *Eigenschaftstheorie*

xxiv German: *Prüfrichtlinien*

xxv German: *Kontaktdichte*

xxvi German: *mit regelmässigen gemeinsamen Gesprächen*

xxvii German: *Entscheidungsbereich*

xxviii German: *Hilf dem Autor, einen besseren Beitrag zum Projekt zu liefern (Gebot 2)*

xxix German: *Achte auf deine Sprache: Es geht um die Sache, nicht um Person (Gebot 3)*

xxx German: *Verteidige dich nicht: es ist ein Review, kein Gericht (Gebot 10)*

xxxi German: *bestehende Verhaltensweisen bewiesen und gerechtfertigt werden*

xxxii German: *Atmosphäre der Öffentlichkeit*

xxxiii German: *kreativitätshindernd*

xxxiv German: *abhängig und unselbständig*

xxxv German: *Gemeinsame Aufgabe*

xxxvi German: *Überschaubare Grösse*

xxxvii German: *gleichwertige aber nicht gleichartige Mitglieder*

xxxviii German: Abrufbare Kenntnisse

xxxix German: *Kommunikationsbereitschaft und Teamwork*

xl French: *base non pertinente*

xli German: *Subjektive Arbeitsanalyse – SAA*

xlii German: *Sinnlosigkeit*

xliii German: *Handlungskompetenz*

xliv German: *Handlungsspielraum*

xlv German: *Ereignisse*

xlvi German: *Die Entscheidung verbindet die Motivation mit der Handlung, und sie scheint gleichzeitig eine Verfestigungswirkung auszuüben, die teils durch die Tendenz des Individuum, zu «seinen Entscheidungen zu stehen», und teils durch das «Bekenntnis zur Gruppe» bedingt ist.*

xlvii German: *primär 'nicht-materielle' Arbeitshandlungen*

xlviii German: *vollständige Tätigkeit*

xlix German: *vollständige Handlung*

l German: *vollständige Aufgabe*

li German: *VDA – Verband der deutschen Automobilindustrien*

lii German: *Abrufbare Qualifikationen*

liii German: *Beschleunigungsmärkte*

liv German: Nachlässigkeit

lv German: *Erkenntnis*

lvi German: *Erfahrung*

lvii German: *Erkenntnis*

lviii German: *Eine Kooperation ist die dauerhafte Zusammenarbeit zwischen zwei oder mehreren rechtlich selbständigen Unternehmen, z.B. intensiver Erfahrungsaustausch, Arbeitsgemeinschaft oder Joint-Venture. Nicht darunter fallen etwa Lieferbeziehungen, Handelsvertreter oder der Kauf/Verkauf von Lizenzen*

The author and publisher welcome any comments and/or corrections which will be of benefit to a next edition of this book.

Please send these to both:

Dr. Eginaldo Shizuo Kamata
Oberwiesenstr. 73E
CH-8050 Zürich
Switzerland

and

Space Technology, Physical Sciences Unit
Kluwer Academic Publishers
P.O. Box 17
3300 AA Dordrecht
The Netherlands

MIX
Papier aus verantwortungsvollen Quellen
Paper from responsible sources
FSC® C105338
www.fsc.org

If you have any concerns about our products,
you can contact us on
ProductSafety@springernature.com

In case Publisher is established outside the EU,
the EU authorized representative is:
Springer Nature Customer Service Center GmbH
Europaplatz 3, 69115 Heidelberg, Germany

Printed by Libri Plureos GmbH
in Hamburg, Germany